Canyon Hiking Guide to the
COLORADO PLATEAU
Third Edition

Michael R. Kelsey

Kelsey Publishing
456 E. 100 N.
Provo, Utah, USA, 84606
Tele & Fax 1-801-373-3327

First Edition April 1986
Second Printing March 1988
Second Edition March 1991
Third Edition January 1995

Distributors for Kelsey Publishing

Primary Distributor All of Michael R. Kelsey's books are sold by this company. If you'd like to order any book, please write to the following address: **Wasatch Book Distribution,** P.O. Box 11776, 268 S., 200 E., Salt Lake City, Utah, USA, 84111, Tele. 1-801-575-6735, or for bookstores, 1-800-786-6715, Fax 1-801-521-8243.

<u>Some of Kelsey's books are sold by the following distributors.</u>
Alpenbooks, 3616 South Road, Building C, Suite 1, Mukilteo, Washington, 98275, Tele. 1-206-290-8587.
Canyon Country Publications, P. O. Box 963, Moab, Utah, 84532, Tele. 1-801-259-6700.
Canyonlands Publications, 4999 East Empire, Unit A, Flagstaff, Arizona, 86004, 1-602-527-0730.
Crown West Books(Library Service), 575 E. 1000 S., Orem, Utah, 84058, Tele. 1-801-224-1455.
Ingram Books, 2323 Delgany, Denver, Colorado, 80216, Tele. 1-800-876-1830.
Northern Arizona News, 1709 North, East Street, Flagstaff, Arizona, 86001, Tele. 1-602-774-6171.
Many Feathers, 2626 West, Indian School Road, Phoenix, Arizona, 85012, Tele. 1-602-266-1043.
Nevada Publications, 4135 Badger Circle, Reno, Nevada, 89509, Tele. 1-702-747-0800.
Mountain 'N Air Books, 7251 Foothill Blvd., Tujunga, California, 91042, Tele. 1-818-951-4150, or for bookstores, 1-800-446-9696
Recreational Equipment, Inc.(R.E.I.), P.O. Box C-88126, Seattle, Washington, 98188, Tele. 1-800-426-4840(or check at any of their local stores).

For the UK and Europe, and the rest of the world contact:
CORDEE, 3a De Montfort Street, Leicester, England, UK, LE1 7HD, Tele. 0533-543579, Fax 0533-471176.

Printed by Bookcrafters, Chelsea, Michigan, USA

All fotos by the author.
All maps, charts, and cross sections drawn by the author.

Front Cover

Front Cover
1. The Corkscrew, Antelope Canyon, Navajo Nation, Arizona.
2. The Black Hole of Lower White Canyon, Utah.
3. Anasazi Ruins in Road Canyon, Utah.
4. Butterfly Canyon, Navajo Nation, Arizona.

Back Cover

Back Cover
5. Buckskin Gulch of the Paria River, Utah.
6. The Grand Canyon, from the South Rim, Arizona.
7. West Clear Creek, Arizona.
8. Little Wild Horse Canyon, San Rafael Swell, Utah.
9. Newspaper Rock, Indian Creek Canyon, Utah.
10. Brimstone Canyon of the Escalante River, Utah.

Table of Contents

Acknowledgments

It's impossible to recall all the many people who helped with information for this book. They are countless, but special thanks should go to the following people: Larry Royer of the Cedar City BLM; Rod Schipper(Skip), the ranger who lived at the BLM Paria River Ranger Station 10 years and the one in charge of the Paria River; Larry Gearhart, formerly of the Hanksville BLM office; Buster Esplin, of St. George and the Arizona Strip's Wildcat Ranch; Fran Barnes of Moab; Bill Booker of the Hanksville BLM office, and Pearl Biddlecome Baker, now deceased. Extra thanks must go to Grand Canyon hiker Harvey Butchart, now of Sun City, Arizona. He donated his hiking maps to the Special Collections Library of the University of Northern Arizona, and they were most helpful in locating routes down into the Grand Canyon and the Little Colorado River Gorge.

Others include Eberhard Schmilinsky, John Logan, Scott Peterson, Jim Ohlman, Grant Reeder, Hani Heller, Buck O'Herin, Marcus Rhinelander, Mike Tanis, Mike Walters, Audrey Almond, and Richard Bartel.

And of course many thanks goes to my mother, Venetta B. Kelsey who looks after my small publishing business when I'm on the road and who helped proof-read this manuscript.

The Author

The author, who was born in 1943, experienced his earliest years of life in eastern Utah's Uinta Basin, namely around the town of Roosevelt. Then the family moved to Provo, where he attended Provo High School, and later Brigham Young University, where he earned a B.S. degree in Sociology. Shortly thereafter he discovered that was the wrong subject, so he attended the University of Utah, where he received his Master of Science degree in Geography, finishing that in June, 1970.

It was then real life began, for on June 9, 1970, he put a pack on his back and started traveling for the first time. Since then he has seen 175 countries, republics, islands, or island groups. All this wandering has resulted in a number of books being written and published by himself: *Climber's and Hiker's Guide to the World's Mountains(3rd Ed.)*, *Utah Mountaineering Guide*, and the *Best Canyon Hikes(2nd Ed.)*; *China on Your Own* and the *Hiking Guide to China's Nine Sacred Mountains(3rd Ed.)*; *Canyon Hiking Guide to the Colorado Plateau(3nd Edition)*; *Hiking and Exploring Utah's San Rafael Swell(2nd Ed.)*; *Hiking and Exploring Utah's Henry Mountains and Robbers Roost(Revised Edition)*; *Hiking and Exploring the Paria River(Updated Edition)*; *Hiking and Climbing in the Great Basin National Park(Wheeler Peak, Nevada)*; *Boater's Guide to Lake Powell--Featuring Hiking, Camping, Geology, History and Archaeology(Updated Edition)*; *Climbing and Exploring Utah's Mt. Timpanogos*; *River Guide to Canyonlands National Park & Vicinity*; and *Hiking, Biking and Exploring Canyonlands National Park & Vicinity*.

He has also helped his mother Venetta B. Kelsey write & publish a book about the town she was born and raised in, *Life on the Black Rock Desert--A History of Clear Lake, Utah*.

Map Symbols

Town or City	⬜⬜	Viewpoint	▲
Buildings or Homes	◧▪	River (Large)	〰
Campgrounds	⛺	Stream or Running Water	〜
Backcountry Campsite	▲	Intermittent Stream	〜
Camp Sites	♠	Potholes	P.H.
Picnic Site	⛺	Escarpment, Canyon Rim	⬛⬛⬛
School	▮	Peak, or Standing Rock	✖•
Church	⛪	Lake	
Hotel, Motel or Lodge		Mine or Quarry	
Ranger Station or Visitor Center		Waterfall or Dryfall	Ⓕ
Guard Station		Spring or Seep	Ⓢ○
Airport or Landing Strip		Grass or Sagebrush	
Radio Tower		Forest (Pines)	
Railway		Forest (Pinyon-Juniper)	
Interstate Highway	70	Forest (Deciduous)	
U.S. Highway	24	Pass	
State Highway	9	Natural Arch	Ⓐ
Road, Maintained		Pictograph	PIC
Road, Not Maintained		Petroglyph	PET
Mile Posts	3 4	Indian Ruins	Ⓡ•
Trails		Narrows	Ⓝ
Routes (No Trail)	••••	Undercut	UC
Car-Park, Trailhead	Ⓟ	Stock Ponds or Tanks	
Geology Cross Section		Corrals	

Abbreviations

Canyon	C. or Can.	Campground	CG.
Lake	L.	Wilderness Study Area	WSA
River	R.	Mine	M.
Creek	Ck.	4 Wheel Drive Vehicle or Road	4WD
Peak	Pk.	2 Wheel Drive Vehicle or Road	2WD
Waterfall, Dryfall, Formation	F.	High Clearance Vehicle or Road	HCV
Kilometer(s)	km, kms	Off Road Vehicle	ORV
Mile Posts	mp or M.P.	All Terrain Vehicle	ATV
Sandstone	S.S.	Spring	Sp.
Pothole	PH.	Dugway and Meters	D60m

United States Geological Survey	USGS
National Park Service	NPS
Bureau of Land Management	BLM
High Water Mark(Lake Powell & Lake Mead only)	HWM

Metric Conversion Table

1 Centimeter = 0.39 Inch   1 Mile = 1.609 Kilometers   1 Pound = 453 Grams
1 Inch = 2.54 Centimeters   100 Miles = 161 Kilometers   1 Quart(US) = 0.946 Liter
1 Meter = 39.37 Inches   100 Kilometers = 62.1 Miles   1 Gallon(US) = 3.785 Liters
1 Foot = 0.3048 Meter   1 Liter = 1.056 Quarts(US)   1 Acre = 0.405 Hectare
1 Kilometer = 0.621 Mile   1 Kilogram = 2.205 Pounds   1 Hectare = 2.471 Acres

METERS TO FEET (Meters x 3.2808 = Feet)

100 m = 328 ft.	2500 m = 8202 ft.	5000 m = 16404 ft.	7500 m = 24606 ft.
500 m = 1640 ft.	3000 m = 9842 ft.	5500 m = 18044 ft.	8000 m = 26246 ft.
1000 m = 3281 ft.	3500 m = 11483 ft.	6000 m = 19686 ft.	8500 m = 27887 ft.
1500 m = 4921 ft.	4000 m = 13124 ft.	6500 m = 21325 ft.	9000 m = 29527 ft.
2000 m = 6562 ft.	4500 m = 14764 ft.	7000 m = 22966 ft.	

FEET TO METERS (Feet ÷ 3.2808 = Meters)

1000 ft. = 305 m	9000 ft. = 2743 m	16000 ft. = 4877 m	23000 ft. = 7010 m
2000 ft. = 610 m	10000 ft. = 3048 m	17000 ft. = 5182 m	24000 ft. = 7315 m
3000 ft. = 914 m	11000 ft. = 3353 m	18000 ft. = 5486 m	25000 ft. = 7620 m
4000 ft. = 1219 m	12000 ft. = 3658 m	19000 ft. = 5791 m	26000 ft. = 7925 m
5000 ft. = 1524 m	13000 ft. = 3962 m	20000 ft. = 6096 m	27000 ft. = 8230 m
6000 ft. = 1829 m	14000 ft. = 4268 m	21000 ft. = 6401 m	28000 ft. = 8535 m
7000 ft. = 2134 m	15000 ft. = 4572 m	22000 ft. = 6706 m	29000 ft. = 8839 m
8000 ft. = 2438 m			30000 ft. = 9144 m

CENTIMETERS / INCHES

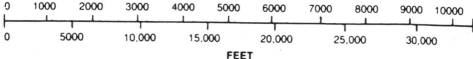

METERS / FEET

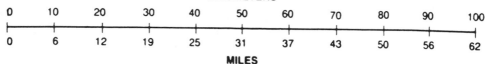

KILOMETERS / MILES

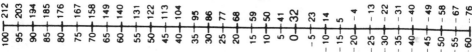

FAHRENHEIT / CENTIGRADE

Introduction

The Colorado Plateau

This is a hiking guide to the canyons of the Colorado Plateau. The Colorado Plateau is a large physiographic region which roughly covers the southeastern half of Utah, the northern half of Arizona, the western fifth of Colorado and a small area in the northwestern corner of New Mexico. It basically includes the middle section of the Colorado River drainage.

Geology--Why the Colorado Plateau is so Unique?

The thing that makes the Colorado Plateau so unique is the geology, and/or the rocks. Most of the rock formations on the Plateau are horizontal or flat-lying, and very much like a layered cake. Each rock formation is of a different color and composition. It's the differential erosion of these very flat-lying rocks which make this region so interesting. For example, the Grand Canyon is made up of cliffs and terraces making a wide open and scenic canyon. In other parts of the Plateau you find deep, dark, narrow canyons some of which may be 50 to 100 meters deep and only a meter or two in width. These narrow slot-type canyons are among the strangest and most interesting and fotogenic features on earth. Many of the best hikes in this book are in slot-type canyons. In short, there is simply no other place on earth quite like it.

Another effect of erosion on these horizontal strata are the alcoves or cave-like depressions in canyon walls. In the past, many of these have provided shelter for indigenous peoples. Thousands of these ruins are scattered throughout the Colorado Plateau which are generally called *cliff dwellings*. These sites were inhabited by what the Navajos call the *Ancient Ones*, or Anasazi. Some of the hikes in this book are to canyons which feature these simple rock shelters.

There are almost no true mountains on the Plateau, except for 8 laccoliths which form the Henry, La Sal, Abajo, Rico, La Plata, Ute, Carrizo and Navajo Mountains. These uplifts were caused by molten magma deep within the earth trying to penetrate the crust and deforming it upward. The magma slowly cooled, then erosion exposed the central core of these mountains, which is normally some kind of granitic-type rock.

Emergency Kit for Your Car

Before leaving home one of the first things you should be aware of when visiting this region is that some places are very isolated and remote and a long way from any garage. It's recommended you take a good reliable vehicle--one you can depend on. Always have a full tank of fuel--and in some cases extra fuel. Also take extra water, food, tools, battery jumper cable, a tow rope or chain, a shovel, tire pump, extra oil and any other item you think might come in handy in an emergency in a far away place. In recent years and with improved roads, there is more traffic on the back roads of the Colorado Plateau, especially on weekends in the spring, summer and fall, but you'll want to go as well prepared as possible. A little time spent planning your trip may prevent a bad experience in the long run.

Best Times to Travel and Hike--Weather and Climate

Most of us have our vacations in summer, but generally speaking the best time to visit the Colorado Plateau is in the cooler months of spring or fall. The reason is, most of the canyons are in areas of low to moderate elevations. Most hikes range from about 1200 to 1800 meters altitude. Few are over 2000 meters, and only along the Colorado River in the western Grand Canyon does the elevation drop to as low as 373 meters(the high water mark of Lake Mead). Therefore summer months get pretty warm and uncomfortable. Spring or fall offers something in between the winter ski season and summer mountaineering trips.

But there are exceptions to every rule. On many hikes, you'll be wading along a small stream or in a slot canyon with a few potholes. In these cases you can still enjoy hiking in summer because you can always cool off in the water or shade. In some, notably West Clear Creek, Wet Beaver Creek and Lower White Canyon, you'll be swimming part of the time, so you can only do these hikes in the warmest summer months. In the text, each hike has the best time stated.

The author generally prefers the time from about mid-March through April(generally the very best time) and May, then again in late September and October. The good things about hiking in spring are longer days than in the fall, and lack of insects. The fall season doesn't have any insects either, but the days are getting very short by late October. Remember, October 21 has the same amount of daylight hours as does February 21; and September 21, the same amount of daylight as on March 21. If you're camping in late December, you've got to have a tent on the ground by about 5pm, then sit there for 5 or 6 hours before bedtime. If you want an early morning start you've got to get up in the dark!

The month with the heaviest precipitation on the Colorado Plateau is August. Next would be July or September, then October. The driest month is June. One important thing you want to remember here is the monsoon season. Generally speaking, it begins in about mid-July and continues through mid-September. However, since this is an arid or semi-arid desert area, the wetter season here isn't the same as the rainy season in the humid tropics. There are still many sunny days, even in August. But when heavy rains come they can turn a dry wash into a raging torrent. When you get close to the area you'll be hiking in, always tune in on local radio stations for the latest weather forecasts. If you're heading for one of the slot canyons, **take precautions**. In one flash flood in the Zion Narrows of southwestern Utah, five hikers were swept away and drowned. This was probably the worst accident of its kind ever to happen on the Colorado Plateau.

The Insect Season

Over the years the author has found that insects begin to appear about the last week in May and continue until mid-summer in most cases. Probably the most bothersome insects are the large **deer** or **horse flies**. These pests are often found along some generally dry washes but near seeps with tamarack, and sometimes with cottonwood trees. They are also generally found in areas where cattle come down to waterholes. The author remembers them being very bad along the Escalante River, The Gulch and in Davis and Lavender Canyons. He hiked these in early summer. In slot-type canyons, such as the Buckskin Gulch, Antelope, Lower White Canyon and Kaibito Creek, there seem to be no insects of any kind no matter what the season.

In early summer you may also be bothered by small **midges, gnats,** or **no-see-ums**. These rascals get in your hair and bite your scalp. June seems to be the worst month for these, and by mid-summer, they seem to disappear. Wearing a hat, especially one with a curtain sewn on around the back, will be a deterrent. These insects aren't everywhere, but they did bother the author along the shore of upper Lake Powell once, and up on Gray's Pasture(Island in the Sky) in Canyonlands National Park at another time.

Mosquitoes are almost never a problem. They are usually found along well-watered valleys or in swampy places, something the Colorado Plateau has little of. What few mosquitoes there are, seem to come out only at night and are almost never seen in the daylight hours. Make sure your tent has mosquito netting and these insects should never be a problem. Incidentally, the author never carries insect repellent.

Another possible nuisance is the ordinary **house fly**. Flies come out in daylight hours and are usually seen only in places where camps have been made. If campers would simply leave their campsites clean, no one would ever see flies.

Drinking Water

Despite what many people think about this so-called desert, there is water of some kind in almost every canyon in this book. The question then arises, what water is safe to drink as is, and what needs to be treated, filtered or boiled?

When it comes to actually drinking the water, old timers used to say; *"if it's clear and it's a fast flowing stream, then it's normally safe to drink."* Throughout the Colorado Plateau the author always test-drinks water from springs not muddied up by cattle, and sometime from small streams which have no sign of cattle or beaver upstream, before he recommends it to other hikers. He has yet to become sick from doing so.

But it is important to choose your drinking water carefully because it's possible to get an intestinal disorder called **giardiasis**, caused by the microscopic organism, **Giardia Lamblia**. Early day fur trappers used to call this ailment Beaver Fever. Giardia are carried in the feces of humans and some domestic and wild animals, especially cattle and beaver. The cysts of Giardia may contaminate surface water supplies. The symptoms usually include diarrhea, increased gas, loss of appetite, abdominal cramps, and bloating. It is not life threatening, but it can slow you down and make life miserable.

BLM and national park rangers constantly warn hikers about its deathly possibilities(which is a little overdone in this author's opinion), but they are required by the Public Health Service to make such statements to stay away from lawsuits. Since many hikers haven't the experience to determine what water is safe to drink as is, and what is not, here are some tips. If you're on a day-hike, simply *carry your own water*. On backpacking trips, take water directly from a spring source or perhaps from a pothole with clear water which is out of reach of cattle. Or take it from a stream near a spring source free from signs of cattle or beaver. Or boil water for one minute, treat with iodine, or filter. On backpacking or mountain climbing trips the author always carries a small bottle of iodine tablets which can be used in an emergency. In all the author's travels to 175 countries, republics, islands and island groups, and to all corners of the Colorado Plateau, he has used iodine tablets or Clorox bleach only 5 times to purify water while climbing or hiking.

Remember, the most common carriers of Giardia are cattle and beaver, so when you see signs of either of these animals, take precautions. There are cattle grazing in many canyons of the Plateau which is normally their winter range. Beaver also range throughout the region, where ever there's a nice year-round flowing stream with willows and trees.

Here's something of interest to Grand Canyon hikers. One NPS ranger on Lake Powell, and another BLM employee familiar with the Paria River and Lee's Ferry, both told the author they have never found Giardia in either Lake Powell or the Colorado River at Lee's Ferry. The water released from Glen Canyon Dam comes from far below the surface which makes the Colorado very cold, clear and for the most part hygienically drinkable, *it would seem(?)*. The main problem then would be water coming in from the Paria River(which is a very small amount) or the Little Colorado River. If either stream is in flood stage, especially the Little Colorado, then it makes the Colorado in the Grand Canyon muddy. The water quality would also be compromised. As a general rule, if the Colorado is clear, the water would be much safer to drink as is than if it's muddy. *It's recommended you always purify this water before drinking, but in an emergency, you could likely drink it as is and not get sick(?).* Nothing is guaranteed however!

Equipment for Day-Hikes

Here's a list of clothes and other items the author usually takes on day-hikes: A small to medium sized day-pack, a one liter bottle of water(2-3 liters in hot summer weather!), camera and lenses, extra film, short piece of nylon rope or parachute cord, toilet paper, pen & small notebook, map, chapstick, compass, pocket knife, a walking stick(made of a ski pole or aluminum shaft with a camera clamp on top which substitutes as a camera stand or a probing or walking stick on the river hikes), or a tripod, a cap with a "sun shield" or "cancer curtain" sewn on around the back, a pair of long pants(for possible deer flies) and usually a lunch.

In warmer weather, he wears shorts and a "T" shirt; in cooler weather, long pants and a long-sleeved shirt, plus perhaps a jacket and gloves. In cooler weather and with more things to carry, a larger day-pack is required.

Equipment for Overnight Hikes

For those with less experience, here's a list of things the author normally takes on overnight hikes. Your list may vary: A large pack, sleeping bag, Thermal Rest sleeping pad, tent--with rain sheet, small kerosene stove and maybe extra fuel, several lighters(no more matches!), 10 meters of nylon cord, camera, lenses & extra film, walking stick with camera clamp on the top end, one large water jug, a one liter water bottle, a stitching awl and waxed thread, small pliers, canister with odds and ends(bandaids, needle & thread, patching kit for sleeping pad, wire, pens, etc.), maps, notebook, reading book, chapstick, compass, toilet paper, pocket knife, rain cover for pack, small alarm clock, candles for light & reading, tooth brush & paste, face lotion, sunscreen, cap with cancer curtain, long pants & long sleeved shirt, soap, small flashlight, a lightweight mini-umbrella and perhaps a coat and light-weight gloves.

Food usually includes such items as oatmeal or cream of wheat cereal, coffee or chocolate drink, powdered milk, sugar, cookies, crackers, candy, oranges or apples, carrots, Ramen instant noodles, soups, canned tuna fish or sardines, Vienna sausages, peanuts, instant puddings, bread, butter, peanut butter, salt & pepper, plastic eating bowl and cup, spoon, and small cooking pot.

Boots or Shoes

Most of the hikes discussed in this book are along canyon bottoms with sand or gravel and boulders. Therefore, there is usually no need to wear heavy duty mountaineering-type boots. Most people, including the author, wear some kind of simple running shoe. These are light weight and comfortable and seldom if ever cause blisters. Most people have a pair of these shoes already in their closet, so there's no need to go out and buy a new pair. If you decide to buy a pair of boots or shoes for hiking canyons, then the best ones are the light weight nylon and leather hiking boots. These are the perfect boots for the Colorado Plateau because they are relatively inexpensive, light weight, comfortable and normally never cause blisters. The only drawbacks are they're not nearly as rugged as all-leather boots, and you have problems walking through cactus fields.

If you're planning to do any hikes along stream bottoms such as the San Rafael River, West Clear Creek, Deer Creek or other watery hikes, you'll need a boot or shoe suitable for wading. The very best would be the canvas & rubber Converse All Star basketball shoe which is unaffected by water. Most people however, would prefer to take an older pair of running shoes which are already near the end of life anyway. For longer hikes involving several days of wading, be sure and take footwear in good condition, or an extra lightweight pair in your pack.

Take a Good Map

The author has done his best to make the maps in this book as good as possible, but no matter how carefully they are drawn, these sketch maps are no substitute for a real good USGS topographic map. Always have a state highway map with you, and buy and use the USGS or BLM maps of the region you're about to visit. The map needed is listed under each hike. *If you get lost using just the maps in this book, the author assumes no responsibility for your predicament!*

Off Road Vehicles(ORV's and ATV's)

In recent years with increased traffic of all kinds on the Plateau, there's also been a dramatic increase in the number of off road vehicles. Since these are called "off road vehicles", naturally the owners want to test drive them "off the road". This indiscriminate use and the destruction of public lands has caused a backlash from other people(tree huggers or environmentalists) who want to protect the land, especially lands that are as special as the Colorado Plateau. This is the primary reason why there's been a move in recent years to protect the more scenic regions by making them into wilderness areas. Just one thought for ORV owners; *it's you who have been using the public lands as test tracks for your fun machines, and who have caused so much land in recent years to be locked up into wilderness areas.* Slowly but surely the Forest Service and BLM are closing off areas formerly destroyed by ORV traffic. *So please, just stay on existing roads and there'll be no problem!*

Mountain Bikes

In the second half of the 1980's, there's also been an increase in the use of mountain bikes. Fortunately, these can seldom if ever be ridden off a road or trail, so they aren't as destructive as ORV's. You simply can't ride them in sandy areas or in most dry creek beds. However, if a lot of people start trying to ride them in areas where they're forbidden, then they too will be in the same category as ORV owners. The Colorado Plateau is criss-crossed with old mining or survey roads which offer plenty of mtn. biking opportunities. *Please keep bikes on established roads or tracks.*

In the late 1980's, the author finally bought a mountain bike which extends the range of his car. He uses it only on 4WD-type roads where his car's undercarriage could be damaged. Incidentally, the author still drives a 1981 VW Rabbit Diesel, which has been equipped with tires 10% larger than normal. Initially the tires would sometimes rub on the wheel wells, so he then had to install gas shocks and cargo coil springs on all four wheels. Now he is able to go on many roads normally frequented by 4WD's and HCV's only, because he's now about 7 or 8 cms(2-3 inches) higher off the ground. His mtn. bike takes over when the car has to be parked.

For those hikers with cars only, a mtn. bike is a cheap way to extend the range of the family car. There are many places in this book where a mtn. bike will save a lot of walking, because you can park your car a few kms from the end of a 4WD road, and use the bike to reach the trailhead. In some places you can lock up a bike to a tree or fence at one end of a canyon and drive to the other end; then do the hike and bike back to the car. This will eliminate backtracking or a lot of road-walking, or the need for two cars. Store the mtn. bike inside your car if you can, rather than on the roof. One bike on the roof will increase fuel consumption by about 20%.

Some Driving Tips for Back Roads

As you leave the paved highways and head out on gravel or dirt roads, here are some things to keep in mind. First, you should stop and lower the air pressure in your tires. The author normally runs his car on the highways with 40 lbs. in his tires. When he leaves the pavement, he lowers it to about 20-25 lbs. This does three things. First, it gives you a smoother and softer ride. Second, it helps prevent sharp stones from puncturing your tires. And third, it gives you better traction, whether it be sand, snow or mud. This means you'll have to carry at all times a tire gauge, and pump of some kind to re-inflate your tires when you get back on pavement. This can be a hand pump, or electric, which is run off the car battery. The electric ones work faster, which means you're more likely to lower the tire pressure in the first place.

If you're driving in sandy areas, lower the pressure in your tires even more, and you can drive to places you couldn't ordinarily, even in a car. This is what drivers of dune buggies do, and they never get stuck. Simply lower your tire pressure to somewhere between 5 and 10 lbs in your drive tires. This puts more rubber to the road, thus increasing traction. Each person will have to experiment a little with this technique in order to gain confidence; *in the meantime, always carry a shovel and a tow rope!*

Also keep in mind when you're driving sandy roads, a little rain helps. It makes the sand more firm and you don't sink in as far. But, if you're in areas with roads made of clay, you'd better get out and onto pavement if a storm is coming. Clay-based roads become very slick when wet. If you get in such a situation and can't move, simply wait an hour or two, and you can normally drive away OK. If heavy rains come, then you may have to wait overnight for the road to firm-up. This is one of the best reasons for always carrying more water and food than you think you'll need!

Hiking in National Parks or Monuments

Hikers in our national parks and monuments have special rules to follow. Each park is a little different, but basically they all have the same general rules. If you're going on a day-hike, you need not have a hiking or camping permit. However, if you're going to be hiking and camping overnight, then you will need a backcountry camping permit. To get this permit, go to the visitor center in the park to be visited. They will give you a free permit and you must carry it with you when hiking. You usually have to camp where the permit states, but that policy varies from park to park. In the case of Grand Canyon, you must write to the Backcountry Reservations Office, Grand Canyon Village, South Rim, Arizona. Write after October 1, for reservations through the next 12 months. The Grand Canyon is obviously the most crowded of all, so write early if you can. In most other parks just go to the visitor center when you arrive and get the camping permit on the spot. For further information, write to or call the national park or monument of your choice and ask for hiking and backpacking information.

Here are some of the general rules for hiking or backcountry camping in national parks. You must camp in designated areas only and use a stove rather than build an open fire. Camp fires are normally banned. Dogs are not allowed while hiking or backpacking. In addition to these rules, there are some common widely accepted practices listed below in *Respect the Land*.

Hiking on Lands of Native Americans

In you're going hiking on Native American lands, be aware of several things. First, keep in mind this is nearly the same as being on private property, although there may or may not be any fences, and maybe no one has a title or deed to the land. Hiking in the Western Grand Canyon on the South Rim you'll likely be on Hualapai Lands. There you must get a camping and use permit at Peach Springs. This cost $7 a day in 1994. It gives you a day and night on their land. When you go down into Havasu Canyon to the village of Supai, you'll pay $9 entrance fee during summer; $7 in winter. Also $12 a night for camping in summer; $8 a night in winter. Pay at the tourist office in the village of Supai in the bottom of the canyon. These dollar amounts may increase by the time you arrive.

Even though most of us don't like the idea of having to pay to go hiking, it could in the long run help get you rescued if something went wrong. And if you did it all legally, maybe you wouldn't have to pay! Also, hiking on lands of Native Americans gives you a chance to see how the other side lives, plus an opportunity to be in some real wild and isolated regions without the tourist traffic one sees on the trails in the heart of the Grand Canyon or other national parks.

For special rules concerning hiking and backcountry camping on the Navajo Nation, see *Introduction to Hiking on Lands of the Navajo Nation*, at the beginning of the Arizona part of this book.

Preserving Archaeology Sites

In some canyons of the Colorado Plateau, there are many ancient cultural sites such as Anasazi or

Fremont cliff dwellings, petroglyph and pictograph panels, and flint chip sites. The author has marked some he found on the hiking maps.

However, if too many people visit these sites, damage will occur; not so much by vandals or pot hunters, but simply by careless visitors. It's highly unlikely you will ever discover any ruins which have not already been plundered, recorded and studied. It's also true that the simple cliff dwellings you'll see will never contribute anything more to our present understanding of the Anasazi people. But regardless of how simple the sites may be, it's important to prevent any further damage to them and make as little impact as possible. Many more interested people will follow in your footsteps, so it's important to leave the sites as unchanged as possible.

Here's a list of things you can do to help preserve Anasazi ruins. First, don't climb onto the walls or any part of the structure. Some may seem solid, but in time, all walls will tumble down. Why not just observe, take pictures, and leave it as is for the next visitor to enjoy. Second, if the ruins are under an overhang approachable via a talus slope, try to get there from the side instead of scrambling straight up the talus. Undercutting ruins is a big problem for some sites and most damage occurs simply by thoughtless individuals. Third, if mother nature calls and you have to use the toilet, please don't do it in or near the ruins. Defecate as far from these sites as possible, and bury it; or in some cases carry it out in plastic bags or a metal box. In recent years, this is becoming a big problem on some Lake Powell and Grand Gulch sites.

And fourth, keep in mind it is against Federal law to damage any ancient artifacts. Part of the Federal law states, *No person may excavate, remove, damage, or otherwise alter or deface any archaeological resource located on public lands or Indian lands unless such activity is pursuant to a permit issued.....* What this is saying is, picking up a potsherd or corn cob and taking it home is illegal. As is putting your initials on a wall next to some petroglyphs. This writer is proud to say that in his collection of rocks and other travel memorabilia, there's not one piece of pottery or corncob. Hopefully everyone who reads and uses this book will follow this example.

Hiking Times

As you read this book and do the hikes, be aware that the author is a full-time hiker-climber-traveler. If he is hiking a canyon or climbing a mountain he has, in addition to enjoying the great outdoors and the scenery, a business-type motive in mind; he is gathering information for this or some other book. So normally he travels faster than the average person. Keep this in mind as you read the description of each hike.

Under each hike in this book there are 10 sub-titles, breaking the page down into specific information categories. Two of these are *Hike Length and Time Needed*, and *Author's Experience*. In some cases the actual length of the hike is stated, however, terrain varies greatly so the number of kilometers(kms) to walk sometimes has little meaning. The time it actually takes is more important. An attempt has been made to calculate the *Time Needed* for the average hiker. Instead of putting the time needed in hours, it's sometimes put in terms of half a day or all-day. A half-day hike will take maybe 4 hours, round-trip; a long half-day hike is about 5-6 hours; an easy day-hike, 6-7 hours; a full day means about 8 hours; and a long all day hike might take 9 or 10 hours or longer.

Under each hike the author's experience is given, stating exactly where he went and the actual time it took. People who have only a few days each year to see and enjoy the Colorado Plateau may want twice as much time as he took.

Respect the Land

Some people are becoming alarmed at the slow destruction of parts of the Colorado Plateau and want to lock it up into wilderness areas. The main reason for this movement is the overuse and abuse by ORV's and 4WD's. Another reason is the amount of trash left behind by a few thoughtless individuals. Around some more heavily used campsites and along some roads one can see the signs of the times; the aluminum soda pop and beer cans. There are very few sites with garbage collection service, so it's up to all of us to pick up our own garbage and in some cases, the trash of our less-concerned neighbors, and dispose of it properly. The author always arrives home with a sack full of aluminum cans. Hopefully, others will do the same.

Although the Colorado Plateau may look rough and mostly barren, it supports a fragile desert ecology that can easily be damaged by careless visitors. It's a great place for outdoor fun, but keep a few things in mind.

1.   Protect fragile soils and vegetation by keeping motor vehicles and mtn. bikes on designated roads and trails.

2.   If you camp some place other than at a campground with toilets, bury all solid human body waste at least 30 meters from water sources.

3.   Carry out what you carry in. While packing up, police the area yourself and pick up trash left by others.

4.   Use existing fire pits and clean them after you're through. If you make a fire, let it burn to ashes rather than burying charred stubs.

5.   Leave prehistoric and historic artifacts as you find them. It's the law.

Metrics Spoken Here

As you can see from reading thus far, the metric system of measurement is used almost exclusively in this book. It's not meant to confuse people, but for some it surely will. Instead, the reason it's used here is that when the day comes for the USA to join the rest of the world and change over to metrics, the author won't have to change his books. The author feels that day is soon coming.

In 1975, the US Congress passed a resolution to begin the process of changing over to the metric system. They did this because the USA, Burma, and Brunei were the only countries on earth still using the antiquated British System. This progressive move ended with the Reagan Administration in 1981.

Use the Metric Conversion Table on page 6 for help in the conversion process. It's easy to learn and use once you get started. Just keep a few things in mind; one mile is just over 1.5 kms, 2 miles is about 3 kms, and 6 miles is equal to 10 kms. Also, 2000 meters is about 6600 feet, 100 meters is about the same as 100 yards. A liter and a quart are roughly the same, and a US gallon jug is 3.78 liters. One pound is 453 grams, and one kilogram is about 2.2 pounds.

Don't Blame Me!

Something happened in the summer of 1993, which has prompted this writer to add another segment to this book. It has to do with the responsibility hikers, as well as guide book writers have, in regards to who is to blame when someone gets lost while hiking or has to be rescued from some canyon. The following is a letter the author sent to several newspapers in Utah some time after the event explaining his point of view, something was wasn't allowed on Salt Lake City's Channel 4 TV station at the time. The letter was dated March 21, 1994.

This one falls into the category of "a day late and a dollar short", but I'd still like to have the last word on something that happened last year which I think the public should know about. It has to do with TV news reporting.

Sometime in early July, 1993, some 4 or 5 people went to the San Rafael Swell to do some hiking along the San Rafael River in what is called the Upper Black Box. This is a deep, narrow canyon that offers both challenges and fine scenery. The hike lasted longer than expected and they had to spend the night in the canyon. Someone became worried and called the Sheriff, who sent a rescue squad out to find them. They were found OK the next day.

*A while later, one of those who had to be rescued called Channel 4 TV in Salt Lake City, and reporter Debbie Dujanavik got on the story which was aired on July 13, 1993. In that feature, two of the lost hikers blamed me and my guidebook, **Hiking and Exploring Utah's San Rafael Swell--2nd Edition**, for all their troubles.*

I was in Africa mountain climbing at the time, and only learned about it just before coming home in October. After I returned, I called around to find out what had happened, then called Dujanavik to see if it was possible for Channel 4 to allow me to air my opinions on the matter.

When TV reporters do a story, they normally present both sides of an argument or issue. On this occasion that wasn't done, because I was halfway around the world. Dujanavik did contact my mother, but she of course knew nothing about the San Rafael Swell.

Later in my interview, I told Dujanavik there was basically only one thing I wanted the public to know and which would qualify as the "other point of view". That was this; those hikers left Provo, Utah, on a Saturday(?) morning and drove two vehicles to the San Rafael Swell. They left one at the bottom end of the canyon, then went back to the trailhead. By the time they started walking it was around noon--that's in the middle of the day!

Now in the 2nd Edition of my guidebook, which they had read, I have clearly stated 3 warnings concerning the length and difficulty of the hike into the Upper Black Box. They are: this is "a long and tiresome hike"; "a very long all day hike"; and this is "for experienced and tough hikers only".

At three different times during the interview, I stated, and in three different ways, that these hikers had ignored, or not read carefully, the warnings in my book and had started hiking in the middle of the day. Because they were embarrassed by having to be rescued, they blamed me and my guidebook for their own stupidity and mistakes. When the story was featured that night(November 3, 1993) on Channel 4, Dujanavik completely left out the only thing I wanted to say!

Dujanavik also did some fancy cutting and splicing of the tape to make it sound like I hadn't even completed the hike--she was obviously trying to discredit me and my guidebook. However, I had completed the hike, but on the second trip, opted to walk the canyon rim instead of swimming through a log jam in the river at the very bottom end of the gorge. I did that to keep my cameras safely out of the water.

Besides wanting to set the record straight, I'd also like to remind readers and TV viewers that reporters have their own biases and opinions and sometimes they try to make them sound like facts; or to dramatize a point for a more impressive story. I feel Dujanavik, who is definitely not an outdoor person, failed totally in her presentation because she knowingly and purposely left out half of the facts. Had she let the public know all the details, and especially the fact that those people had started "a long and tiresome hike" in the middle of the day, it would have totally invalidated her story! She left out half the facts so it would conform to her own biased opinions, which came about because of her failure to properly research the other half of the issue.

For those who might be thinking of hiking the Upper Black Box, start hiking early in the morning and you'll have no trouble.

Incidentally, for those who might have a First Edition of the San Rafael Swell book, and are contemplating a hike through the Upper Black Box, you're advised to at least look at a copy of the Second Edition, or use this book's update on that fine hike before you go.

Readers and hikers should keep a few things in mind. This writer has done his best to collect information and present it to the reader as best and accurate as he can. He has drawn the maps as carefully as possible, and of course encourages everyone to buy the USGS topo maps suggested for each hike in this book. He has tried to inform hikers that canyon bottoms change with every flash flood, and that many of the hikes in this book are in isolated wilderness regions. He has tired to tell hikers that some canyons are for experienced and tough hikers only. Some hikes here are easy, but others are definitely not a *Sunday picnic in the park!* So for those who will somehow get lost, stranded, or have to spend an extra night in some canyon and be rescued, all I can say is, *I've done my best.* The rest is up to you, *so don't blame me for your mistakes.*

Before doing any hike, always stop at a national park visitor center or BLM office in the area, and get the latest information on road, trail, water, flood or weather conditions before venturing out. In some cases, the information in this book can be outdated the minute it goes to press! This is especially true in slot canyons where conditions can change dramatically with every flash flood.

Fotography in Deep, Dark Canyons

If you've talked to anyone who has tried taking pictures in some of the deep, dark slot canyons of the Colorado Plateau, and especially narrow canyons like Buckskin Gulch, Antelope Canyon, Kaibito Creek, Starting Water Wash and Lower White Canyon, they'll likely tell you the fotos just didn't turn out right. Since most of these hikers have tried taking pictures in such places only once or twice in their lives, it's easy to understand how they failed. Here are some tips on how to make fotos better on your first trip into a slot-type canyon

Some of the more common problems are these. Perhaps the most common thing to happen is that the fotos turn out far too dark. This is because there's very little light in the narrow canyon. Usually this means the camera just doesn't have the capabilities to take a good foto in dark places.

Another common problem is the pictures are often fuzzy or blurred. This is caused by a combination of low light, a slow shutter speed, and no tripod; or the lens is just out of focus.

Still another problem is, half of a foto may be very bright or totally washed out, while the other half is dark, sometimes black and shows almost nothing. This is caused by taking a picture with half of the area in bright sunshine; the other half in the shade or shadows. These are the more common problems. Below are some suggestions on how you can correct the situation.

Camera and Lens

The most common type of camera used by hikers today is the 35 mm, with a through-the-lens metering system. Others may use the instamatic type, but these are of the poorest quality, and will produce poor quality fotos.

The recommended equipment to take is a 35 mm camera, with two lenses; one the normal 50 mm which comes with the camera, and a moderately wide angle lens, such as a 24 mm, 28 mm or 35 mm. A lens wider than about 24 mm, will distort the foto more than most people want. If only one lens could be taken into a narrow canyon, the author would likely choose a 35 mm. A serious fotographer will take at least two lenses, perhaps more.

Most of these 35 mm cameras have a shutter speed range from one full second up to 1/1000 of a second. In the darker canyons, you'll likely be using low shutter speeds, from 1/30 to 1/15 of a second, or lower, depending on your film speed and how dark it is.

The author now carries a Pentax K-1000, a totally mechanical camera, with a through-the-lens metering system, and a screw-on self-timer. He would prefer a built-in timer, because he uses it often.

Mechanical cameras, of which there are very few on the market today, are much more sturdy and can withstand more bumps and abuse, than the newer electronic models. Mechanical cameras are also more likely to work after they've been dropped in water than the electronic ones. They are also easier and less expensive to repair. With electronic cameras, you may have a developing minor short in the system, but which the repairman can't find or fix. The mechanical camera is easily diagnosed and repairs made quickly. The mechanical Pentax K-1000 cameras are usually less expensive as well.

The author carries at all times a 50 mm lens, a 28 mm, and a 70-210 mm zoom. The zoom lens is seldom used in the canyons, but was used often when fotographing more open areas in this book. In the past, zoom lenses have been pretty poor quality, but the newer ones are much improved. There are new models out now which range from 28 mm to 70 mm, or thereabouts. This would be a good lens to have, except that it would be rather slow; that is, the F stop would be no better than about 2.8, and likely closer to 3.5, which makes it difficult to take good fotos in dark places. A lens with an F stop of 1.4 or 1.7, allows much more light onto the film, than a 2.8 or 3.5 lens. Try to have a lens of at least a F 2.0, or faster, if you can.

Film Type

Besides using a fast lens(F1.4 or F1.7), you can also compensate for the darkness by using a faster film. Film speeds range from a slow 25, 50, or 64 ASA, up to fast 400 or 1000 ASA. If you're out on the ski slopes on a bright sunny day, you'd want a film with a slow speed, such as 25 or 50 ASA. But if you're in darker places such as Buckskin Gulch, you'll want one of the faster films such as 200, 400 or 1000 ASA. The author now uses 100 ASA color slide film only, and it works moderately well in bright sunny scenes, as well as in the darkest canyons.

Another technique you can use, is to "push" the film. For example, if you use 200 ASA Ektachrome(color slide film), you can set your camera ASA setting on 400 or 800 ASA, which is one or two full stops ahead. This higher setting(say two full stops) will change your shutter speed from 1/15 of a second, to 1/60 of a second. In this example, instead of using a tripod, you can hand hold the camera, and still get good results.

When it comes time to develop the film, you'll have to take it to a lab which specializes in film finishing, and tell them you "pushed" the film either one or two stops. They will then leave it in the developer for a longer period of time, thus compensating for the adjustment you made with the ASA setting on the camera.

You can use the "pushing" technique on Ektachrome 200 or 400 ASA film(or any color slide film which uses the E-6 processing method, such as Agfachrome 200 ASA), or Black + White 400 ASA film. If using the B + W 400 film, you can push it as far as 1600 ASA, or two stops, which would virtually eliminate the need for a tripod. Lab people have also told the author this technique does not work for color print film, such as the new 1000 ASA Kodacolor print film. But that's plenty fast anyway(and seems grainy). Before you try this method of "pushing the ASA", better call a local lab first, to get everything straight.

Tripod or Camera Stand

Carrying a tripod or camera stand down a long canyon hike, is asking a lot from some people, but it will pay off with better results. Actually, if you have a fairly fast lens(F1.4 or F1.7), a fast film(200, 400 or 1000 ASA), and compensate by pushing the film(one roll of film in the darkest narrows only) one or two full stops, you can easily get by without the tripod. But here's an alternative to a bulky tripod.

Make a walking stick out of an aluminum rod of some kind. Cut off one end, and have someone weld it to form a "T". Or better still, modify an old ski pole which is ready-made. You can use it to probe for possible deep holes in a stream, and can stick it in the mud or sand and with the aid of a small camera clamp on top, use it as a tripod. This is what the author often uses, since he is alone more than 99% of the time. One thing to remember though, when you're using it in a very dark part of the narrows, with the shutter speed at perhaps 1/15 of a second or lower, be sure to lean it against the canyon wall. Otherwise, you'll get a swaying motion in the single leg, which will make your foto blurred.

If your camera indicates the shutter speed is below 1/60 of a second, use a stand of some kind--either a rock, tripod, or walking stick. Sometimes you can lean against a wall and get moderately good results by hand-holding it at 1/30 of a second. Also, don't forget to set the focus carefully. The lower the F-stop, the smaller the field of focus will be. Over the years the author has learned exactly where to place the needle in the viewfinder to get the correct amount of light, so the only thing left is to take several shots at different focus settings. This is the best move he has made in a long time. Now at least some of his slides are well focused as well as having the correct amount of light.

More Tips

When in a place like Antelope Canyon, never take a foto where there's a streak of sunlight in your subject area. If you do, part of the picture will be washed out with too much light; the other part will be dark or totally black. Instead, take a picture where the sunlight is being bounced off the upper walls and diffused down into the slot bottom below. Another way would be to wait for a cloud to cover the sun, then the light is diffused, thus eliminating the harsh contrasts between sunny places and shadows. Maybe the best time to take fotos in the slot canyons is in mid-morning or mid-afternoon. This way you can easily find places where the sun isn't shining down into the narrows, but instead is shining on part of an upper wall. The light then is being bounced down into the dark corners.

If you should slip and fall into water, or somehow drop your camera in a stream, here are the steps to take. Immediately take out the camera battery. Quickly roll the film back into the cassette, and remove it. Open the camera and shake or blow out any water. Allow it to sit in the sun to dry, turning it occasionally to help evaporate any water inside. If you're near your car, start the engine, turn on the heater, and hang the camera in front of a heater vent. The warmer the camera gets, the better. This helps the water evaporate more quickly. The quicker the water evaporates, the less corrosion there will be on the electrical system; and the less rust there will be on metal parts.

If your camera is under water for just a second or less, like when you slip down and get right back up, there will likely not be any water inside. In this case, by following the above steps, you will likely be fotographing again in half an hour, especially if the water is clear, no sand has gotten into the camera, and if the sun is warm. The author has had several of these little accidents, and with each of his cameras. The last several times, no repair work was needed because he did the right things to get the camera dry fast.

Important BLM Offices

Before doing any hike, always stop at one of these BLM offices and get the latest information on road, trail, water, flood or weather conditions before venturing out. In some cases, the information in this book can be outdated the minute it goes to press! This is especially true in slot canyons where conditions can change dramatically with every flash flood.

Utah
Cedar City, 176 East, D. L. Sargent Drive, 84720, Tele. 1-801-586-2401.
St. George, 225 North, Bluff Street, 84770, Tele. 1-801-673-4654.
Escalante, Highway 12, West side of town, 84726, Tele. 1-801-826-4291.
Kanab, 320 North, First East, P.O. Box 459, 84741, Tele. 1-801-644-2672.
Richfield, 150 East, 900 North, P.O. Box 768, Tele. 1-801-896-8221.
Hanksville, P.O. Box 99, 84734, Tele. 1-801-542-3461.
Moab, Sand Flats Road, P.O. Box M, 84532, Tele. 1-801-259-8193.
Moab, 82 East, Dogwood, P.O. Box 970, 84532, Tele. 1-801-6111.
Price, 900 North, 7th East, 84501, Tele. 1-801-637-4584.
Monticello, 435 North, Main Street, 84535, 1-801-587-2141.
Vernal, 170 South, 500 East, P.O. Box F, 84078, 1-801-789-1362.

Colorado
Grand Junction, 2815 H Road, 81506, Tele. 1-303-244-3000.
Montrose, 2505 South, Townsend, 81401, Tele. 1-303-249-6047.
Durango, Federal Building, 701 Camino del Rio, 81301, Tele. 1-303-247-1082.
Doloras, 275021 Highway 184, 81323, Tele. 1-303-882-1825.

New Mexico
Grants, 620 East, Santa Fe Avenue, 87020, Tele. 1-505-285-5406.

Arizona
St. George, Utah (Arizona Strip District), 390 North, 3050 East, 84770, Tele. 1-801-673-3545.
Kingman, 2475 Beverly Avenue, 86401, Tele. 1-602-757-3161.

The best known feature of Great West Canyon is this narrow spot called The Subway.

The narrows of Butterfly Canyon, located southeast of Page, Arizona.

The old Hanson sheep bridge across Swazys Leap in the lower Black Box of
Utah's San Rafael Swell.

Index Map of Hikes--Colorado Plateau

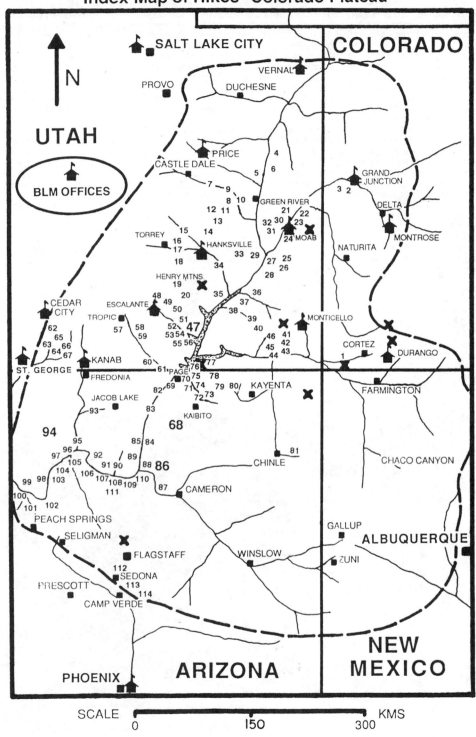

Sand Canyon and Rock Creek, Colorado

Location and Access Sand and Rock Creek Canyons are located in extreme southwestern Colorado. These are both short drainages which flow south into McElmo Creek. To get there, drive south out of Cortez on Highway 160-666. After about 3 kms, and between mile posts 35 & 36, turn west onto McElmo Road. Head in the direction of the dry ice plant, a small cemetery and the Battle Rock School as shown on the map. The small cemetery near the mouth of Sand Canyon is about 22 kms from Cortez. The McElmo Road is paved to Battle Rock, then it's gravel. About one km past the cemetery is a slickrock parking area on the right or north side of the road. This is immediately south of Castle Rock. Park there. You should see some kind of trail register 20-30 meters from the highway. In the fall of 1993, archeologists were working on ruins at Castle Rock, so the parking situation may change when you arrive. This is now the trailhead for both Rock Creek and Sand Canyons. The roads into both Sand and Rock Creek Canyons are presently closed to public access.

Trail or Route Conditions From the car-park, walk north along the west side of Castle Rock. There are signs and stone cairns marking the hiker's trail. About 150 meters north of Castle Rock the trail splits. Turn right to enter Sand Canyon. You'll first go east, then north. Along the way will be a number of short canyons on the left. Some of these have Anasazi ruins in small south-facing Entrada Sandstone alcoves. You can walk all the way up-canyon and circle around to the east side of the drainage while staying on the same bench. You could go all the way around to Graveyard Canyon, which apparently has more ruins(?). To get into Rock Creek Canyon, walk north from Castle Rock but turn left at the trail junction. There are some ruins in one minor side canyon to the right, then more ruins up at the head of the Entrada Sandstone part of Rock Creek.

Elevations Cemetery, 1675 meters; most ruins in both canyons, about 1775 meters.

Hike Length and Time Needed From the highway to the most-northerly ruins in Sand Canyon is only about 5 kms. To visit all of them will take more than half a day for most people. The distance isn't great, but the ruins are interesting. The east fork of Rock Creek has fewer ruins, but the time needed will be about the same. Of the two, Sand Canyon is more interesting.

Water There's no water in either canyon, so carry it with you and have plenty in your car.

Maps USGS or BLM map Cortez(1:100,000), or Battle Rock(1:24,000).

Main Attractions Anasazi ruins. The author counted 10 in Sand Canyon, 5 in the east fork of Rock Creek. There are also several around Castle Rock. Most sites are in Entrada Sandstone alcoves, but some are out in the open. You'll be able to see more ruins to the north as you drive along McElmo Road, but they're on private property. Take binoculars.

Ideal Time to Hike Spring or fall; or maybe some winter warm spells. Summers are pretty warm.

Hiking Boots Any dry weather boots or shoes.

Author's Experience The author hiked into both of these canyon on one hot August day; about 5 hours in Sand Canyon, and 3 in Rock Creek. In late November, 1993, he made a quick trip into part of Sand Canyon in 2 1/2 hours round-trip.

Anasazi ruins. This one is in the east-side drainage of Rock Creek Canyon.

Map 1, Sand Canyon and Rock Creek, Colorado

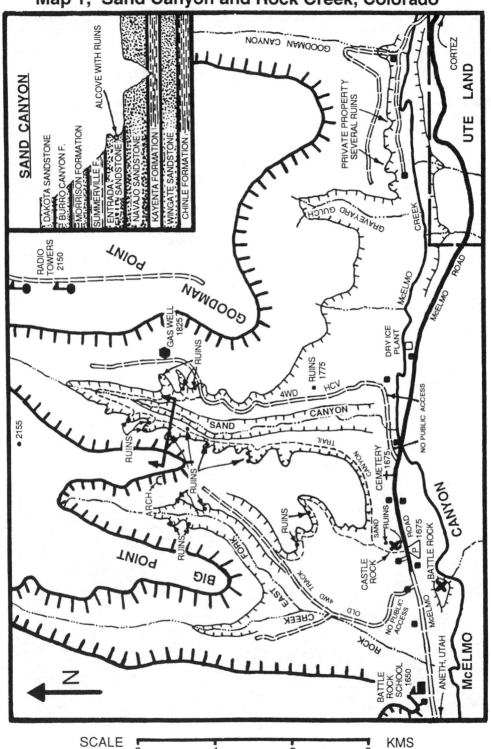

SAND CANYON

ALCOVE WITH RUINS

DAKOTA SANDSTONE
BURRO CANYON F.
MORRISON FORMATION
SUMMERVILLE F.
ENTRADA SANDSTONE
NAVAJO SANDSTONE
KAYENTA FORMATION
WINGATE SANDSTONE
CHINLE FORMATION

GOODMAN CANYON

CORTEZ

UTE LAND

PRIVATE PROPERTY SEVERAL RUINS

GRAVEYARD GULCH

CREEK

McELMO

McELMO ROAD

RADIO TOWERS 2150

GOODMAN POINT

GAS WELL 1825

RUINS

RUINS 1775

DRY ICE PLANT

NO PUBLIC ACCESS

4WD

HCV

SAND

CANYON

• 2155

RUINS

TRAIL

CEMETERY 1675

CANYON

NO PUBLIC ACCESS

ARCH

RUINS

RUINS

SAND

RUINS

ROAD 1675

BIG POINT

RUINS

FORK

RUINS

CASTLE ROCK

CANYON

EAST

4WD TRACK

P

BATTLE ROCK

CREEK

ROCK

OLD

McELMO

NO PUBLIC ACCESS

ANETH, UTAH

McELMO

N

BATTLE ROCK SCHOOL 1650

SCALE   0   1   2   3   KMS

19

Ute and Monument Canyons, Colorado

Location and Access Ute and Monument Canyons are located in western Colorado and in Colorado National Monument, which lies immediately west of Grand Junction. Access to Grand Junction or Fruita is easy as I-70 runs right through both towns. In the center of Grand Junction look for the beginning of State Road 340. It runs west from the old downtown area, crosses a bridge on the Colorado River, then continues west toward Glade Park. About 11 kms from Grand Junction, a good paved road turns right or north, and runs into and through the national monument. It connects Grand Junction and the town of Fruita, located immediately northwest of this park. Drive from either of these towns and park at the Ute Canyon, Liberty Cap, or Monument Canyon Trailheads. The road in Colorado N. M. is open year-round.

Trail or Route Conditions For Ute Canyon, let's start at Liberty Cap Trailhead. The beginning of this trail is actually an old road until it drops off the Entrada Sandstone near Liberty Cap Rock. Motor vehicles are no longer allowed to use it. There's a good trail beyond the end of the old road, which drops off the Wingate Cliffs further along. This trail continues on to the suburb of Redlands, but you'll want to make a right turn at the bottom of the cliffs and walk up-canyon as shown. There's no trail in the bottom of Ute Canyon but walking is easy. Near the upper end of the canyon the Ute Canyon Trail zig zags up through a rock slide covering the Wingate Sandstone cliff to the Kayenta Bench, where the park road is located. From there you'll have to road-walk or use a mtn. bike to get back to your car. Now Monument Canyon. Just after you begin walking from the trailhead, one part of the trail goes to the *Coke Ovens;* but the main trail descends into Monument Canyon, then contours around the base of the Wingate Cliffs on top of the Chinle Formation. It soon heads in a northwest direction, then turns northeast and runs along the bottom of the canyon. It ends(or begins) at a lower trailhead near the main Fruita-Redlands Highway.

Elevations Liberty Cap Trailhead, 1975 meters; Ute Canyon Trailhead, 1981; bottom end of Ute Canyon, 1700; Monument Canyon Trailhead, 1859; park visitor center, 1764 meters.

Hike Length and Time Needed From Liberty Cap Trailhead to Liberty Cap Rock is about 9 kms. From the Ute Canyon Trailhead to Liberty Cap via the canyon, is about the same. This makes the loop-hike about 23 kms, which includes the distance along the highway back to your car(use a mtn. bike here). Most people can do this loop in one easy day-hike. The length of the Monument Canyon Trail is about 7.5 kms, one way. If one were to begin at either end, walk to the other trailhead and return, it would be a short day-hike for most people.

Water Take your own water. Both canyons are dry, but with occasional seeps in wetter times.

Maps USGS or BLM map Grand Junction(1:100,000), or Colorado National Monument(1:24,000).

Main Attractions Year-round access, scenic views into red rock canyons & Grand Junction, and contrasting red rocks above the black Precambrian granites in lower Monument Canyon.

Ideal Time to Hike Spring, fall or perhaps winter warm spells, or morning-hike in summer.

Hiking Boots Any dry weather boots or shoes.

Author's Experience He walked the Liberty Cap--Ute Canyon loop in about 4 1/2 hours. He also walked about halfway down Monument Canyon and back in less than 2 hours.

Independence Rock, as seen from a highway viewpoint.

Map 2, Ute and Monument Canyons, Colorado

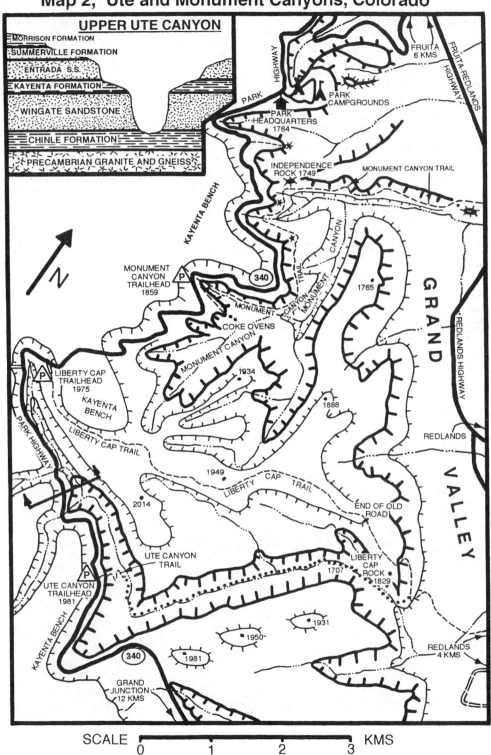

UPPER UTE CANYON

MORRISON FORMATION
SUMMERVILLE FORMATION
ENTRADA S.S.
KAYENTA FORMATION
WINGATE SANDSTONE
CHINLE FORMATION
PRECAMBRIAN GRANITE AND GNEISS

FRUITA 6 KMS

FRUITA-REDLANDS HIGHWAY

PARK HIGHWAY

PARK CAMPGROUNDS

PARK HEADQUARTERS 1764

INDEPENDENCE ROCK 1749

MONUMENT CANYON TRAIL

KAYENTA BENCH

N

MONUMENT CANYON TRAILHEAD 1859

P

340

MONUMENT CANYON TRAIL

MONUMENT CANYON

1765

GRAND

COKE OVENS

MONUMENT

MONUMENT CANYON

1934

1888

LIBERTY CAP TRAILHEAD 1975

P

KAYENTA BENCH

REDLANDS HIGHWAY

LIBERTY CAP TRAIL

1949

REDLANDS

PARK HIGHWAY

2014

LIBERTY CAP TRAIL

END OF OLD ROAD

VALLEY

UTE CANYON TRAIL

LIBERTY CAP ROCK 1829

1707

UTE CANYON TRAILHEAD 1981

P

1931

KAYENTA BENCH

1950

REDLANDS 4 KMS

340

1981

GRAND JUNCTION 12 KMS

SCALE   0   1   2   3   KMS

Knowles and Mee Canyons, Colorado

Location and Access These two canyons are located about 25 kms due west of Grand Junction. Drive southwest out of Grand Junction on State Road 340 in the direction of Colorado National Monument and Glade Park. In the center of Glade Park is an intersection with a small store & gas station. From the store drive west 10 kms to 11 1/2 Road, then go north 2 kms to BS Road and turn left, or west. Drive past one cattle guard, a stock tank, 3 Quonset huts, another stock tank, and park at the bottom of Twentyeight Hole Wash. With a 4WD one could get closer, but this route is generally good for any car. The BS Road begins about 1 km north of Glade Park Store, which is an alternate route to the car-park. Total driving distance from Grand Junction to the trailhead is about 40 kms, most of which is on paved roads.

Trail or Route Conditions As you enter and exit these canyons you'll walk on old 4WD or HCV roads. Follow the arrows on the map from the car-park at 1860 meters. This description makes a loop down Knowles and up Mee. From the car-park, walk north up a shallow drainage to the road, as shown. Just after the road drops off the Entrada Sandstone in Knowles, you must walk around an Entrada bluff and locate a route down through the Wingate Sandstone into the canyon bottom. Look for greenery and a spring. Also, look for a lone standing rock and a deer trail down. In the canyon bottom simply follow the drainage, which is generally very easy walking. After Knowles and as you walk along the Colorado River, you'll find cliffs at one point, but there's a route through, perhaps 60 meters above the river. To exit in upper Mee Canyon, start about 250 meters below the big undercut cave, and look for a route up through the Wingate to the Kayenta Bench on the east. Then contour south 2 to 3 kms before exiting up through the Entrada. The geology cross section shows the area of the big cave or undercut. Once on top of the mesa, find the road and walk west back to your car. Before starting this hike make sure you take a compass and one of the maps listed below.

Elevations The car-park about 1860 meters; the Colorado River about 1350 meters.

Hike Length and Time Needed To walk down Knowles, along the Colorado River and up Mee Canyon, plan to take three days. The loop-hike is about 48-50 kms. Some really strong and fast hikers can do it in 2 long days.

Water The author was there about 4 days after good rains and found a fair amount of water. Mee Canyon has more seeps and springs, while the lower end of Knowles is nearly dry.

Maps USGS or BLM map Grand Junction(1:100,000), or Ruby Canyon and Sieber Canyon(1:24,000).

Main Attractions Both canyons are included in the Black Ridge Canyons WSA, so solitude and scenery are great. Also, at the head of Mee Canyon are several huge, deep undercut caves.

Ideal Time to Hike Spring or fall, but it can be done in summer, or warm spells in winter.

Hiking Boots Any dry weather boots or shoes.

Author's Experience The author started one July afternoon, then walked one full day and another half day to finish it off. Total walk-time was about 17 1/2 hours.

This is the huge undercut in Upper Mee Canyon, as seen from the slopes above.

Map 3, Knowles and Mee Canyons, Colorado

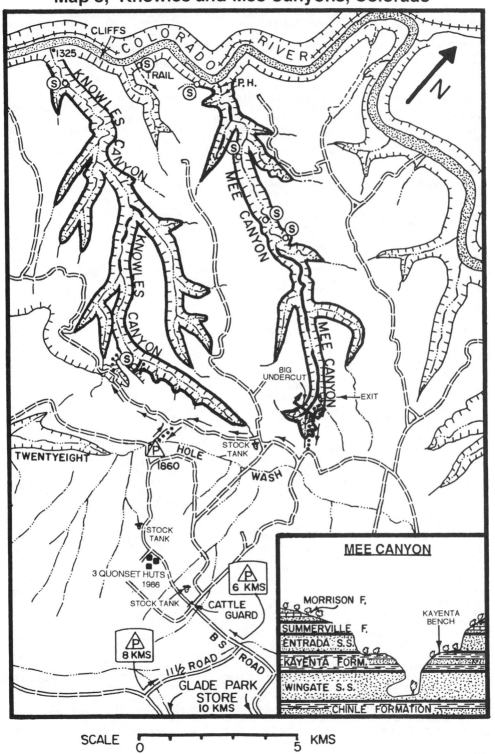

CLIFFS

COLORADO RIVER

●1325

Ⓢ

Ⓢ

TRAIL

Ⓢ

P.H.

Ⓢ

KNOWLES CANYON

Ⓢ

MEE CANYON

N

KNOWLES CANYON

Ⓢ

MEE CANYON

BIG UNDERCUT

←EXIT

Ⓢ

STOCK TANK

TWENTYEIGHT

P HOLE

1860

WASH

STOCK TANK

3 QUONSET HUTS 1986

STOCK TANK

P 6 KMS

CATTLE GUARD

P 8 KMS

BS ROAD

11½ ROAD

GLADE PARK STORE 10 KMS

MEE CANYON

MORRISON F.

KAYENTA BENCH

SUMMERVILLE F.

ENTRADA S.S.

KAYENTA FORM.

WINGATE S.S.

CHINLE FORMATION

SCALE

0   5   KMS

23

Desolation Canyon, Utah

Location and Access Desolation Canyon of the Green River begins, for all practical purposes, at the mouth of Sand Wash in the Uinta Basin. This dry creek is found south of Ouray and is the first canyon north of the mouth of Nine Mile Creek or Canyon. From Sand Wash, the Green River then flows almost due south to Green River, Utah. It's just above this town on Interstate Highway 70 that the river flows out of the canyon and into flat land for a few kms. Most people refer to this entire gorge as Desolation Canyon, but in reality, the region below Three Fords Canyon and Km 56(Mile 35), is actually Gray Canyon. The rocks in that part are gray in color, thus the name.

Most rafters or boaters floating through the canyon begin at Sand Wash. To get there, drive to a point about 3 kms west of Myton near mile post 105 on Highway 40. There a paved road runs south to Pleasant Valley. After three kms, turn left at a major junction and drive past some oil wells, old gilsonite mines and finally to Sand Wash(If you were to turn right at the major junction, you would end up in Nine Mile Canyon and Wellington). Just follow the signs to "Sand Wash" on first a paved, then a good dirt and gravel road. The distance from Highway 40 is 54 kms(34 miles). At Sand Wash is a seasonal ranger station, a boat launch ramp, and the remains of an old abandoned ferry boat site. There are several historic old cabins in the area, and you should plan to spend a little time there. Sand Wash is located at Km 153.3 or Mile 95.8. The railroad bridge at Green River is Km and Mile 0.

On the map you'll notice letters and numbers like D40m, or D20m. This means that you will have to walk up on a bench or Dugway about 40 or 20 meters above the river.

Elevations Sand Wash, 1407 meters altitude; Green River town is 1240 meters.

Maps Don't attempt this long and tiring hike without good maps. The author recommends the USGS or BLM metric maps Seep Ridge, Huntington, and Price(1:100,000). Or Nutters Hole, Flat Canyon, Firewater Canyon, Range Creek and Gunnison Butte(1:62,500). The Utah Travel Council map Northeastern Utah will help you get to and down the river, and the Southeastern Utah map shows the lower part of the canyon. Both are 1:250,000 scale.

Water For the entire hike you'll have the Green River at your left hand side(this hike follows the west bank), so there's no shortage of water on this trip. But you should plan to take a water filter, or some kind of water purification tablets, or Clorox bleach, as did the author. To make sure the river water was drinkable, he used 15 drops of Clorox for each liter of water. The best solution however, is to use Iodine tablets. These have become easily available in recent years. Also, there are several small streams and a few springs where you can get water good enough to drink as is.

Nine Mile Creek is a perennial stream, but there are a number of ranches upstream. Purify this water! At K120(M75) there are two springs up a small box canyon. Flat Creek is another perennial stream, but in dry years, you may have to walk up-canyon a ways to find water. Steer Ridge Canyon has springs, but they are a long way up the canyon. Rock Creek is a perennial trout stream and just marvelous. Chandler Creek is marvelous too, but it's on the east side of the river. At Trail Canyon there's a spring. The next water is at Range Creek with a year-round flow. The author drank it without purification, but there are often cattle up-canyon. Further down, Coal and Rattlesnake Creeks usually have year-round flows, but they are on the other side, as is the undrinkable water from Poverty Creek. The last water entering this canyon is the Price River, but forget it! Green River water is better. About 16 kms from Green River town is the Ekker Ranch feed lot. From this point south and to the town, you'll be able to get water from ranch houses along the road.

Main Attractions This canyon is one of the more remote regions in the continental USA. On the east side of the river is the Hill Creek extension of the Uintah & Ouray Indian Reservation(down to Coal Creek), and on the west is the West Tavaputs Plateau, the Roan or Brown Cliffs and the Beckwith Plateau. The river cuts through the Book Cliffs near the bottom end of the canyon. There are several proposed WSA's along both sides too. Wilderness solitude is a big attraction, but in recent years about 7000 rafters float the canyon each year. Boating season runs from about May 1 through the end of September, but there's another busy time each year during the deer hunt. This begins the Saturday closest to October 20, and usually runs for 11 days. During the three months of summer, you'll have competition for some select campsites. However, the number of rafters going down is regulated by the BLM River Rangers and the numbers should remain constant. As yet, hikers are not included in some of the rules and regulations governing passage through this canyon. To be aware of the problems you might write to the BLM office in Price, Utah, or contact the river ranger stationed at Sand Wash(his home base is the Price office). All the rules have to do with preserving the wilderness character of the canyon.

So solitude is an important attraction, as are the dozens(if not hundreds) of petroglyphs. Most of these are very well preserved. The author found panels at the mouth of Rock House Canyon, and the first canyon immediate to the east. About 2 kms north of Lighthouse Rock is a single big horn sheep on one rock, and just south of the mouth of Three Canyon is another panel with etchings of early white men and Indians. There are two panels on boulders just north of the mouth of Price River. The best panel along the river is at the mouth of Flat Creek Canyon. This one is nearly as good as Newspaper Rock(found west of The Needles--Canyonlands National Park), but it has a different style.

There are three rincons or abandoned meanders along the way. The first is the Tabyago Rincon, at K139(M87); next is the Gold Hole Rincon, near K130(M81); and Three Canyon Rincon, at K78(M49). Another attraction is the ruins of the Rock Creek Ranch. This is an outdoor museum, so kindly leave everything in place. On the east side of the river is the McPherson Ranch. It too is abandoned and interesting. Near the bottom of the canyon is the Swazy(Swasey) Cabin.

Hiking Boots There's no wading here, so wear some tough dry weather hiking boots in good condition. This hike is 153 kms or nearly 96 miles long, so a good pair of boots is very important.

Ideal Time to Hike Early fall is best, but it could be done in summer after the high water has passed. Don't

Map 4, Desolation Canyon--North, Utah

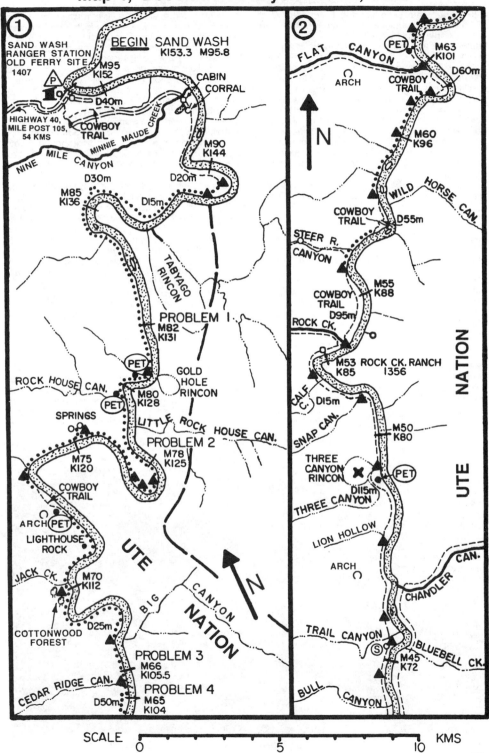

① SAND WASH RANGER STATION OLD FERRY SITE 1407

BEGIN SAND WASH KI53.3 M95.8

M95 KI52

CABIN CORRAL

D40m

HIGHWAY 40, MILE POST 105, 54 KMS

COWBOY TRAIL

MINNIE MAUDE CREEK

NINE MILE CANYON

M90 KI44

D30m

D20m

M85 KI36

DI5m

TABYAGO RINCON

PROBLEM 1

M82 KI31

PET

GOLD HOLE RINCON

ROCK HOUSE CAN.

M80 KI28

PET

SPRINGS

LITTLE ROCK HOUSE CAN.

PROBLEM 2

M75 KI20

M78 KI25

COWBOY TRAIL

ARCH PET

LIGHTHOUSE ROCK

UTE NATION

JACK CK.

M70 KII2

BIG CANYON NATION

COTTONWOOD FOREST

D25m

PROBLEM 3

M66 KI05.5

CEDAR RIDGE CAN.

PROBLEM 4

D50m

M65 KI04

② FLAT CANYON

PET

M63 KI0I

ARCH

D60m

COWBOY TRAIL

N

M60 K96

WILD HORSE CAN.

COWBOY TRAIL

D55m

STEER R. CANYON

M55 K88

COWBOY TRAIL

D95m

ROCK CK.

M53 ROCK CK. RANCH 1356

K85

CALF C.

DI5m

SNAP CAN.

M50 K80

THREE CANYON RINCON

PET

DI5m

THREE CANYON

LION HOLLOW

ARCH

CHANDLER CAN.

TRAIL CANYON

S

BLUEBELL CK.

M45 K72

BULL CANYON

NATION

UTE

SCALE

0   5   10   KMS

make this hike during the late spring-early summer runoff(generally from mid-May to mid-June), as there are 4 possible problem areas for hikers. October is a great time with fall colors and very few rafters.

Hike Length and Time Needed From Sand Wash to the railway bridge at Green River town is 153.3 kms or 95.8 miles. With luck you can get a ride for the last 16 kms, from the Ekker feed lot to Green River. Many people would want 8 to 9 days, but it can be done in as little as 5 days. This faster hike would mean hurrying though, to the discouragement of some. If you're not a really strong hiker, consider asking a rafter to drop off a sack of food halfway down river. This would make your pack a bit lighter and more bearable.

Author's Experience The author was driven to Sand Wash, had a look around, and began hiking in the late afternoon between the two weekends of the deer hunt in late October, 1985. The trip took 6 days, but the first and last days were both short. The four days in between were about 9 1/2 hours each. The total walk-time was 47 1/2 hours. He managed to hitch a ride for the last 7 kms. This works out to about five, 9 1/2 hour days. A long hike, but a great trip. He hitch hiked back home to Provo that same afternoon.

Trail or Route Conditions There's an old road along the river down from Sand Wash, but don't use it, because you'll be cliffed-out after about one km. Instead, walk back up the road to the west 300 meters from the river, cross over the dike, and look for a cowboy trail running upon the bench. This trail(and road) gets you to the mouth of Nine Mile where more man-made trails continue for a ways, then hardly a trail at all until Flat Canyon and K101(M63). It's this part of the canyon that may present problems for the hiker on the west bank.

Just above K136(M85), you'll have to go up a small draw to avoid a cliff, but this is no problem. The first real problem is at K131(M82). There's a cliff there but you can walk under it OK when the river is low. You could also backtrack a bit, and go above the cliffs. Or you could also use an air mattress to float your pack around the cliff in calm waters during spring runoff. Problem 2 is at K125(M78). During low water you can walk under these cliffs. But to find a route above this place looks discouraging. It would be a long detour. If you were there at high water, you could float your pack on an air mattress and swim yourself. Problem 3 at K105.5(M66) is about the same as Problem 2, in that you can walk between a large boulder and the cliff at low water time, but not with higher water. To detour around this cliff looks like a very long and difficult hike, if not practically impossible. Problem 4 at K104(M65) is different. Here you cannot walk around this cliff in either high or low water. Instead, you must walk up the sandy bank to a cat-walk or ledge, then climb 3 to 4 vertical meters at a place where loose shale-type rock is separating from the main wall. The author did this first without his pack; then after the scouting was finished, he managed to take his pack up. It's best not to do this part alone(as was the author's case) unless you've developed a bit of confidence on rock. It's recommended you take along a short piece of rope, maybe 10 meters of parachute cord, to get your pack up and over this pitch. After contouring about 100 meters, drop back down to the river and breathe easy, as this is the last difficult place for hikers.

Below Flat Canyon is an old cattle or cowboy trail for most of the way. In several places you'll have to look for stone cairns marking the route up and over some cliffs. Across the river from the mouth of Wild Horse Canyon, you'll walk on a cairned trail, which disappears in a slide. From that point, don't go up, but remain low until you again find the trail. Just after Wire Fence Canyon, the trail heads west and up onto bench land. You should first follow the trail, but drop back down to the river at Three Fords Canyon West. At about K53(M33), you should go up and contour south for half a km, then you'll find another trail taking you to Range Creek. Take the trail from the mouth of Range Creek up on the Bluecastle Bench, but further on look for a cairned route and trail back down to the river just above Rabbit Valley. From there on down there's a trail, though it fades at times. From the mouth of Price River onward, there's an old and unused road to Gunnison Butte, then it's a road-walk or hitching back into town.

This is part of the petroglyph panel at the mouth of Flat Creek Canyon.

Map 5, Desolation Canyon--South, Utah

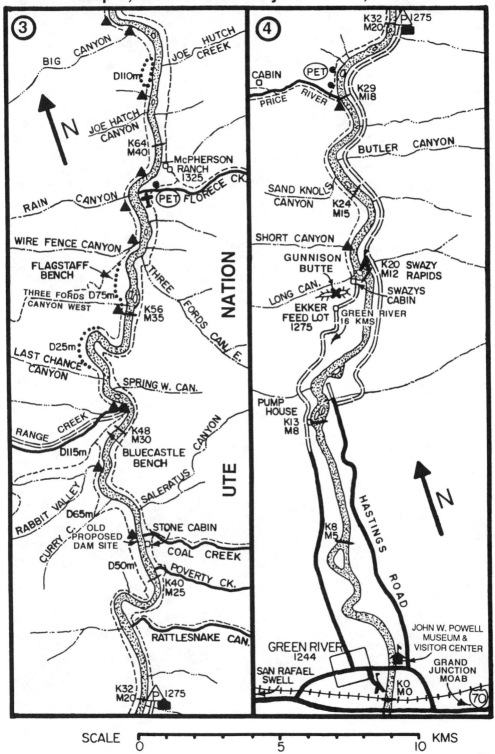

③ BIG CANYON
JOE HUTCH CREEK
D110m
JOE HATCH CANYON
K64 M40
McPHERSON RANCH 1325
RAIN CANYON
PET FLORECE CK.
WIRE FENCE CANYON
FLAGSTAFF BENCH
THREE FORDS CANYON WEST D75m
K56 M35
THREE FORDS CANYON E.
D25m
LAST CHANCE CANYON
SPRING W. CAN.
RANGE CREEK
K48 M30
D115m
BLUECASTLE BENCH
SALERATUS CANYON
RABBIT VALLEY
D65m
CURRY C.
OLD PROPOSED DAM SITE
STONE CABIN
COAL CREEK
D50m
POVERTY CK.
K40 M25
RATTLESNAKE CAN.
K32 M20 1275

④ K32 M20 1275
CABIN
PET
K29 M18
PRICE RIVER
BUTLER CANYON
SAND KNOLLS CANYON
K24 M15
SHORT CANYON
GUNNISON BUTTE
LONG CAN.
K20 M12 SWAZY RAPIDS
SWAZYS CABIN
EKKER FEED LOT 1275
GREEN RIVER 16 KMS
PUMP HOUSE
K13 M8
HASTINGS ROAD
K8 M5
JOHN W. POWELL MUSEUM & VISITOR CENTER
GREEN RIVER 1244
SAN RAFAEL SWELL
GRAND JUNCTION MOAB
K0 M0
70

NATION
UTE

N

SCALE   0   5   10   KMS

The best grove of cottonwood trees on the entire hike is at the mouth of Jack Canyon.

The author rests after finally climbing above the cliff at Problem 4 in Desolation Canyon.

Typical fall scene in the middle portion of Desolation Canyon

The workshop of the old and now abandoned Rock Creek Ranch is a museum piece.

Florence and Coal Creek Canyons, Utah

Location and Access These two canyons flow from the east into the Green River and Desolation Canyon about 30 to 40 kms north of Green River, Utah. To get to this region, drive east out of Green River, cross the river bridge, and turn north on the paved Hastings Road. This is the one and only road running north along the east side of the Green River and into lower Desolation(actually Gray Canyon). The road ends at Nefertiti Rapids, 32 kms north of town(at mile 20 on the USGS river survey charts). One thing to keep in mind while hiking in this area is that all the area north of Coal Creek and east of the Green River is Ute Tribal Land.

Trail or Route Conditions From the trailhead, an old wagon road and cowboy trail begins and runs north along the east side of the Green River. It ends at the abandoned McPherson Ranch, where another old abandoned road from the north ends. Hikers should stay on this good trail which is still used by cowboys and cattle. When you arrive at Florence Creek, be sure to check out the McPherson Ranch before heading east. Next to the older McPherson Ranch buildings is a newer structure which was built by the Ute Tribe. It was built as a dude ranch, but when floods wiped out the road in Chandler Canyon, both were abandoned. From the ranch, there's an old trail running up Florence Creek a ways, but it slowly fades into deer and elk trails. In places in upper Florence, you'll walk through 3 meter-high sagebrush. The high part between Florence and Coal has several dirt roads, then a trail down along some springtime waterfalls into Coal Creek Canyon. Along upper Coal Creek you can use elk and/or cattle trails or you can just walk in the mostly dry creek bed. Take and use long pants in the upper portions of both these canyons, as there are some thorny wild rose bushes.

Elevations The trailhead is just under 1300 meters; McPherson Ranch about 1325; and the divide between Florence and Coal Creek Canyons about 2800 meters.

Hike Length and Time Needed The loop-hike up Desolation Canyon and Florence Creek, then down Coal Creek, and back to the trailhead is about 100 kms. This will take most people from 4 to 6 days which covers some of the wildest and most remote canyons in the Lower 48.

Water Green River water is good after treatment. Florence Creek water is crystal clear and normally should be good to drink as is. The upper end of Coal Creek normally has year-round flowing water, as does some of the lower parts. In dry years, there may be less water than the author has shown on this map. He made his hike in mid-May, with deep and melting snows on the pass.

Maps USGS or BLM maps Huntington and Westwater(1:100,000), or Range Creek, Moonwater Point and Gunnison Butte(1:62,500).

Main Attractions Wilderness semi-alpine and desert scenes, historic ranch relics, one Fremont ruin in upper Florence Creek Canyon, and a good chance to see elk and bear.

Ideal Time to Hike For the loop-hike, late May through October, but mid-summers are warm.

Hiking Boots Any dry weather boots or shoes(there's snow on top in late spring and in the late fall).

Author's Experience The author made the loop-trip in 3 1/2 days, or 28 hours total walk-time. He used an ice ax going over the divide for this mid-May hike and saw elk and bear along the way.

The stone ranch home of the abandoned McPherson Ranch at the mouth of Florence Creek.

Map 6, Florence and Coal Creek Canyons, Utah

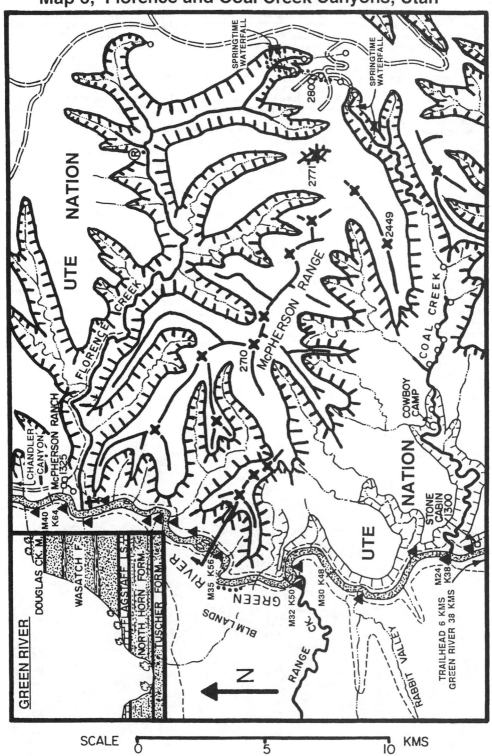

Upper San Rafael River Gorge, Utah

Location and Access This canyon is found in the northern part of the San Rafael Swell southeast of Huntington. This entire mapped region is north of Interstate 70. To get to the area, you can leave I-70 at mile post and Exit 129, and drive north to the San Rafael River and campground. Or drive south out of Price on Highway 10. Between mile post 56 and 57, or between 49 and 50, turn southeast toward Cleveland and the Buckhorn Draw. Or continue south on Highway 10 to a point just north of Castle Dale and turn east between mile posts 39 and 40. After a few kms on either road, you'll come to Buckhorn Flat and Buckhorn Well which is a 4-way junction. At the well is a large metal tank, small pump house and a watering trough. Turn right at that point, then after 200 meters, turn right again and drive 10 kms down Fuller Bottom Draw on a good sandy road to the San Rafael River where you can park and/or camp. For more information of the area, see the author's book, *Hiking and Exploring Utah's San Rafael Swell, 2nd Edition.*

Trail or Route Conditions This hike begins at the end of Fuller Bottom Draw and ends at the San Rafael Campground at the junction of the San Rafael River and Buckhorn Wash. In the beginning, you'll be walking in a shallow valley with a meandering small stream, but soon the river enters a gorge and the walls get higher and higher. At the deepest point, the Little Grand Canyon(as some call it) is nearly 500 meters deep. Along the way you'll be crossing the stream many times but will also be walking on cow trails much of the way. The deepest part of the gorge is in Sids Mtn. Wilderness Study Area, but some ORV's continue to run up and down the canyon bottom illegally. Along the way you will pass at least 4 panels of petroglyphs or pictographs as shown on the map.

Elevations Trailhead 1625 meters; The Wedge Overlook, 2025; San Rafael Campground, 1557 meters.

Hike Length and Time Needed From Fuller Bottom Draw to the campground is about 28-30 kms. This can be done in one day, but camping one night can be enjoyable. To do this one-way hike you'll need two cars(or maybe a mtn. bike could be used as a shuttle). With just one car you'll have to walk into the gorge from either end and return.

Water Always carry some in your car and in your pack because the San Rafael River water must be boiled or treated. Virgin Spring at the bottom end of Virgin Spring Canyon has the only drinkable-as-is water along the way. Don't expect to find drinking water at the San Rafael Campground!

Maps USGS or BLM map Huntington(1:100,000), or Wilsonville SE and Red Plateau SW(1:24,000).

Main Attractions A deep rugged wilderness gorge with petroglyphs and pictographs.

Ideal Time to Hike Spring or fall, but if the snowpack is heavy in the mountains, then the spring runoff could make the river a little deep for wading. Normally high water comes in late May. In October water levels are low, plus you'll have fall colors.

Hiking Boots Wading boots or shoes.

Author's Experience Because he's always alone, the author has walked into the middle of the gorge from both ends many times, as well as having seen the canyon from The Wedge Overlook on several occasions. He has also located the cabins on Sids Mountain; to get there, see his book on the San Rafael Swell.

The new and the old San Rafael River Bridges at the very bottom end of the Upper San Rafael River Gorge.

Map 7, Upper San Rafael River Gorge, Utah

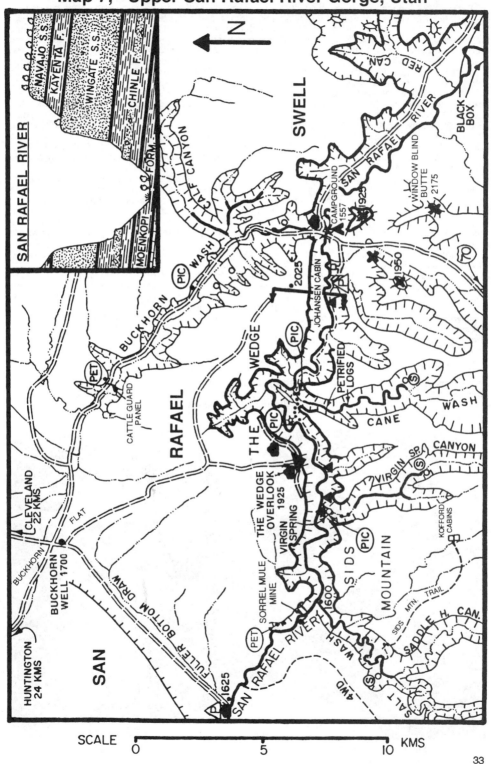

SAN RAFAEL RIVER

NAVAJO S.S.
KAYENTA F.
WINGATE S.S.
CHINLE F.
MOENKOPI FORM.

N

RED CAN.

SWELL

SAN RAFAEL RIVER

BLACK BOX

CALF CANYON

CAMPGROUND 1557

1925

WINDOW BLIND BUTTE 2175

1950

BUCKHORN WASH

PIC

2025

JOHANSEN CABIN

RAFAEL

THE WEDGE

PIC

PETRIFIED LOGS

S

PET

CATTLE GUARD PANEL

PIC

CANE WASH

VIRGIN SP.
S

VIRGIN CANYON

CLEVELAND 22 KMS

FLAT

THE WEDGE OVERLOOK 1925

VIRGIN SPRING

KOFFORD CABINS

SIDS MOUNTAIN

PIC

BUCKHORN WELL 1700

BUCKHORN

PET

SORREL MULE MINE

1600

SIDS MTN. TRAIL

SADDLE H. CAN.

FULLER BOTTOM DRAW

SAN RAFAEL RIVER

WASH

HUNTINGTON 24 KMS

SAN

1625

4WD

SALT

S

SCALE

0   5   10   KMS

33

Introduction to the Black Boxes, Utah

During 1988, the author received a couple of letters from hikers who had a tough time getting through the Upper Black Box of the San Rafael River. As it turned out, sometime between late 1986 and mid-1988, there was a big storm which sent a flash flood through the canyon. That flood left a huge pile of logs in the bottom end of the Upper Black Box. As a result, this log jam created a major delay and inconvenience at the end of a long and tiresome hike. In the spring of 1989, the author returned to the area and discovered several exits or entry points one can use to get around the log jam. In about 1992, another flood took the log jam away, but of course it could return any time. This edition does a better job of explaining this situation in greater detail. For more information on the San Rafael Swell, see the author's book, *Hiking and Exploring Utah's San Rafael Swell, 2nd Edition*.

Location and Access The normal way for people from northern Utah to reach the Upper Black Box, is to drive south from Price on Highway 10 in the direction of Huntington and Castle Dale. Between mile posts 56 and 57, or between 49 and 50, turn southeast and drive to Cleveland, then turn south towards Buckhorn Wash and the San Rafael Campground on the San Rafael River. Another alternate route to the campground is to drive Highway 10 to between mile posts 39 and 40, and turn east toward the same Buckhorn Wash. You can also drive I-70 to mile post and Exit 129, then turn north and drive down Cottonwood Draw to the San Rafael Campground and the two bridges.

From the north side of the San Rafael River bridges, turn east onto the Mexican Mtn. Road. Drive about 16 kms until you arrive at a point where you have a small butte on the left or north, and a small old stock pond with tamaracks on the right or south. You can park and camp there, or you could drive a newly created track running southeast as shown on the Upper Black Box map. The author believes it's best to park on the road, and walk southeast to the river. If it's a really long hike you want, you could begin further upstream at what some people call the Lockhart Box. But that part of the gorge isn't as impressive as the lower end.

To get to the Lower Black Box, either drive south from the San Rafael Campground and up Cottonwood Draw; or leave I-70 at mile post and Exit 129 and drive north. About 9 kms from Exit 129, you'll come to Sinkhole Flat, as shown on **Map 8, Access Map & Roads to the Black Boxes.** From the north end of the flat, turn east for 3 kms, then turn northeast for another 5 kms. At the Jackass Benches, turn right and drive 6 kms. At that point you can drive with a HCV to near the lower end of the Lower Black Box. There are two rough spots at the beginning of this road, but it's pretty good after that.

If you continue around to the north side of the Jackass Benches, you can get on another road which used to lead to the northern or upper end of the Lower Black Box. This road is good up to one rough spot which is at the WSA boundary. There the BLM has, or at least tried, to block off the road. However, some people are ignoring the blockade and driving all the way to the bench above Swazys Leap and the Hanson sheep bridge--but they run the risk of being cited by BLM rangers.

For people with nice cars, you could also park at the rest area between mile posts 140 and 141 on I-70, and walk cross-country from there across the upper part of Black Dragon Canyon to the lower end of the Lower Black Box. You could also drive to the mouth of Black Dragon Canyon from near mile post 145 on I-70, and get in from there. The author used this route when he hiked all the way through the Lower Black Box.

From on top of the log jam looking downstream at the mass of floating logs.

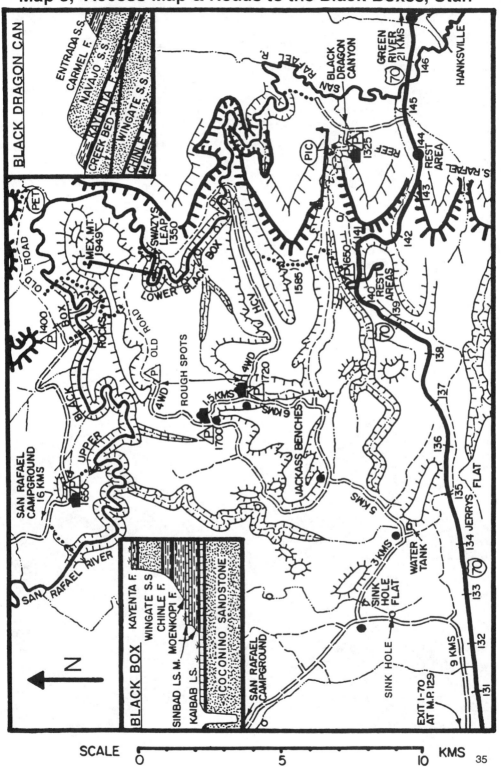

BLACK DRAGON CAN

Entrada S.S.
Carmel F.
Navajo S.S.
Kayenta F.
Creek Bed
Wingate S.S.
Chinle F.
M. F.

PET

OLD ROAD

MEX MT. 1949

SWAZYS LEAP 1350

LOWER BLACK BOX

BLACK

ROCKS

1400

BOX

OLD ROAD

PIC

1325

REEF

BLACK DRAGON CANYON

SAN RAFAEL

GREEN RIVER 21 KMS

70

146

HANKSVILLE

145

144 REST AREA

143

S. RAFAEL

142

1585

HCV

1650

141

REST AREAS

140

139

138

137

ROUGH SPOTS

4WD

4WD 720

1.5 KMS

5 KMS

1700

JACKASS BENCHES

5 KMS

136

135

134 JERRYS FLAT

SAN RAFAEL CAMPGROUND 16 KMS

1650

UPPER

SAN RAFAEL RIVER

WATER TANK

3 KMS

SINK HOLE FLAT

133

70

BLACK BOX

KAYENTA F.
WINGATE S.S
CHINLE F.
SINBAD LS. M. MOENKOPI F.
KAIBAB LS.

COCONINO SANDSTONE

SAN RAFAEL CAMPGROUND

SINK HOLE

9 KMS

132

EXIT I-70 AT M.P. 129

131

N

Upper Black Box, San Rafael River, Utah

Trail or Route Conditions When hiking the Upper Black Box one must walk in the river a lot(up to 50% of the time during years with high water levels). To do this hike successfully, you must have an *inner tube, life jacket, or day-pack lined with plastic bags* for each person, as there are many deep holes along the way. From the car-park at 1650 meters, walk southeast for about one km. When you reach the canyon rim, you should see the river below and a shallow drainage coming in on the right. Look for stone cairns marking an old stock trail down into the Box. About 9/10 of the way through, you'll come to a rockfall which has dammed the river. At that point you'll need a *short rope* to get yourself, cameras, packs, etc., over the short dropoff in a dry condition. If you have nothing to damage in the water, then you could slide down 3 meters, but there may be rocks under the murky water! You could also swim through a narrow opening on the right side avoiding a jump. If you have to, you could exit at that point by climbing a steep ravine up the north wall, but that's for more experienced climbers or hikers only. Downstream from the rockfall there are 3 other entry/exit places and possibly a log jam as shown on the small insert map. The 2nd exit is easy, but it would put you on the south side of the canyon. The 3rd exit is also fairly easy and this is the recommended way out if the log jam is too difficult. It begins where there's a second and smaller rockfall in the river. Just beyond the 3rd exit, is another easy way out, but on the south side of the gorge. About 300 meters below the 4th exit, is a narrowing and a possible(?) log jam. If the log jam is there, you can easily climb up on top of it, but there are 10-15 meters of floating logs on the other side you'll have to swim through or crawl over. This is not so difficult unless you're trying to keep a camera dry. Not far below the log jam area you'll walk out the bottom end of the Box. A second car or mtn. bike left there would eliminate a long road-walk.

Elevations Car-park, 1650 meters; bottom of the Upper Black Box, 1400 meters.

Hike Length and Time Needed From the entry car-park to the bottom of the Box it's about 14 kms; or about 28 kms round-trip. You can shorten the trip by leaving at the 3rd exit; and/or by having a shuttle with 2 cars, or a car and a mtn. bike. Remember, this will be a long all-day hike, so arrive at the trailhead the night before, camp, then start hiking early the next morning.

Water Carry water in your car and in your pack. River water must be treated before drinking.

Maps USGS or BLM maps Huntington and San Rafael Desert(1:100,000), or Red Plateau SE & Beckwith Peak SW(1:24,000), and The Wickiup(1:62,500).

Main Attractions A deep narrow canyon, where you'll need to swim in several places and have a short rope to get all the way through. This is for experienced and tougher hikers only.

Ideal Time to Hike Because of wading, warm or hot weather; from late May through mid-September.

Hiking Boots Wading boots or shoes.

Author's Experience On his first trip, he walked, waded and swam down to the rockfall, and had to exit(no rope and alone), then made it back to his car in 7 hours. On the second trip, he entered the canyon at the rockfall and climbed up and out at all 4 entry/exits as shown. He was worried about getting his cameras wet on the downstream side of the log jam, so he left the canyon at the 3rd exit finishing the hike along the canyon rim.

This is the middle section of the Upper Black Box in the San Rafael Swell.

Map 9, Upper Black Box, San Rafael River, Utah

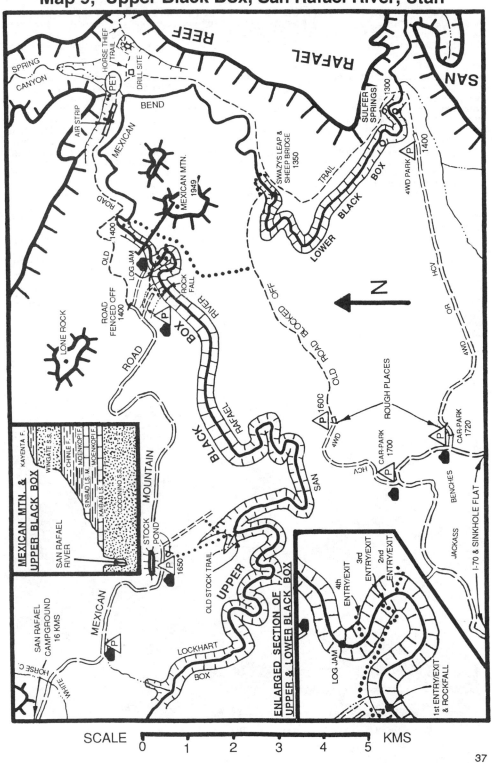

SCALE 0 1 2 3 4 5 KMS

Lower Black Box and Swazys Leap, Utah

Trail or Route Conditions In order to successfully get through the Lower Black Box you will need an *inner tube, life jacket,, or a day-pack lined with plastic bags,* because you'll pass through many deep swimming holes especially during years with heavy runoff. One way there is to walk to the end of the old road which ends on the ledge just above Swazys(Swasey's) Leap, then head for the upper end of the narrows and get down into the river just as it begins to dive into the Lower Black Box. Soon you'll walk and/or float under the old Hanson sheep bridge at the narrowest place in the gorge called Swazys or Sid's Leap. It's about 3 meters wide at the top and 17 meters above the river. Better not walk across this bridge, it's pretty rickety these days. As you head on down the canyon, you will find several minor rockfalls and places with deep channels. You must swim or float through these places. This canyon is similar to the Upper Black Box, only it's a bit shorter and without major obstacles. It's also a bit more narrow. When you reach the bottom end, turn left and walk back to the sheep bridge along an old stock trail on the east side of the gorge. If you drive or ride a mtn. bike to the lower end of the canyon, then walk the trail to the upper end of the Lower Black Box first, then walk down-canyon in the water. Don't attempt to walk upstream against the current. With the road closed to the upper end of the Box, this second route in is the shortest of the two route possibilities. It's best to have at least one experienced hiker along on this trip, in case the river channel changes. Also, take along a *short rope* in case of an emergency. Stay out of this gorge if the weather looks bad.

Elevations Swazys or Sid's Leap, 1350 meters; the bottom end of the Lower Black Box, about 1300; the two Jackass Bench Car-parks, about 1700 meters.

Hike Length and Time Needed The distance through the Lower Black Box is about 8 kms. If you must leave your car at one of the Jackass Bench car-parks and make the round-trip hike from there, it'll be 28-30 kms--a very long all day hike. This is about the same distance as if you were to start at the rest stop on I-70 or at the mouth of Black Dragon Canyon, two other possible routes in. It's best to take a shovel and try to drive to the lower end of the Box. There are two or three bad spots along this road. This would be a great place to try out a mtn. bike if you have a normal car. Driven with care, most cars should have no trouble getting to and around the Jackass Benches, at least during dry conditions.

Water Always carry water in your car and in your pack, but there are several minor freshwater seeps within the Box. The springs at the lower end are sulfur tainted and undrinkable. Treat river water before drinking.

Maps USGS or BLM map San Rafael Desert(1:100,000), or Tidwell Bottoms(1:62,500).

Main Attractions A deep, dark canyon with an interesting old historic sheep bridge.

Ideal Time to Hike In warm or hot weather(late May through September).

Hiking Boots Wading boots or shoes.

Author's Experience The author has been there several times, but when he did the complete Lower Black Box hike, he did it from Black Dragon Canyon with a total walk-time of 9 1/2 hours.

This foto shows the bridge over Swazys Leap from the river, about 17 meters below.

Map 10, Lower Black Box and Swazys Leap, Utah

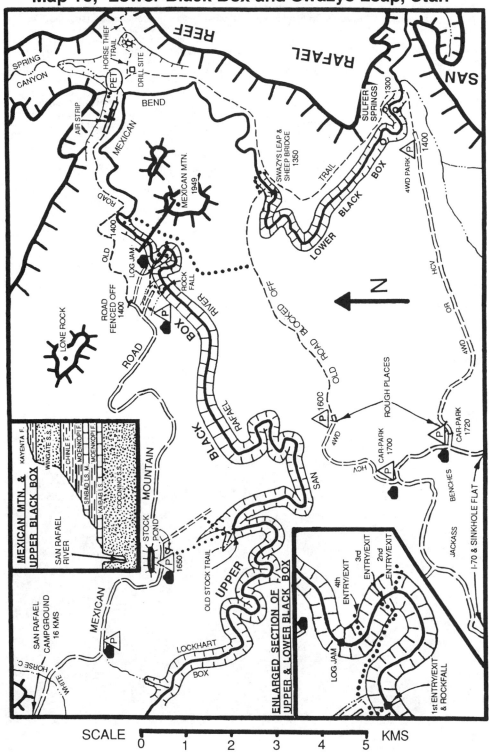

SCALE 0 1 2 3 4 5 KMS

San Rafael Reef, Utah

Location and Access A kind of "reef" or ring of steep tilted cliffs surround the entire San Rafael Swell, but the place where it's most spectacular is on the Swells' eastern side just south of Interstate Highway 70. This area is about 30 kms west of Green River, Utah. To get there, drive west out of Green River on I-70, or east out of Richfield and Salina, or north from Hanksville. People with nice cars can park right on I-70 at the lower rest stop located at mile post 144. For those who don't mind a little dust and some rough roads, you can get close to the middle parts of the Reef by using Highway 24. It used to be you could use the road running west from the Hatt Ranch, but they'd rather you not use that route any more. Now there are two all-public lands access routes to the middle and southern end of the Reef from Highway 24. One is right at mile post 148. Drive west from there to Iron Wash. Most of the time cars can now get across OK. Beyond the wash it's a good road to the mouth of Straight Wash, where there's a good campsite in the cottonwood trees. The best way to get to the middle part of the Reef and Three Fingers Canyon, is to drive west from between mile posts 153 and 154(take the most northerly of two roads), and drive west to the corral, then north a ways, then west again. Car drivers should always carry a shovel to help get over the occasional rough gully. These dry washes change after every heavy storm. For more information on the Swell, see the author's book, *Hiking and Exploring Utah's San Rafael Swell, 2nd Edition.*

Trail or Route Conditions One suggested hike is to park at the rest area on I-70, and walk west up Spotted Wolf Canyon along side the freeway, then south along the Chinle-Moenkopi Valley which is west of the highest cliffs. Day-hikers can get to about Three Fingers Canyon, then walk north along the eastern base of the Reef back to I-70. If beginning at Three Fingers, you could walk south inside the Chinle-Moenkopi Valley to Straight Wash, then return via the eastern base again. It's easy walking along the eastern front of the Reef, as well as in the Chinle-Moenkopi Valley located behind the Navajo, Kayenta and Wingate fins. The walk from I-70 to Straight Wash in this back-valley, is one of the most spectacular hikes around. When terrain permits, climb to the top of the Reef for fine views.

Elevations Eastern base of the Reef, 1350 meters; Chinle-Moenkopi Valley passes, up to about 1600.

Hike Length and Time Needed From I-70 south to Three Fingers Canyon and back is an all day hike, as is the hike from Three Fingers to Straight Wash and back.

Water Always carry plenty in your car. There may also be pothole or seep water in some canyons in the period right after rains.

Maps USGS or BLM map San Rafael Desert(1:100,000), or Tidwell Bottoms(1:62,500).

Main Attractions Amazing geology of the San Rafael Reef, old mines and petroglyphs.

Ideal Time to Hike Spring, fall or in mild winter weather. Summers here are extra warm.

Hiking Boots Any dry weather boots or shoes.

Author's Experience Years ago the author drove from Hatt's Ranch to the middle of the Reef and camped; he then walked along the Chinle-Moenkopi Valley to Straight Wash, then back along the eastern face to his car in 8 hours walk-time. On another trip he walked from 1-70 to Three Fingers Canyon via the Chinle-Moenkopi Valley, and back along the front of the Reef in 6 hours.

By doing a little climbing, one has a fine view of the San Rafael Reef.

Map 11, San Rafael Reef, Utah

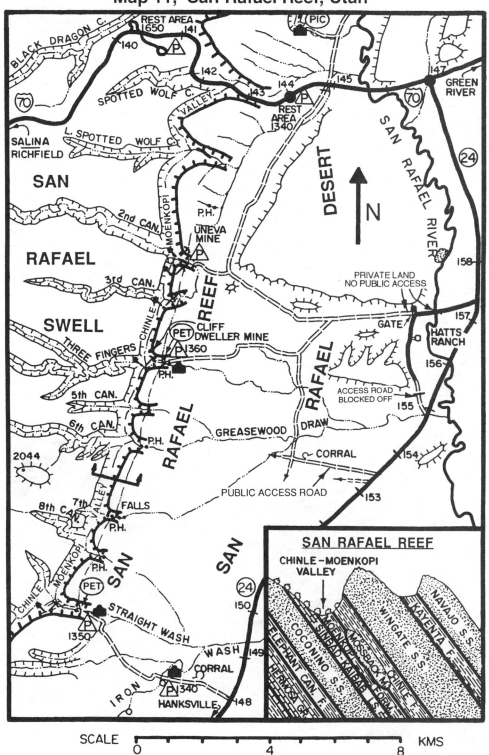

SCALE   0   4   8   KMS

41

Eardley Canyon and Straight Wash, Utah

Location and Access This canyon is located in the eastern part of the San Rafael Swell between Interstate 70 and Highway 24 and about 30 kms west of Green River. One can get to the upper parts of the canyon by exiting I-70 at mile post and Exit 129. First head west 1 1/2 kms, then turn southeast and cross Indian Flat. Or exit 1-70 through an ordinary ranch gate near mile post 133, and drive southeast to the landing strip between Red and Hyde Draws called Cliff Dweller Flat. An easier way might be to leave I-70 at Exit 129, and use the road heading for the San Rafael River, then turn south and drive under the freeway to reach Cliff Dweller Flat. There's also a good road running down into the upper part of Red Draw from the freeway and the mile post 133 ranch gate exit. To reach the bottom end of Eardley, drive south from I-70 on Highway 24 about 20 kms to mile post 148, then turn west on a good dirt road. Any vehicle can make it to Iron Wash, and in normal conditions, get across OK. Car drivers take a shovel just in case. The road beyond is good. For more information on the area's history and more hikes, see the author's book, *Hiking and Exploring Utah's San Rafael Swell, 2nd Edition.*

Trail or Route Conditions From the car-park at the mouth of Straight Wash, there's the faded remains of an old mining exploration road running into the canyon, but as one gets into Eardley, there is no trail(Eardley Canyon actually begins, or ends, just west of the Reef). The bottom of this canyon is totally untouched and a real gem. One hundred meters inside Eardley is an Olympic-sized pothole which cannot be passed. You must therefore walk up the slope on either side of the bottom narrows, 'till a descent route can be found up above. The author entered, then exited, just below and just above the geology cross section arrow shown on the map. From the rincon up-canyon, there are no route problems with a number of exit possibilities. Perhaps the best part of Eardley lies in the upper end. Enter at Red Draw or Charley Holes. There are several routes down into the Coconino Narrows. Just follow the route symbols on the map, and compare this one to a USGS topo map. One of the best parts of this hike is walking through the Reef along Straight Wash, then to the almost hidden mouth of Eardley. The other good parts are the narrow sections below Red Draw; and just where the dryfall is at the head of Crawford Draw..

Elevations Elevations range from 1350 meters at Straight Wash, to 1980 in the upper canyon.

Hike Length and Time Needed To walk the entire length of Eardley may take two days, and you'd need 2 cars. One-day hikes are best however, from either end of the canyon and into the best parts.

Water Always carry plenty of water in your car and in your pack.

Maps USGS or BLM map San Rafael Desert(1:100,000), or Tidwell Bottoms(1:62,500).

Main Attractions The San Rafael Reef, huge potholes and narrows in the upper and lower end of the canyon, and solitude in a canyon that can't be seen from the highway.

Ideal Time to Hike Spring or fall. Summers are very warm down low, but cooler in the upper canyon. During some winter warm spells you could hike into the lower end of the canyon.

Hiking Boots Any dry weather boots or shoes.

Author's Experience The author camped on Iron Wash, then walked into Straight Wash and eventually into Eardley on the north side, exiting just south of the rincon. This hike took 8 hours, round-trip. He also spent 5 hours in the upper canyon and Red Draw on another hike. On his last trip he found the route into upper Crawford Draw, walked through 2/3's of it, then returned, all in 3 hours.

This slot is in the middle part of Crawford Draw, which is an upper tributary of Eardley Canyon.

Map 12, Eardley Canyon and Straight Wash, Utah

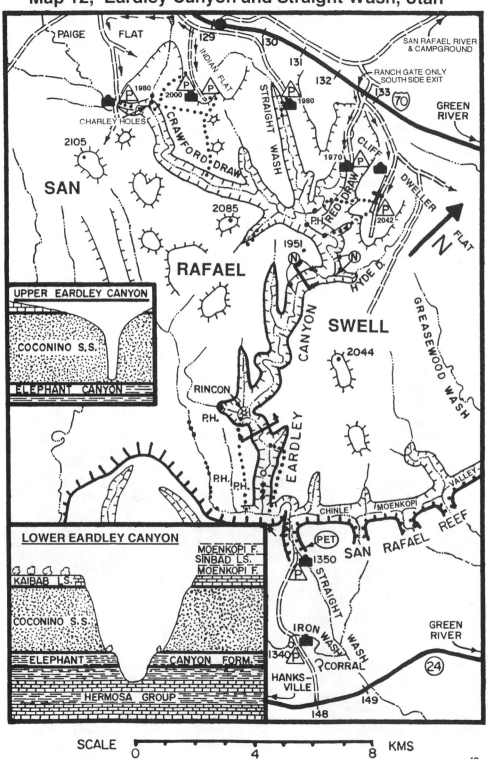

PAIGE FLAT

INDIAN FLAT

129
30
131
132
133
70

SAN RAFAEL RIVER & CAMPGROUND

RANCH GATE ONLY SOUTH SIDE EXIT

GREEN RIVER

P 1980
2000 P
P
STRAIGHT WASH
P 1980

CHARLEY HOLES

CRAWFORD DRAW

2105

SAN

2085

1970 P
CLIFF DWELLER
FLAT
N

P.H. RED DRAW
P 2042

1951

RAFAEL

HYDE D.

SWELL

GREASEWOOD WASH

2044

UPPER EARDLEY CANYON

COCONINO S.S.

ELEPHANT CANYON

CANYON

EARDLEY

RINCON

P.H.

P.H. P.H.

CHINLE MOENKOPI

VALLEY

SAN RAFAEL REEF

LOWER EARDLEY CANYON

MOENKOPI F.
SINBAD LS.
MOENKOPI F.

KAIBAB LS.

COCONINO S.S.

ELEPHANT CANYON FORM.

HERMOSA GROUP

PET

1350 P

STRAIGHT WASH

IRON WASH

1340 P CORRAL

HANKS- VILLE

GREEN RIVER

24

148 149

SCALE 0 4 8 KMS

Crack and Chute Canyons, Utah

Location and Access These two canyons are found due north of Hanksville and southwest of Green River, Utah. They're also just north and northwest of Goblin Valley State Park, and along the southeastern part of the San Rafael Reef. To get there, turn off Highway 24(which connects I-70 with Hanksville) between mile posts 136 and 137 and drive west on the good paved Temple Mtn. Road. After about 8 kms you can turn south toward Goblin Valley on a well-maintained road. About 1 km before arriving at Goblin Valley, turn to the right or southwest and drive to the bottom of the Wild Horse drainage and park at the 1460 meter car-park as shown on the map. One problem with parking there is the deep sand which is at its worst during long dry spells. Therefore, it's recommended that those with regular cars head for the west, or inner side, of the Reef. To get there, drive west on the Temple Mtn. Road all the way to the old Temple Mtn. townsite, then about half a km past it, turn to the left or southwest on to the Chute Canyon Road. Drive this for about 13 kms to where it ends near the Erma Mine. There you'll see a metal shack and a rock house called "Morgan's Cabin". If you just want to hike Crack Canyon, then drive only 8 or 9 kms along this same road and park. There are good camp sites near Morgan's Cabin and near the beginning of the Chute Canyon Road.

Trail or Route Conditions If you start on the Goblin Valley side of the Reef, walk up Chute, then road-walk northeast to the head of Crack Canyon, then down Crack and back to your car. Some ORV's are using Chute Canyon, but thank goodness they can't get up or down Crack! There are three sets of good narrows in Crack making it one of the best slot canyons around. For people driving cars, doing these canyons from the top end is by far the best. One reader wrote and complained about a big dropoff in Crack Canyon, but the author failed to see it as an obstacle. Climbing over minor dryfalls, dropoffs or ledges is just part of the canyon hiking experience. Don't enter this canyon if bad weather is threatening. For more information on the area, see the author's book, *Hiking and Exploring Utah's San Rafael Swell, 2nd Edition*.

Elevations Head of Crack Canyon, 1650 meters; head of Chute Canyon, 1550; bottom of both canyons, about 1460 meters.

Hike Length and Time Needed To make a circle route of both canyons will be to walk about 25 kms(a long all-day hike), but hikers will be most interested in Crack Canyon. You can hike most of the way down from the Chute Canyon Road and return, in about 3 or 4 hours, or half a day.

Water Carry water in your car and in your pack, as there are no springs around. If you need more water, drive to Goblin Valley State Park to replenish your supply. Goblin Valley is a fee-use area.

Maps USGS or BLM map San Rafael Desert(1:100,000), or Wild Horse and Temple Mtn.(1:62,500).

Main Attractions High cliffs, deep canyons, good narrows, and old mines and cabins.

Ideal Time to Hike Spring or fall, but some winter warm spells can be pleasant. Summers are hot.

Hiking Boots Any dry weather boots or shoes.

Author's Experience He once camped at the 1460 meter site, then went up Chute and down Crack in about 5 1/2 hours round-trip. Another time he walked down from the road past the narrows in Crack Canyon and back, in less than 2 hours.

This is part of the narrows of Crack Canyon as it cuts through the San Rafael Reef.

Map 13, Crack and Chute Canyons, Utah

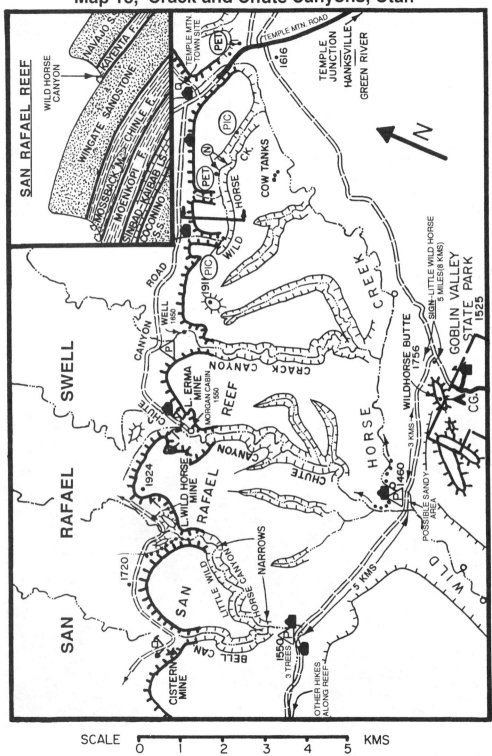

SAN RAFAEL REEF

SAN RAFAEL SWELL

WILD HORSE CANYON

NAVAJO S.S.
KAYENTA F.
WINGATE SANDSTONE
CHINLE F.
MOSSBACK M.
MOENKOPI F.
SINBAD - KAIBAB LS.
COCONINO S.S.

TEMPLE MTN. ROAD
TEMPLE MTN. TOWN SITE
PET
1616
TEMPLE JUNCTION
HANKSVILLE
GREEN RIVER

N

PIC
N
PET
WILD HORSE CK.
COW TANKS
CREEK

191 PIC
WELL 1650
CANYON ROAD
CRACK CANYON
L. ERMA MINE 1550
MORGAN CABIN
1924
CHUTE
REEF
RAFAEL CANYON
CHUTE CANYON
HORSE
WILDHORSE BUTTE 1756
SIGN-LITTLE WILD HORSE 5 MILES(8 KMS)
GOBLIN VALLEY STATE PARK 1525
CG.
3 KMS
1460
POSSIBLE SANDY AREA
WILD

L. WILD HORSE MINE
1720
SAN
LITTLE WILD HORSE CANYON
NARROWS
BELL CAN
1550
3 TREES
5 KMS
OTHER HIKES ALONG REEF
CISTERN MINE

SCALE   0   1   2   3   4   5   KMS

45

Bell and Little Wild Horse Canyons, Utah

Location and Access These two popular canyons are located due west of Goblin Valley State Park and in the southeastern part of the San Rafael Swell. To get there, drive along Highway 24 between Hanksville and I-70. Between mile posts 136 and 137, turn west onto the paved Temple Mtn. Road signposted for Goblin Valley. Drive this road for 8 kms then turn south toward Goblin Valley. After another 11 kms, or about one km before arriving at Goblin Park, turn right or southwest, and drive another good road down through the bottom of Wild Horse Creek. In this dry creek bed area, you may find it sandy; but lately, this road has been maintained better than before and today there's lots of traffic, so it's not the sand trap it used to be. After Wild Horse Creek, continue west another 5 kms. You'll go over a low divide, then drop down into Little Wild Horse dry wash with 3 small cottonwood trees next to a parking area. Some people are driving up the creek bed a ways and camping under shade trees.

Trail or Route Conditions It's recommended you walk up-canyon past some minor dryfalls, then at the junction of the two canyons, veer left into Bell(if you just want to do Little Wild Horse, pay attention to this part which is about 10-15 minutes past the upper parking area; some people miss the turnoff). Walk through Bell, then when you reach the west side of the Reef, road-walk to the right or northeast to the upper part of Little Wild Horse. Then return down this canyon to your car. Little Wild Horse is probably the best of the two canyons. It has about 2 kms of narrows, of which one km averages one to two meters in width. In one section, it would be difficult to take a frame pack through because of its narrowness. It's loaded with potholes, but one hiker told the author he went through a day after a big storm and the pothole water was no more than thigh-deep. Obviously, you don't want to be there during a big storm, or if bad weather is threatening. For more information on the area see the author's book, *Hiking and Exploring Utah's San Rafael Swell, 2nd Edition.*

Elevations Car-park at Little Wild Horse, 1550 meters; high point inside the Swell, 1720 meters.

Hike Length and Time Needed From the car-park, up Bell, down Little Wild Horse, and back to one's car is about 13 kms. This loop-hike can be done in about half a day, but most people would want to spend all day in these canyons.

Water Always carry plenty of water in your car and pack. If you need more, stop at Goblin Valley State Park, which is a fee-use area.

Maps USGS or BLM map San Rafael Desert(1:100,000), or Wild Horse(1:62,500).

Main Attractions One of the best little narrow canyons on the Colorado Plateau and a fun hike for the whole family. You might also check out the Cistern Mine area.

Ideal Time to Hike Spring or fall, but some winter warm spells can be pleasant. Summers are hot, but in a shaded canyon, hiking is still possible. Weekends are busy, so try to be there during the middle of the week.

Hiking Boots Any dry weather boots or shoes, except right after a storm, then waders.

Author's Experience The author camped at the mouth of Little Wild Horse Canyon, then walked up Bell and down Little Wild Horse, in about 4 hours. On a second trip he did the same hike but in the reverse direction in under 4 hours. He also was in Little Wild Horse during one Easter weekend, and found lots of people!

This is typical of the one-km-long narrows in Little Wild Horse Canyon.

Map 14, Bell and Little Wild Horse Canyons, Utah

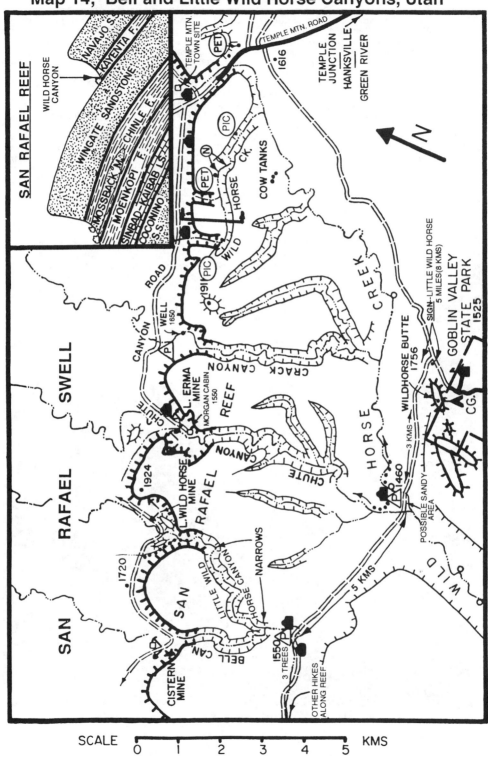

SAN RAFAEL REEF

WILD HORSE CANYON

NAVAJO S.
KAYENTA F.
WINGATE SANDSTONE
CHINLE F.
MOSSBACK M.
MOENKOPI F.
SINBAD LS.
KAIBAB LS.
COCONINO S.S.

TEMPLE MTN. TOWN SITE
PET
TEMPLE MTN. ROAD
1616
TEMPLE JUNCTION
HANKSVILLE
GREEN RIVER

N

PIC
WILD HORSE CK.
COW TANKS
PET
N

191 PIC
CANYON WELL
ROAD
1650
P

SAN RAFAEL SWELL

L. ERMA MINE
1550
MORGAN CABIN
CHUTE
1924
CHUTE CANYON
CRACK CANYON

WILDHORSE BUTTE
1756

SIGN-LITTLE WILD HORSE
5 MILES (8 KMS)
GOBLIN VALLEY STATE PARK
1525
CG.

HORSE CREEK

L. WILD HORSE MINE
REEF CANYON
1720

NARROWS
LITTLE WILD HORSE CANYON
SAN RAFAEL
3 KMS
P 1460
POSSIBLE SANDY AREA

1550
3 TREES
BELL CAN.
CISTERN MINE
5 KMS
WILD
OTHER HIKES ALONG REEF

SCALE 0 1 2 3 4 5 KMS

The Chute--of Muddy Creek, Utah

Location and Access The Chute of Muddy Creek is located in the extreme southern end of the San Rafael Swell not far north of Hanksville. The Chute is found on the inside of the Reef and in between Tomsich Butte and the Delta(Hidden Splendor) Mine. Muddy Creek in this section cuts down into the Coconino Sandstone, thus making a very narrow and deep trench. It does the same thing here as the San Rafael River does in the north end of the Swell as it flows through the Black Boxes. These two canyons are very similar except there's a lot less water and no obstacles in Muddy Creek(but changes can occur with each flood!). To get to this canyon, exit I-70 at mile post and Exit 129 and drive south to Tomsich Butte; or leave Highway 24 at Temple Junction, which is between mile posts 136 and 137. From there drive west to the middle of the Swell and turn south towards Tomsich Butte. You could also begin the hike at the Delta Mine. However, most of the best parts of The Chute are easier to reach from Tomsich Butte.

Trail or Route Conditions If you begin at the Delta Mine, then simply get into the stream and walk up-canyon. From Tomsich Butte, turn left or south and drive as far as you can. That will be just before the stream crossing. Park there and walk along the very old and faded mining road for about 2 kms as shown on the map, then it's into the water and walking downstream. In the part labeled *Coconino Narrows*, you'll be in the water about 50% of the time; while in the area of the log jam, you'll be in water 90% of the time(during dry years in mid-summer, there may be no water). The water is normally less than ankle deep and there are no obstructions as one finds in the Black Boxes of the San Rafael River. For more information on the area, see the author's book, *Hiking and Exploring Utah's San Rafael Swell, 2nd Edition*.

Elevations The creek bed at Tomsich Butte, 1554 meters; bottom end of The Chute, 1432 meters.

Hike Length and Time Needed From Tomsich Butte to the Delta Mine is about 25 kms one way. This can be done in one day, but you'd need two cars, or a car and a mtn. bike. It's perhaps best to begin and end the hike at one car-park. Plan on a full day's hike.

Water Take all the water you'll need in your car and in your pack. Better treat Muddy Creek water.

Maps USGS or BLM map San Rafael Desert(1:100,000), or Wild Horse(1:62,500).

Main Attractions A deep narrow canyon. Maybe the best narrows hike in the San Rafael Swell.

Ideal Time to Hike Spring through fall, but if it's too early or late in the season the water can be very cold. Summers are OK, because you're in the water and there's shade. Winter is out of the question, unless it's the middle of a drought. In years with heavy snowfall in the mountains, kayakers sometimes make a run through this canyon from sometime late in May to early June. <u>Don't go into this gorge with an unsettled or rainy weather forecast!</u>

Hiking Boots Wading boots or shoes.

Author's Experience The author camped at the Tomsich Butte car-park and walked downstream on April 19, 1986. At first his feet felt like blocks of ice, then a little later they warmed up OK. He walked to within 4 kms of Chimney Canyon and returned in 8 hours.

The Chute of Muddy Creek. This foto shows the log jam high above the creek.

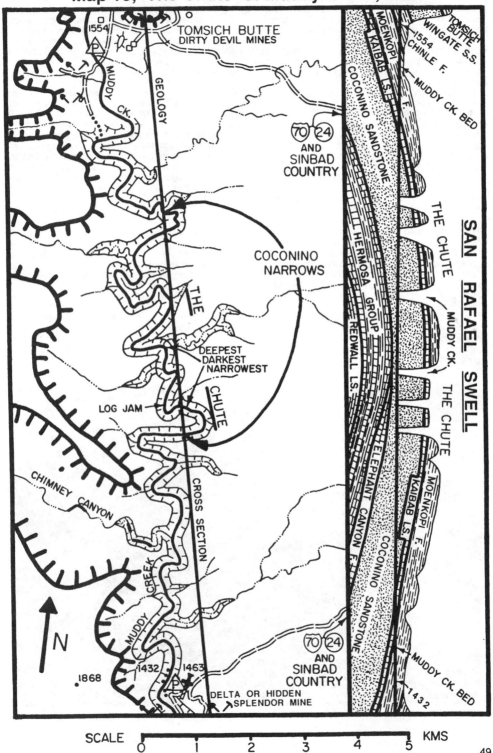

1554
TOMSICH BUTTE
DIRTY DEVIL MINES

P

MUDDY CK.

GEOLOGY

70 24
AND
SINBAD
COUNTRY

COCONINO
NARROWS

THE

DEEPEST
DARKEST
NARROWEST

LOG JAM

CHUTE

CHIMNEY CANYON

CROSS SECTION

MUDDY CREEK

N

.1868

1432   1463

P

DELTA OR HIDDEN
SPLENDOR MINE

TOMSICH
BUTTE
WINGATE S.S.
1554
CHINLE F.
MUDDY CK. BED

MOENKOPI F.

KAIBAB LS.

COCONINO SANDSTONE

HERMOSA GROUP

REDWALL LS.

THE CHUTE

MUDDY CK.

THE CHUTE

SAN RAFAEL SWELL

ELEPHANT CANYON

COCONINO SANDSTONE

KAIBAB LS.

MOENKOPI F.

70 24
AND
SINBAD
COUNTRY

MUDDY CK. BED

1432

SCALE   0   1   2   3   4   5   KMS

49

Spring Canyon, Utah

Location and Access The hike featured here is in Spring Canyon, located in the northern part of Capitol Reef National Park. It's the major canyon just to the north of Highway 24, Torrey and Fruita. This drainage begins on the east side of Thousand Lake Mountain and heads southeast into the Fremont River. To reach the bottom end of the canyon, drive along Highway 24 east of Fruita and park visitor center, to between mile posts 83 and 84. To enter the canyon at its mid-point, park at the Chimney Rock Trailhead, located about 5 kms west or uphill, from the visitor center and between mile posts 76 and 77. To enter the head of the canyon, drive to the Cooks Mesa Trailhead which is about 1 km west of the park boundary and very near mile post 73.

Trail or Route Conditions You might consider making a loop-hike of either the upper half or the lower half of this canyon. One possibility is to start at Cooks Mesa Trailhead. Drive about 200 meters off the road to the north and park. Then look for a stone cairn at the beginning of a trail running north and up-slope to the top of the Shinarump Bench just below the Wingate Cliffs. Follow this same bench trail northwest to "W" Pass. It's easy to follow until you reach the petrified trees, then it fades. Continue northwest on the same bench. Later, and just under the pass, you'll see cairns again. Go over "W" Pass and into the head of Spring Canyon. You then walk in the dry creek bed all the way to the Fremont River; or make an exit up a side canyon running to the southwest and end up at the Chimney Rock Trailhead. Road-walk, hitch a ride or use a mtn. bike to return to your car. If you start at the Chimney Rock Trailhead, follow a good trail up over a pass, then down a side drainage into Spring Canyon. You'll exit the gorge at the Fremont River and Highway 24.

Elevations Cooks Mesa Trailhead, 2000 meters; "W" Pass, 2375; Chimney Rock Trailhead, 1850; bottom of Spring Canyon, 1575 meters.

Hike Length and Time Needed From Cooks Mesa Trailhead to the Fremont River is about 32 kms. That's one very long day-hike, or an easier two day backpacking trip(with two cars, or a mtn. bike at one end). If you enter the upper canyon and exit at Chimney Rock, it's about 27 kms including the road-walk or bike ride back to your car. This will be an easier, but all-day hike none-the-less.

Water Take some with you. There is a little running water in the canyon, but with cattle there in the cooler months, take it directly from a spring or seep source, or purify it first.

Maps USGS or BLM map Loa(1:100,000), or Torrey and Fruita(1:62,500).

Main Attractions A deep red rock canyon, many petrified logs, and solitude.

Ideal Time to Hike Spring or fall, or hike early in the mornings during summer. The upper end of this drainage is at a higher elevation, therefore cooler and better for summer hiking.

Hiking Boots Any dry weather boots or shoes.

Author's Experience The author camped at the Cooks Mesa Trailhead, then made it over the "W" Pass. He later exited at Chimney Rock, and hitched a ride back to his car, all in just over 9 hours.

Just one of hundreds of petrified trees found on Cooks Mesa and the Shinarump Bench.

Map 16, Spring Canyon, Utah

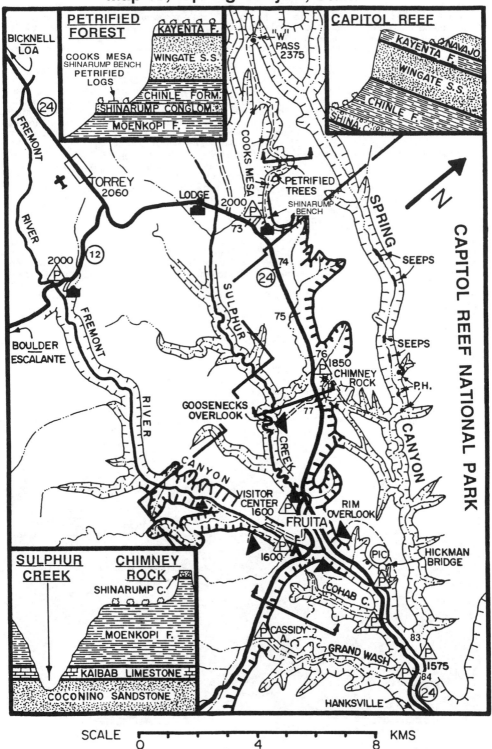

PETRIFIED FOREST

KAYENTA F.

WINGATE S.S.

COOKS MESA
SHINARUMP BENCH
PETRIFIED LOGS

CHINLE FORM.

SHINARUMP CONGLOM.

MOENKOPI F.

CAPITOL REEF

NAVAJO
KAYENTA F.

WINGATE S.S.

CHINLE F.

SHINA. C.

BICKNELL
LOA

FREMONT

RIVER

24

TORREY
2060

LODGE

2000

73

"W" PASS
2375

COOKS MESA

PETRIFIED TREES

SHINARUMP BENCH

SPRING

N

CAPITOL REEF NATIONAL PARK

2000
P
12

BOULDER
ESCALANTE

FREMONT

RIVER

SULPHUR

74

24

75

76
1850
CHIMNEY ROCK

SEEPS

SEEPS

P.H.

GOOSENECKS OVERLOOK

CREEK

77

CANYON

CANYON

VISITOR CENTER
1600

FRUITA

RIM OVERLOOK

PIC

HICKMAN BRIDGE

P

SULPHUR CREEK

CHIMNEY ROCK

SHINARUMP C.

MOENKOPI F.

KAIBAB LIMESTONE

COCONINO SANDSTONE

1600

CASSIDY A.

COHAB C.

GRAND WASH

P

P

83

P

1575

84

HANKSVILLE

24

SCALE   0   4   8   KMS

51

Sulphur Creek, Utah

Location and Access Sulphur Creek makes another interesting hike in the northern part of Capitol Reef National Park. This creek begins in the area just north of Torrey and emerges from a deep canyon just west of the park visitor center at Fruita. To hike the full length of the gorge, start at a point northeast of Torrey on Highway 24 between mile posts 72 and 73. A shorter alternative is to do just the bottom half of the gorge, which is more interesting anyway. To do this hike, park at the Chimney Rock Trailhead between mile posts 76 and 77.

Trail or Route Conditions If you start between mile post 72 and 73, get right in the shallow creek bed and walk. It's uninteresting at first, then it deepens. The lower half of the canyon is "V" shaped and deep. In the area below the Goosenecks Overlook, there are several small waterfalls, but these can be circumvented easily. You will exit the canyon just above the park visitor center. If you park at the Chimney Rock Trailhead, walk south across the Highway and into a shallow wash. This drainage deepens quickly and joins the main canyon just above the Goosenecks. Before or after your hike, you might drive to the Goosenecks Overlook. This side road begins just east of the Chimney Rock Trailhead and very near mile post 77. After two kms you'll arrive at the overlook, where you can look straight down about 200 to 250 meters. *Don't do this hike if bad weather is threatening.* The hike through Fremont River Canyon is not featured in this edition, but it's still a pretty good hike. You can walk down-canyon on hiker-made trails from Highway 12 in the upper part; or hike up from the campground at the bottom end of the gorge. The lower end of the canyon is the best part.

Elevations The upper creek entry point is about 2000 meters altitude, while the Chimney Rock Trailhead is about 1850. The visitor center is about 1600 meters.

Hike Length and Time Needed To hike the full length of the canyon is to walk about 13 kms. This will take some people almost an entire day to complete. If you start at the Chimney Rock Trailhead and hike to the visitor center, it's about 8 kms. To do just this bottom end will be an easy half-day hike for the average person. If you have but one car, and need a ride back to your vehicle, ask someone at the visitor center about people going that way. Many park employees go home to Torrey at about 4:30 pm. You can also use a mountain bike to return to your car; but it's a steep ride back up to the head of the canyon.

Water Although there is a year-round flow in Sulphur Creek, it's not the safest or best tasting water in the world, so take your own.

Maps USGS or BLM map Loa(1:100,000), or Torrey(1:62,500).

Main Attractions A deep canyon, moderately good narrows, several waterfalls, easy access.

Ideal Time to Hike Because you'll be wading so much, do this one in warm weather, from about early May through late September or early October.

Hiking Boots Wading boots or shoes.

Author's Experience The author camped at the Cooks Mesa Trailhead, and hiked down-canyon and up to Chimney Rock; then immediately back down into and through the lower canyon to the visitor center, all in about 5 hours. He hitched a ride back to his car with a park employee.

One of several waterfalls in Sulphur Creek in Capitol Reef National Park.

Map 17, Sulphur Creek, Utah

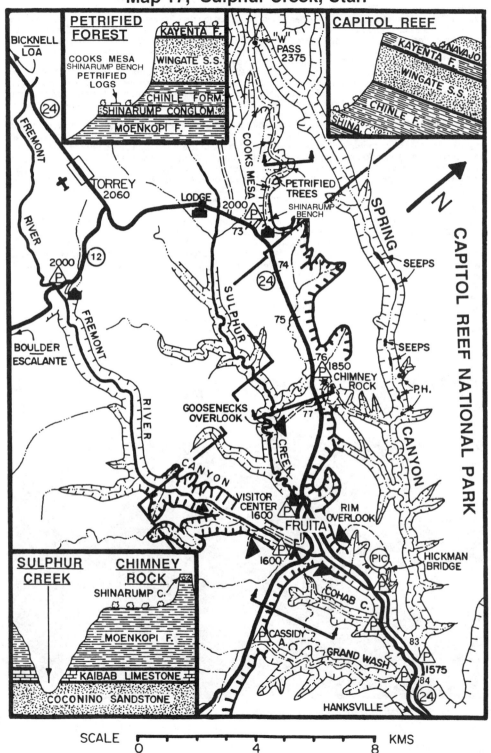

PETRIFIED FOREST

KAYENTA F.
WINGATE S.S.
COOKS MESA
SHINARUMP BENCH
PETRIFIED LOGS
CHINLE FORM.
SHINARUMP CONGLOM.
MOENKOPI F.

CAPITOL REEF

NAVAJO
KAYENTA F.
WINGATE S.S.
CHINLE F.
SHIN. C.

BICKNELL
LOA
24
FREMONT
RIVER
TORREY 2060
LODGE
2000
73
P
"W" PASS 2375
COOKS MESA
PETRIFIED TREES
SHINARUMP BENCH
SPRING
N
SEEPS
2000
P
12
FREMONT
RIVER
BOULDER
ESCALANTE
SULPHUR
74
75
24
76
1850
P
CHIMNEY ROCK
SEEPS
P.H.
77
CANYON
CAPITOL REEF NATIONAL PARK
GOOSENECKS OVERLOOK
CREEK
CANYON
VISITOR CENTER 1600
FRUITA
P
RIM OVERLOOK
HICKMAN BRIDGE
P
1600
P
PIC
P

SULPHUR CREEK   **CHIMNEY ROCK**

SHINARUMP C.
MOENKOPI F.
KAIBAB LIMESTONE
COCONINO SANDSTONE

P
CASSIDY A.
COHAB C.
GRAND WASH
83
P
1575
P
84
HANKSVILLE
24

SCALE   0   4   8   KMS

Waterpocket Fold Canyons, Utah

Location and Access The Waterpocket Fold is a geologic feature that forms the backbone of Capitol Reef National Park. This fold runs from Thousand Lake Mountain to Lake Powell in south central Utah. Buckling of the earth's crust, combined with different erosional characteristics of each formation, make the Capitol Reef. The three featured formations are the Wingate Sandstone, Kayenta Formation and Navajo Sandstone. That part of the Fold shown here is in the middle section of the park, between Notom and the Sandy Ranch. To get there, drive northwest from Bullfrog on the partly-paved Bullfrog-Notom Road. Most people however, drive along Highway 24 west of Fruita to between mile posts 88 and 89, then turn south on the Bullfrog-Notom Road. One can also get to these canyons via the now-paved Burr Trail Road from the Escalante-Boulder area. The top of this map is about 10 kms south of Highway 24 and 3 kms south of Notom. The Sandy Ranch is about 100 meters off the bottom of this map. You should see a sign naming each drainage as you drive along the road.

Trail or Route Conditions There are four canyons featured here: Burro Wash, Cottonwood Wash, Fivemile Wash, and Sheets Gulch. To enter each merely walk west or southwest from the road in the dry creek beds. You have to walk 2-3 kms in each to get to the challenging parts, then they all constrict and you'll be confronted with potholes, dropoffs, and tight narrows. You can come down Sheets Gulch, but going up is difficult as you'll have a 3 meter dropoff to climb. The author is familiar only with Burro and Cottonwood, and in each of those he was stopped by pools of water and the wrong kind of hiking boots. This leaves room to explore. There are other good little canyons all along this part of the Waterpocket Fold.

Elevations All car-parks, about 1500 meters; top of fold, about 2000 meters.

Hike Length and Time Needed Each of these canyons is only 8 or 9 kms long. For the most part each is a half-day hike. Some people might do two in one day, as they're all near one another. As you hike into each canyon, remember the narrows are in the Navajo Sandstone. If you can reach the upper end of any, you'll come to the Wingate Sandstone, which forms the western cliff face.

Water There's no permanent running water in any of these canyons, but you will find plenty of potholes or "waterpockets". If there's water in these potholes, it will likely be good to drink(right after a storm), but it's recommended you always carry your own.

Maps USGS or BLM map Loa(1:100,000), or Notom(1:62,500).

Main Attractions Short, but good narrows in each of these little-known canyons.

Ideal Time to Hike Spring or fall, or morning hike in summer. Maybe in some winter warm spells(?).

Hiking Boots Because of the many potholes, better take wading boots or shoes.

Author's Experience The author has read about Sheets Gulch, and has hiked up about halfway into Burro and Cottonwood Washes. These were both done on the same day. If you're there after recent rains, expect to do some wading.

All canyons of the Waterpocket Fold have narrows, but this one in Burro Wash is extra narrow and dark.

Map 18, Waterpocket Fold Canyons, Utah

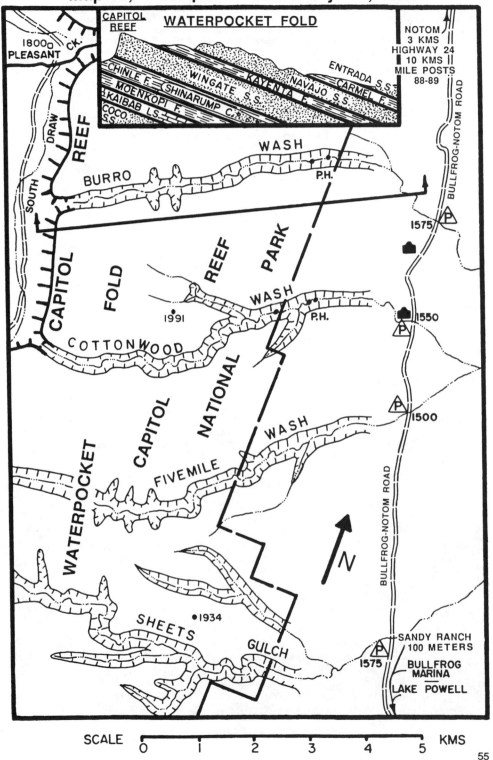

SCALE

0 1 2 3 4 5 KMS

Muley Twist Canyon, Utah

Location and Access Muley Twist Canyon is located in the southern part of Capitol Reef National Park, about halfway between Lake Powell on the south, and Highway 24 and Fruita on the north. There are three ways of reaching the trailhead. Perhaps the most-used route is to drive along Highway 24 east of Fruita to between mile posts 88 and 89, then turn south on the Bullfrog-Notom Road. After driving about 53 kms you'll come to the Burr Trail Road. Turn west and drive up the good zig zag road 3 kms to the pass at the top of the dugway and the trailhead just beyond. The other possibility is to go to Boulder, a small settlement on the south side of the Boulder Mountains and turn east on the now-paved Burr Trail. Drive about 56 kms to this same trailhead. You can also drive northwest from the Bullfrog Marina on Lake Powell along the Bullfrog-Notom Road. Parts of this road not administered by the National Park Service are now paved.

Trail or Route Conditions From the trailhead begin walking south in the dry creek bed. It's shallow at first, then high cliffs begin to rise to the east. About halfway through, you'll come to a sign stating "The Post" and "Halls Creek". This points out the Cut-off Trail which is used by many because it makes a shorter hike out of what is far too long of a day for most people. If you use this trail it will take you to a place called The Post on the Bullfrog-Notom Road. If you continue down-canyon, you'll find at least three very deep undercuts. The biggest of these is labeled Cowboy Camp on the map. It's full of Cowboy Glyphs dating back to the 1920's. Not far below this alcove the canyon turns east and soon meets Halls Creek. Turn north and follow an old wagon road back to the Bullfrog-Notom Road and The Post, then road-walk, drive a second car, or mtn. bike, back to the trailhead.

Elevations The trailhead, 1740 meters; bottom end of canyon, about 1440 meters.

Hike Length and Time Needed From the trailhead to Halls Creek is about 19 kms; Muley Twist Canyon to The Post is 8 kms; from The Post to the trailhead via the road is 7 kms. A total distance of 34 kms. That's one very long day-hike. Or you could use the Cut-off Trail which is 4 kms, making the round-trip only about 17 kms. However, Cowboy Camp, one of the best sites in the canyon, is in the bottom end. Another possibility; park at The Post and do a loop-hike of the lower part of the canyon. The walking is easy at all points along the way.

Water Take water with you and have plenty in your car. The author found several small seeps, but that was after a wet spell. Normally there is no reliable water in the canyon.

Maps USGS and BLM maps Escalante and Hite Crossing(1:100,000), or Wagon Box Mesa and Mt. Pennell(1:62,500).

Main Attractions Three huge undercuts, one of which is 55 to 60 meters deep(Cowboy Camp). Also, large Navajo Sandstone bluffs and domes.

Ideal Time to Hike Spring, fall or winter dry spells, or early mornings in summer.

Hiking Boots Any dry weather boots or shoes.

Author's Experience The author camped just out of sight of the road near the trailhead, then hiked this entire 34 km loop in about 7 1/2 hours.

This huge undercut, which the author calls Cowboy Camp, is 55 to 60 meters deep.

Map 19, Muley Twist Canyon, Utah

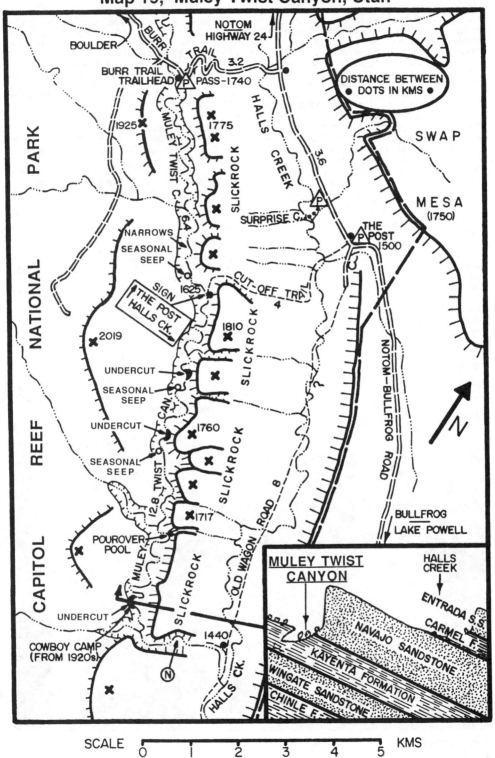

BOULDER

BURR TRAIL

NOTOM HIGHWAY 24

BURR TRAIL TRAILHEAD P PASS-1740

3.2

DISTANCE BETWEEN DOTS IN KMS

SWAP

1925 ✗

MULEY TWIST C.

1775 ✗

HALLS CREEK

3.6

MESA (1750)

SLICKROCK

✗

6.4

NARROWS
SEASONAL SEEP

SURPRISE C. P

THE POST 1500 P

SIGN
THE POST
HALLS CK.

1625

CUT-OFF TR.

4

✗ 2019

1810 ✗

SLICKROCK

NOTOM-BULLFROG ROAD

UNDERCUT
SEASONAL SEEP

HALLS C. CAN.

✗

UNDERCUT

1760 ✗

SLICKROCK

SEASONAL SEEP

12.8 TWIST C.

✗

OLD WAGON ROAD 8

BULLFROG LAKE POWELL

✗ 1717

POUROVER POOL ✗

MULEY

1717

SLICKROCK

MULEY TWIST CANYON

HALLS CREEK

ENTRADA S.S.
CARMEL F.

NAVAJO SANDSTONE

UNDERCUT

CAPITOL

REEF

NATIONAL

PARK

SLICKROCK

KAYENTA FORMATION

WINGATE SANDSTONE

CHINLE F.

COWBOY CAMP (FROM 1920s)

1440

N

HALLS CK.

✗

N

SCALE 0 1 2 3 4 5 KMS

57

Lower Halls Creek Narrows, Utah

Location and Access The narrow slot canyon in the lower end of Halls Creek is located at the extreme southern end of Capitol Reef National Park and just northwest of Bullfrog Marina on Lake Powell. It's also just west of the Bullfrog-Notom Road. Get to this hike from Bullfrog by driving north on Highway 276 about 10 kms, then turn northwest and drive about 25 more kms on the partly-paved Bullfrog-Notom Road in the direction of Capitol Reef and/or Notom. You will then see a sign at a side road which will say something about the Halls Creek or Capitol Reef Overlook. Drive this good secondary road for 4 or 5 kms to the trailhead. If you're coming from the north, leave Highway 24 between mile posts 88 and 89 and drive south to the same turnoff. From Boulder, drive east along the now-paved Burr Trail Road. Part of the Bullfrog-Notom Road is made of gravel and clay, and can be slick in places during heavy rains. Be sure to have a USGS map of the region to help locate the right turnoff.

Trail or Route Conditions From the trailhead walk north on a good and well-used trail. This path takes you down about 2 kms to the bottom of the valley and Halls Creek. At the bottom you can hike up a side canyon coming down from the west. Walk this moderately rugged canyon bottom to Brimhall Arch, a worth while side-trip. If going to the narrows, walk south along some of several old cow trails. At times you'll be using an old road originally built by Charles Hall who built Halls Ferry in 1881. It's easy walking all the way. When you reach the narrows, the creek bed will veer to the right or west where it cuts deep into the Navajo Sandstone. Walk down through the narrows, then head back north along Halls Road which crosses Hall Divide. From the Divide, walk west for a look down into the slot. Another interesting hike would be to drive south from the trailhead, past the airstrip, and just beyond the spring turn south and park not far north of an overlook of Halls Creek Narrows. The last part of this road is rough, so for those driving a car, take a shovel to smooth out a gully or two; or use a mtn. bike.

Elevations Trailhead, 1600 meters; Hall Divide, 1275; Halls Creek Narrows Overlook, 1550 meters.

Hike Length and Time Needed From the trailhead to the narrows is about 14-15 kms, one way. It'll seem longer than that on a hot summer day. Plan on a long all-day hike round-trip. From the end of the 4WD road to the Halls Creek Overlook will only take a couple of hours round-trip.

Water Always have water in your car and in your pack, but there will always be some water in Halls Creek Narrows year-round. The small seeps or springs come out where the Navajo Sandstone and the Kayenta Formation meet.

Maps USGS or BLM map Hite Crossing(1:100,000), or Hall Mesa(1:62,500).

Main Attractions Good narrows, a fine arch, great scenery and interesting geology.

Ideal Time to Hike Spring or fall are best. Summers are hot. By late May through July there will be many large deer flies along the creek bed which bite bare legs. Wear long pants at that time!

Hiking Boots Dry weather boots or shoes, but you'll do some wading in the narrows, so you might take a second pair of waders for that section.

Author's Experience Once the author hiked up from the lake and through the narrows in about 9 hours, round-trip. Another time he hiked to Brimhall Arch, then to the beginning of the narrows and back to the trailhead, in 6 1/2 hours round-trip. Still later, he drove and walked to the overlook on the eastern rim.

Looking southwest where Halls Creek begins to cut into the Navajo Sandstone to form the narrows.

58

Map 20, Lower Halls Creek Narrows, Utah

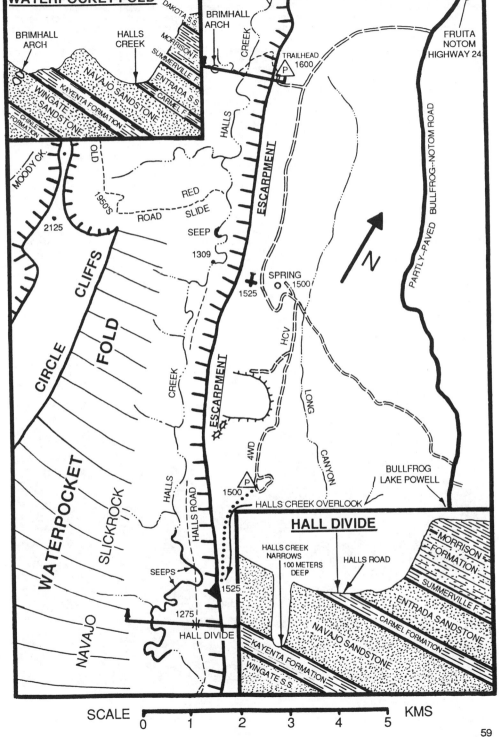

Courthouse Wash, Utah

Location and Access Courthouse Wash is located within Arches National Park, which is about 8 kms due north of Moab. To reach this canyon, turn south from Interstate 70 at Crescent Junction and drive to where the paved road turns west towards Canyonlands National Park and Dead Horse Point(or drive north out of Moab). Park just south of this major junction which is at the upper end of the wash and between mile posts 136 & 137. To enter in the middle part of Courthouse, proceed to the park visitor center as shown. After getting more information and/or maps, drive into Arches and park at the 1260 meter car-park. Still a third access point would be from the bottom or lower end of the canyon. To get there, drive north out of Moab and park next to mile post 129, which is just west of the Colorado River bridge. It's there Courthouse Wash leaves the canyon and enters the river. All three access points are easy to get to with no driving on dirt roads.

Trail or Route Conditions At the upper end of the canyon is a large dryfall. The creek, which is usually dry, drops off the rim, which is made of the Curtis Formation, and into the canyon bottom which is made of the Entrada Sandstone. On the south side of the dryfall is an old cattle trail which allows entry to the upper wash. The upper part of the wash has high walls and is moderately narrow. Further down the wash opens up, and there are side canyons to explore. There are no trails in the wash; you simply walk down the stream bed. It's very easy walking and there's no bushwhacking. The bottom is sandy and smooth. Many people walk barefoot through the water during the summer heat. There are more interesting narrows two or three kms below the park highway.

Elevations Upper trailhead, 1375 meters; midway point, 1260; bottom end, 1225 meters.

Hike Length and Time Needed From top to the bottom this canyon is about 20 kms long; about 12 kms in the upper part, about 8 in the lower end. Most people can do this hike easily in one day, but you'd need either two cars, or one car and a bike. Otherwise you'd be looking at a 13 km walk or hitch hike back to your car along the highway.

Water The author found running water the length of the wash except for the upper 5 kms or so. There are no cattle in the canyon today, so it would likely be good to drink in the upper end. However, nowadays there's going to be human pollution. The bottom end of the wash is especially popular, so it's best to carry your own drinking water.

Maps USGS or BLM map Moab(1:100,000), or Arches National Park(1:50,000).

Main Attractions The Courthouse Towers and a cool wade in a sandy wash in summer.

Ideal Time to Hike Temperatures are more comfortable in spring and fall, but with the wading, it can be enjoyable even during the heat of summer. Winter hiking? Maybe.

Hiking Boots Wading boots or shoes.

Author's Experience The author walked from the Canyonlands turnoff to the bottom in less than 4 hours, but as usual he was in a hurry. He then walked part way back to his car before getting a ride. A road or mtn. bike locked up at the bottom of the canyon would provide easy transportation back to your car.

Most of Courthouse Wash has a flat sandy bottom just like you see here.

Map 21, Courthouse Wash, Utah

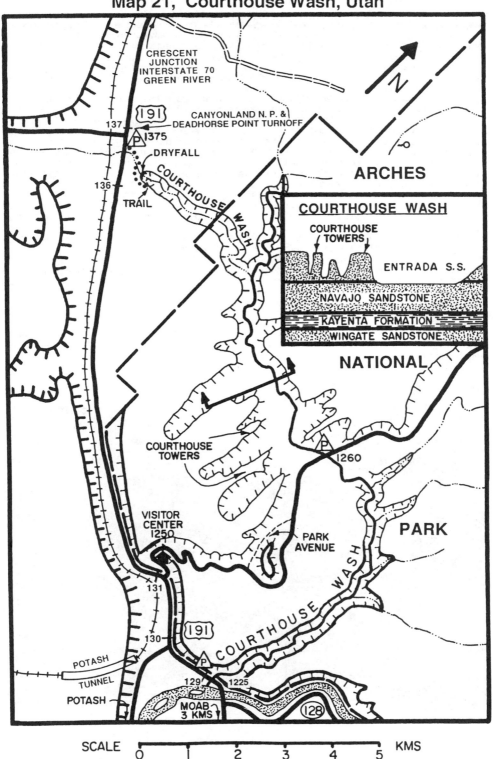

CRESCENT
JUNCTION
INTERSTATE 70
GREEN RIVER

191
137
1375

DRYFALL

136

COURTHOUSE WASH

TRAIL

CANYONLAND N. P. &
DEADHORSE POINT TURNOFF

N

ARCHES

COURTHOUSE WASH

COURTHOUSE
TOWERS

ENTRADA S.S.

NAVAJO SANDSTONE

KAYENTA FORMATION

WINGATE SANDSTONE

NATIONAL

COURTHOUSE
TOWERS

P
1260

PARK

VISITOR
CENTER
1250

PARK
AVENUE

COURTHOUSE WASH

131

191

130

POTASH

TUNNEL

POTASH

P

129   1225

MOAB
3 KMS

128

SCALE   0   1   2   3   4   5   KMS

Professor Creek(Mary Jane Canyon), Utah

Location and Access Professor Creek, sometimes called Mary Jane Canyon, is located due east of Moab and just south of Utah State Highway 128 and the Colorado River. Get there by driving north out of Moab for 3 kms, then turn east on Highway 128 and drive to a turnoff between mile posts 18 and 19. Turn south at the sign "Ranch Road--Dead End". This road runs into the Professor Valley. You could also get there by exiting Interstate 70 at Cisco, not far west of the Utah-Colorado state line, and drive south on Highway 128. About 3 1/2 kms up Professor Valley is a turnoff to the south in a grove of cottonwood trees. You can park and/or camp there. This last 3 1/2 km section of road has a good gravel surface and is considered all-weather.

Trail or Route Conditions From the trailhead and campsite at 1325 meters, walk south on one of several cow trails heading into the canyon. It's a broad valley at first, then the canyon narrows and the cow trails slowly fade. Further up, one must wade right in the creek, but it's a small shallow stream and the water is barely ankle deep. The narrow section of this little canyon is 2-3 kms long. In the upper end you'll eventually reach a chokestone as shown in the foto. It makes a waterfall you can't get around. If you want to go further into the canyon, you'll have to back-track a ways, and locate a route out of the slot, then walk along side, but above the narrows to the south. This is slow walking. Further up there are more narrows and another chokestone. The energetic hiker could walk and climb to the south and reach the top of Fisher Mesa for some excellent views of the valley.

Elevations Trailhead, 1325 meters; first waterfall 1450; top of Fisher Mesa, 2100 meters.

Hike Length and Time Needed From the trailhead to the first waterfall is about 7 kms. This waterfall is in the best part of the canyon so it seems a bit fruitless to try to go further. One can do this hike in about half a day.

Water The water in Professor Creek comes from springs three or four kms above the first falls so it should be good to drink. There seems to be no cattle in the upper part of this drainage.

Maps USGS or BLM map Moab(1:100,000), or Castle Valley(1:62,500).

Main Attractions A short hike with moderately good narrows and easy access. Just below the first falls, the walls close in to about 2 1/2 or 3 meters. If you have the energy to hike to the mesa top, the view of Castle Rock and other nearby monoliths is very good.

Ideal Time to Hike Spring through fall. Summers are hot, but you can cool off in the water.

Hiking Boots Wading boots or shoes.

Author's Experience The author hiked up to the waterfall, then back-tracked to an exit point. He then climbed upon the east side of the canyon narrows and walked up-valley. Finally he got back down into the stream bed, where he found the second waterfall. The walking in this upper valley outside the narrows is difficult and slow, so it's not recommended.

This chokestone and resultant waterfall impedes traffic in Professor Creek.

Map 22, Professor Creek(Many Jane Canyon), Utah

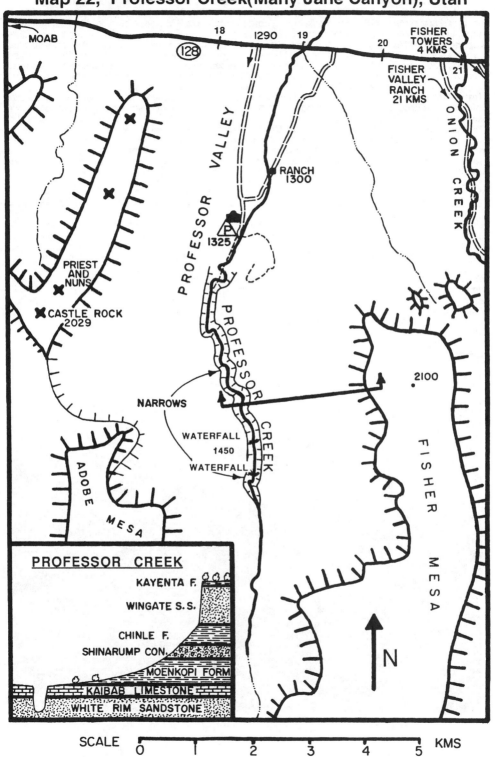

MOAB

128

18
1290
19
20

FISHER
TOWERS
4 KMS

21

FISHER
VALLEY
RANCH
21 KMS

ONION CREEK

PROFESSOR VALLEY

RANCH
1300

P
1325

PRIEST
AND
NUNS

CASTLE ROCK
2029

PROFESSOR CREEK

2100

NARROWS

WATERFALL
1450

WATERFALL

ADOBE MESA

FISHER MESA

PROFESSOR CREEK

KAYENTA F.
WINGATE S.S.
CHINLE F.
SHINARUMP CON.
MOENKOPI FORM.
KAIBAB LIMESTONE
WHITE RIM SANDSTONE

N

SCALE   0   1   2   3   4   5   KMS

63

Negro Bill and Mill Creek Canyons, Utah

Location and Access Both of these canyons are just east of Moab, and are very popular. Access is very easy. To reach Negro Bill Canyon, drive north out of Moab about 3 kms, then turn right or east, and drive along Utah State Highway 128 for 5 kms to the mouth of the creek which is right at mile post 3. Park in the parking lot. For Mill Creek, drive southeast out of Moab on Mill Creek Road, cross the Mill Creek Bridge, then a short ways after that turn left or east on the road going into the mouth of Mill Creek Canyon. This is called Power House Lane. From the trailhead at 1350 meters, you can enter either Mill Creek or the North Fork of Mill Creek. One could also use the Sand Flats Road to gain access to the upper part of each canyon. Either of the BLM offices in Moab can lead you in the right direction if you need help.

Trail or Route Conditions There's a well-used trail into Negro Bill Canyon. The old road is now blocked off to all motorized vehicles and mtn. bikes. There's a year-round flowing stream throughout. The lower end of this canyon is especially popular up to Morning Glory Arch, which is in a side canyon. Just stay on the most-used trail. In the North Fork of Mill Creek, there's also a trail, but it gradually fades the further up you go. Near the upper end of both Negro Bill and North Fork canyons, and where the route symbol is shown on the map(1725), there is no trail. At that point you can leave either canyon and walk over the saddle between and enter the other. There's more than one way down into either canyon from the mesa and the Sand Flats Road. If you're making a loop-hike of both canyons, you'll have to route-find at that point. A loop-hike is a very long day-hike for most people, therefore not recommended unless you have two cars, or a car and a mtn. bike to make a shuttle.

Elevations Mouth of Negro Bill, 1250 meters; mouth of Mill Creek, 1350; and the pass between the two drainages, about 1725 meters.

Hike Length and Time Needed The length of each canyon is about 13 kms. To make a loop-hike, for example, to walk up Negro Bill, then down Rill and the North Fork, will take some people two days, but it can be done in one long day. Or one can just hike up either canyon on day-hikes. The lower parts of both canyons are the most scenic and popular. Most people do the hike up to Morning Glory Arch in about 2 hours, round-trip.

Water Most parts of all the canyons on this map have year-round flowing water which is generally good for drinking, but there are cattle there at times. There's an excellent spring right under Morning Glory Arch.

Maps USGS or BLM map Moab(1:100,000), or Moab and Castle Valley(1:62,500).

Main Attractions The best parts of either canyon are the Navajo Sandstone walls near the bottom of each. Also, Morning Glory Arch in Negro Bill is worth a side-trip.

Ideal Time to Hike Spring or fall. Summers are hot, but just tolerable with all the running water and wading pools around.

Hiking Boots Wading boots or shoes.

Author's Experience The author parked at Negro Bill Canyon, walked up and over to Rill Creek, then down to the mouth of Mill Creek. That was about 9 hours. He then took half an hour to hitch hike back to his car. On another trip, he just went up to Morning Glory and returned in 2 hours.

At the upper end of a side drainage in Negro Bill Canyon is Morning Glory Arch.

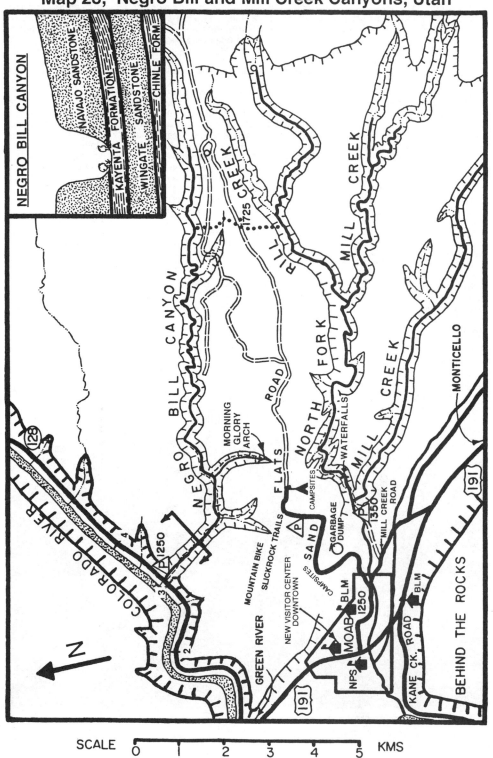

NEGRO BILL CANYON

NAVAJO SANDSTONE

KAYENTA FORMATION

WINGATE SANDSTONE

CHINLE FORM.

NEGRO BILL CANYON

BILL CREEK

1725 CREEK

MILL CREEK

NORTH FORK

MILL CREEK

MONTICELLO

MORNING GLORY ARCH

ROAD

FLATS

WATERFALLS

CAMPSITES

128

1250

MOUNTAIN BIKE SLICKROCK TRAILS

GARBAGE DUMP

P

SAND

1350

MILL CREEK ROAD

191

COLORADO RIVER

GREEN RIVER

NEW VISITOR CENTER DOWNTOWN

CAMPSITES

BLM

1250

BLM

BEHIND THE ROCKS

MOAB

KANE CK. ROAD

NPS

191

N

SCALE   0   1   2   3   4   5   KMS

Pritchett and Hunters Canyons, Utah

Location and Access The area of Pritchett and Hunters Canyons, sometimes known as "Behind The Rocks", is found just to the southwest of Moab. To get there, drive to the south end of town, and turn west on Kane Creek Road. This takes you to the Colorado River, then southwest and eventually into Kane Springs Canyon. At the end of the paved road turn left to find a car-park and campsite just inside the mouth of Pritchett Canyon. Or continue driving south into Kane Springs Canyon. At the mouth of Hunters is another place to park and camp. Since this whole area is close to Moab, it's a favorite place for locals to use 4WD's and motor cycles. If you dislike the racket made by these vehicles, avoid the area in late afternoons or on weekends. However, in the 1990's, there seems to be fewer motor bikes, and lots more mtn. bikes. The lower end of Kane Springs Canyon is a popular area for tourists to camp, hike and mtn. bike.

Trail or Route Conditions Pritchett Canyon has a 4WD road, but many people walk into the area and to Pritchett Natural Bridge. This route is now becoming popular with mtn. bikers. If you're hiking or mtn. biking in this canyon, you can take any number of side trips into and through the fins of the Navajo Sandstone. There are hundreds of possibilities. There's a hiker's trail in Hunters Canyon which is deep and moderately narrow. At the upper end of the drainage you'll come to a dryfall. Beneath it is a spring and the beginning of the year-round flowing stream. When you arrive at the falls, back-track about 100 meters and climb an old Indian trail up the south wall. This will put you in the upper drainage and at the end of a road which you can use to reach Pritchett Canyon. With two cars, or a car and a mtn. bike, you can make a loop-hike of both canyons.

Elevations Both car-parks, about 1250 meters; upper Pritchett Canyon road, about 1450 meters.

Hike Length and Time Needed To walk up Pritchett, over the pass and down Hunters, then road-walk(or bike) back to your car, would be about 20 kms. From the mouth of Pritchett to Pritchett Natural Bridge is about 7 kms, or 14 round-trip. Either of these hikes will take the average person a full day. Perhaps the best hiking is in the area called "Behind The Rocks". This is a maze of bluffs and fins made of Navajo Sandstone, with many short but narrow side canyons.

Water There's a spring in Hunters Canyon just below the Wingate dryfall and it flows year-round. Otherwise, take water in your car and in your pack.

Maps USGS or BLM maps Moab and La Sal(1:100,000), or Moab and Hatch Point(1:62,500).

Main Attractions Navajo fins and bluffs, Pritchett Natural Bridge, and petroglyph and pictograph panels along both sides of the Colorado River. There's also one panel in lower Pritchett Canyon. Also, very easy access, with Moab just a few minutes away.

Ideal Time to Hike Spring or fall, or some winter warm spells. Summers are hot.

Hiking Boots Any dry weather boots or shoes.

Author's Experience On one trip, the author camped at the mouth of Hunters Canyon, then next morning he hiked up Hunters, down Pritchett, and along the road back to his car, all in 6 hours.

These are the fins and bluffs of the area known as "Behind the Rocks". La Sal Mountains in the background.

Map 24, Pritchett and Hunters Canyons, Utah

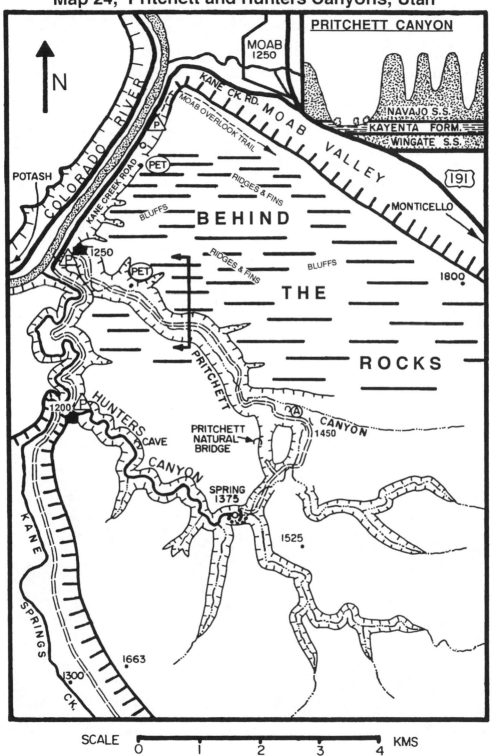

PRITCHETT CANYON

MOAB
1250

NAVAJO S.S.
KAYENTA FORM.
WINGATE S.S.

N

KANE CK. RD. MOAB VALLEY

MOAB OVERLOOK TRAIL

191

COLORADO RIVER

KANE CREEK ROAD

P

PET

POTASH

RIDGES & FINS

BLUFFS

BEHIND

MONTICELLO

1800

P 1250

PET

RIDGES & FINS

BLUFFS

THE

PRITCHETT

ROCKS

1200 P

HUNTERS

CAVE

CANYON

PRITCHETT
NATURAL
BRIDGE

A CANYON
1450

SPRING
1375

1525

KANE

SPRINGS

1663

1300

CK.

SCALE   0   1   2   3   4   KMS

67

Hatch Wash, Utah

Location and Access Hatch Wash lies about 30-35 kms due south of Moab and due west of La Sal Junction. One could enter this canyon by driving out of Moab on the Kane Creek Road, which takes one to the bottom part of the canyon. But the bottom end is choked with tamaracks and not so interesting. To get to the more interesting upper end, drive along Highway 191 between Moab and Monticello. Just south of mile post 93, the paved Needles Overlook Highway begins and runs northwest. Drive this road past the Wind Whistle Campground about 5 or 6 kms and look for the good road running northeast near the head of Little Water Creek. From there drive as far as your vehicle will allow into Threemile Creek and to, or near, the line cabin on lower Little Water. Watch out though, some parts of these roads are sandy. A mtn. bike will extend the range of your car if the road is too rough. There are other area side roads you can take as shown on the map, but coming in from the west side seems best. The road running past Looking Glass Rock and Hatch Rock & Rockland Ranch, is very well-maintained and heavily used.

Trail or Route Conditions There's a very old mining exploration track running into the bottom of Hatch Wash (shown here as a trail), but it's useless as a road today. Otherwise, the wash bottom has only trails of wild cattle. From the line cabin on Little Water, there's an old cow trail running down into the bottom, then it's walking the creek bed down into Hatch. For Threemile, it's best to stay on the north rim of the canyon until you're near Hatch Wash, then route-find down into the canyon. Further down Hatch Wash and near the rincons, there are two old cattle trails leading into or out of the canyon. There's also another route in or out just around the corner from the wreckage of a test rocket. It's possible to come right down the main drainage, but there is a lot of brush in upper Hatch Wash and Hatch Ranch Canyon. There were some beaver ponds in the main drainage in March, 1994, just above where Hatch Ranch Canyon enters.

Elevations The test rocket wreckage(perhaps from the same rocket as found in Harts Draw) is about 1600 meters, while the well at the head of Threemile Creek is 1825 meters.

Hike Length and Time Needed A loop-hike from the line cabin, down Little Water and Hatch to the rincons, then up to the mesa, and by road back to your car is about 24-26 kms. That's a very long day for most. One could shorten the trip by walking part way in and returning the same way. Or if you could leave a mtn. bike near one exit, it would eliminate a road-walk. Another possibility would be to make an overnight trip of it

Water There's running water for the entire length of Hatch Wash. The author drank it, and lives on, but there are some wild cattle and beaver in the drainage, so it's best to just take you own, at least on day-hikes, or filter or purify it first if camping.

Maps USGS or BLM map La Sal(1:100,000), or Hatch Point and La Sal Junction(1:62,500).

Main Attractions The best part of Hatch Wash is from Little Water Creek down to the rincons. You'll find sheer canyon walls, a year-round stream, cottonwood trees, and many good campsites.

Ideal Time to Hike Spring or fall, or maybe during some winter dry spells(?). Summers are warm.

Hiking Boots Any dry weather boots or shoes.

Author's Experience The author camped at the cave house and explored down Threemile Creek, then the main canyon to the rincons, and finally up and back along the mesa top. That took 10 hours! Second day he explored Little Water Creek for a couple of hours. At a later date he attempted to walk down-canyon from the spring in the upper part of Hatch Wash, but was stopped by beaver ponds and brush.

The bottom of upper Hatch Wash is a narrow green paradise in between red canyon walls.

Map 25, Hatch Wash, Utah

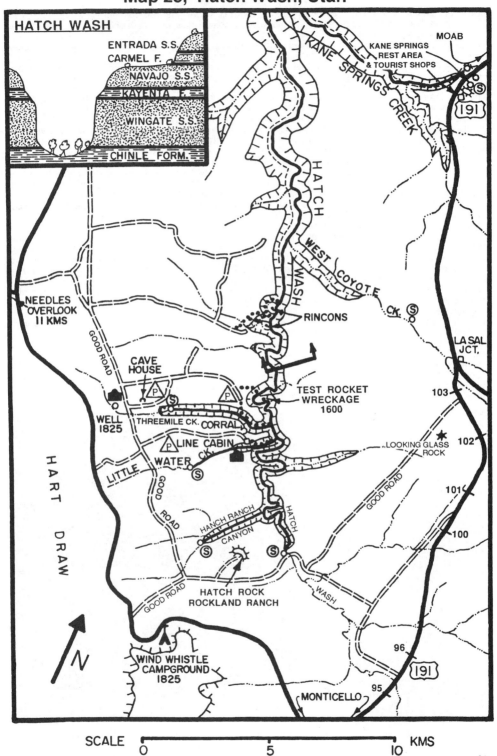

HATCH WASH

ENTRADA S.S.
CARMEL F.
NAVAJO S.S.
KAYENTA F.
WINGATE S.S.

CHINLE FORM.

MOAB

KANE SPRINGS
REST AREA
& TOURIST SHOPS

KANE SPRINGS CREEK

191

HATCH

WEST COYOTE CK.

WASH

RINCONS

LA SAL JCT.

NEEDLES
OVERLOOK
11 KMS

GOOD ROAD

CAVE
HOUSE

P

P

103

TEST ROCKET
WRECKAGE
1600

WELL
1825

THREEMILE CK.

CORRAL

102

LOOKING GLASS
ROCK

P LINE CABIN

WATER

LITTLE

CK.

GOOD ROAD

GOOD ROAD

101

HART DRAW

HANCH RANCH

HATCH

CANYON

100

HATCH ROCK
ROCKLAND RANCH

WASH

GOOD ROAD

96

191

WIND WHISTLE
CAMPGROUND
1825

95

MONTICELLO

SCALE 0   5   10 KMS

N

Harts Draw, Utah

Location and Access Harts Draw is located south of Moab, just east of The Needles District of Canyonlands National Park and Newspaper Rock along Indian Creek, and north of Monticello. Get to Harts Draw by driving along Highway 191, the road linking Moab and Monticello. Between mile posts 86 and 87, turn west on State Highway 211(the paved road running to The Needles) and drive to Photograph Gap. From there turn north on a dirt road running to the head of Hart Spring Draw. This may be the best entry point, but there are others. One can also take the paved road running to The Needles Overlook, and use one of two or three routes into the lower end of Harts Draw. The graded road running between Photograph Gap and The Needles Overlook road is good for all vehicles.

Trail or Route Conditions In the bottom of Harts Draw Canyon are some cattle trails, but for the most part you will walk along side a small stream or in the creek bed itself. In Hart Spring Draw, perhaps the easiest way in, walk down the creek bed which usually has water in it. One can also camp at the Wind Whistle Campground and head southwest to a minor drainage of Bobbys Hole Canyon, and walk down an old abandoned cattle trail. Otherwise, Bobbys Hole Canyon is blocked off by dryfalls. In the canyon just west of the campground, there's a route down, but for roped climbers only. About 7 kms northwest of the campground are two routes into the lower end of Harts. You'll have to look for these. The author used the Indian log route once. A still better route into this same little side canyon is to walk along the rim on the north side and look for an easy route down through the Navajo Sandstone to the Kayenta Bench, then walk east to the head of the canyon and down a rockslide covering the Wingate. This is apparently the route used by the MaComb Expedition in 1859, which was looking for the confluence of the Green and Grand(later Colorado) Rivers.

Elevation The rincons, about 1600 meters; the canyon rim, 1850 to about 1900 meters.

Hike Length and Time Needed From Hart Spring, down Harts Draw and to the trailhead in the side canyon with the pieces of rocket wreckage(Indian log or MaComb Route) is about 23 km. With two cars, or a car and a mtn. bike, you could do it in one day, but since most of us have only one car, the recommended entry & exit points are Hart Spring Draw and the cattle trail into the lower end of Bobbys Hole Canyon One could day-hike or camp overnight.

Water There is year-round running water in the middle and upper parts of Harts Draw. There are also some cattle there at times, so try to take water directly from a spring source, or purify it.

Maps USGS or BLM map La Sal(1:100,000), or Harts Point and Hatch Rock(1:62,500).

Main Attractions Solitude in a little known and deep canyon with moderately good narrows. The part of the canyon most interesting is from the rincons south.

Ideal Time to Hike Spring or fall, or in some winter dry spells. Summers are warm.

Hiking Boots Any dry weather boots or shoes.

Author's Experience The author took two days exploring several routes into and out of this canyon. The first day was spent looking in the northeastern section. On the second day, he left his car near Hart Spring, and hiked down to the end of the flowing water, then back up the same way in 10 long hours. At a later date he hiked the MaComb route into the little side canyon.

This foto shows the droppoff leading down to a side canyon of Harts Draw. On the map this side canyon is labeled "Roped Route Only".

Map 26, Harts Draw, Utah

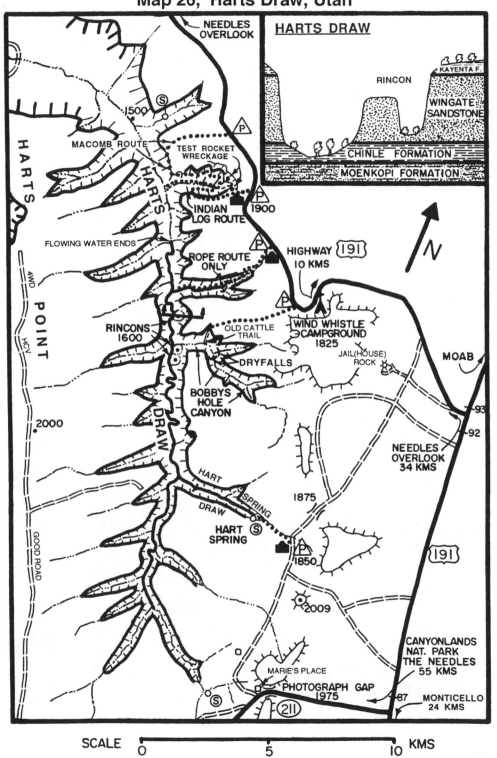

HARTS DRAW

KAYENTA F.
RINCON
WINGATE SANDSTONE
CHINLE FORMATION
MOENKOPI FORMATION

NEEDLES OVERLOOK

1500

MACOMB ROUTE

TEST ROCKET WRECKAGE

HARTS

P

INDIAN LOG ROUTE
1900

FLOWING WATER ENDS

ROPE ROUTE ONLY

P

HIGHWAY 191
10 KMS

P

N

4WD

HCV

POINT

RINCONS 1600

OLD CATTLE TRAIL

WIND WHISTLE CAMPGROUND
1825

MOAB

JAIL (HOUSE) ROCK

DRYFALLS

BOBBYS HOLE CANYON

93

92

2000

DRAW

NEEDLES OVERLOOK 34 KMS

HART

SPRING

DRAW

1875

HART SPRING

S

P

1850

191

GOOD ROAD

2009

CANYONLANDS NAT. PARK THE NEEDLES 55 KMS

MARIE'S PLACE

S

PHOTOGRAPH GAP 1975

211

87

MONTICELLO 24 KMS

SCALE
0   5   10   KMS

Introduction to Hikes in The Needles, Utah

Location and Access This map covers most of The Needles District of Canyonlands National Park. To get there, drive along Highway 191 between Moab and Monticello. Between mile posts 86 and 87, turn west onto the highway running to The Needles. Just after entering the park, stop at the visitor center before going further to get the latest information on trails, camping, etc. From there, drive west to the campground where the paved road ends. Beyond that a good gravel road runs west to Elephant Hill Trailhead. This is where most people with cars begin hiking. The paved road running north from near the campground ends at the Confluence Overlook Trailhead. Get more detailed information & local history from the author's other book, *Hiking, Biking and Exploring Canyonlands N.P. and Vicinity.*

Trail or Route Conditions All trails in this district are well-used and well-marked with sign posts and/or stone cairns. There are many good hikes, but here are three of the most popular. From the Confluence Overlook Trailhead, walk west to view the Confluence of the Green and Colorado Rivers. Ask park rangers about the trail down to the Confluence from the vicinity of the Overlook. From Elephant Hill, take the trail running south toward Chesler Park. There are several junctions along the way, so be sure you take a better map than this. The third hike is also from Elephant Hill. Walk or ride a mtn. bike up the steep 4WD road over Elephant Hill to the west and follow the signs toward Devils Lane, Cyclone Canyon(leave a mtn. bike there) and finally to Lower Red Lake Canyon. Your destination will be Spanish Bottom and the Colorado River. This is just an introduction, so get more details on these and other hikes at the visitor center, or from the book mentioned above.

Elevations Campground, 1580 meters; Colorado River, 1200; Chesler Park, 1710 meters.

Hike Length and Time Needed The one-way hike from Elephant Hill to the Colorado is about 15 kms. This is usually a two day hike, but fast walkers can do in one very long day. A 4WD or mtn. bike will make it shorter, easier and faster. The walk to Chesler Park is about 5 kms. This can be done in about half a day, round-trip. You can extend this hike to other trails and make a full day-hike out of it. The trail to the Confluence Overlook is about 8 kms one-way. It's an easy hike and will take most people 4-6 hours, round-trip; or 3 hours round-trip for fast hikers.

Water At the campground and visitor center. Carry water in your car and in your pack.

Maps USGS or BLM map La Sal(1:100,000), or the plastic Trails Illustrated map Needles & Maze (1:62,500).

Main Attractions Geology--spires and graben valleys. An overlook of the Confluence of the Green & Colorado Rivers. Upper Salt Creek and other nearby canyons are covered on the next map.

Ideal Time to Hike Spring or fall, or some warm dry spells in winter. Summers are really hot.

Hiking Boots Any dry weather boots or shoes.

Author's Experience Here are just three examples of what the author has done. For other hikes and times see his other book on Canyonlands. He once walked from the Confluence Trailhead to the Confluence Overlook, and back to his car in 2 1/2 hours. He once biked to Cyclone Valley, walked to the river and back, then biked back to Elephant Hill in 6 hours. He also hiked into Chesler Park and back in 4 hours.

One of the most popular hikes in The Needles region is into Chesler Park.

Map 27, Introduction to Hikes in The Needles, Utah

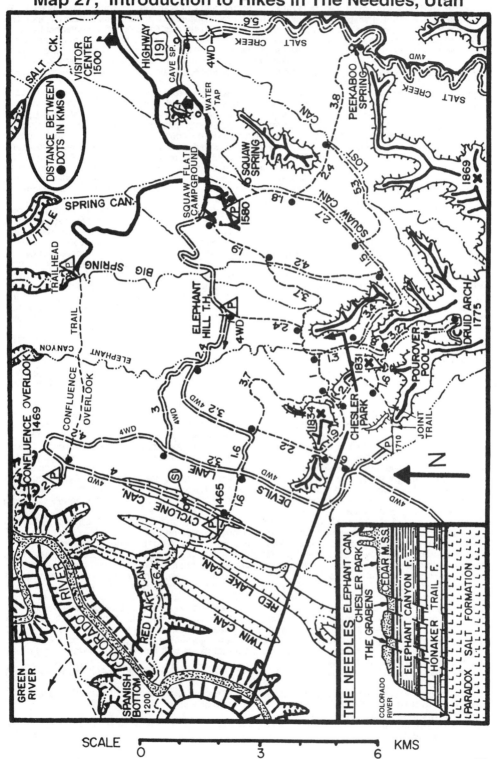

SCALE 0   3   6 KMS

Salt Creek, and Horse, Davis and Lavender Canyons, Utah

Location and Access Featured here are the upper ends of Salt Creek, and Horse, Davis and Lavender Canyons, all of which are located in The Needles District of Canyonlands National Park. The map shown here is from the author's other book, *Hiking, Biking and Exploring Canyonlands National Park and Vicinity*, and the road symbols are a little different; the heavy black dashed lines indicate roads, and the symbols which look like tents are campsites with road access. To get to this region, drive along Highway 191, the main road linking Moab and Monticello. Between mile posts 86 and 87, turn west onto The Needles Highway. This road runs west past Newspaper Rock, down Indian Creek and into The Needles. Just inside the park, stop at the visitor center before going further, to get the latest information on hiking, trails, and the new (1994) all-public-land access road into the lower end of Davis and Lavender Canyons. The book above also has more detailed information & history.

To reach the lower end of Salt Creek and Horse Canyon, drive to a point just south of Cave Spring, which is due south of the visitor center. From there, 4WD, mtn. bike or walk south into the canyons. The beginning part of that road is very sandy and you'll have to push a mtn. bike. The road in upper Horse Canyon is also sandy in places, and a mtn. bike can only be ridden part of the way. If you don't have a 4WD, it might be easier to walk into Horse canyon than to take a mtn. bike. Once into Salt Creek Canyon, the road is fine for mtn. biking, but it could be closed to all vehicles in the near future(?). To reach Davis or Lavender from the bottom end, you must now drive the all-public-land route into both from the Davis Canyon road, which begins next to the bridge over Indian Creek on The Needles Highway. The BLM changed the access roads in 1993-94 to avoid private land.

To get to the upper ends of Lavender and Salt Creek Canyons, leave The Needles Highway about 1 1/2 kms northwest of the Dugout Ranch. Turn left or south at that point and proceed up North Cottonwood Creek Canyon on a newly-built BLM road bypassing the Dugout Ranch. This road takes you to the area around Cathedral Butte and the head of Salt Creek and Lavender Canyons.

Trail and Route Conditions From the end of the 4WD or mtn. bike road in the middle part of Salt Creek, there's a good trail running up-canyon all the way. In the colder half of the year, or when roads to the upper end of the canyon are wet, this it the only way into upper Salt Creek. In Horse Canyon, 4WD's can be taken all the way up, but you can hike along that road and into side canyons. In one side canyon called Trail Fork, there's a developing trail which can be used to reach the flat ridge top above. From there you can use deer trails to reach other tributary forks of upper Salt Creek and Lavender Canyon, and probably Davis as well(?). If you 4WD or mtn. bike up from the bottom of Davis or Lavender Canyons, then you can hike into side canyons from the sandy entry track.

Now for access from the south or upper ends of Salt Creek and Lavender Canyons. To enter the upper end of Salt Creek, park at the trailhead just west of Cathedral Butte. It's signposted and easy to see on the right or west side of the road. You can camp there or on the other side of the road in the tall piñons & cedars. From the trailhead follow the good well-used Bright Angel Trail down in. It passes through Aristocratic Pasture with Kirk's hay rake on the west side under a ledge, and later Kirk's cabin. Lots of great scenery and Anasazi ruins in this upper end.

Davis Canyon is hard to get into from the south, but there is at least one easy route down into Lavender from a point at the west end of a large sagebrush clearing marked 2063 meters on this map. This is about 1 1/2 kms east of Cathedral Butte. From the upper or western side of that sagebrush flat, walk northeast on an old sandy track to the edge of Lavender Canyon and route-find down in. There is an old bulldozed track cutting down through the upper rim, then you'll have to look for deer trails going down from there. At the bottom, there's another old vehicle track in the middle of the drainage you can walk on. Once down into Lavender, you can explorer several of the upper tributaries, either on a day-hike, or an overnight backpacking trip. Two other interesting hikes are to the top of Cathedral Butte and Bridger Jack Mesa. For more details on water and routes in or out of these canyons, consult park rangers at the visitor center, and the author's other book listed above. It has more details on hiking routes, and history of the area.

Elevations Lower Salt Creek, 1500 meters; Kirk's cabin, 1817; Upper Salt Creek Trailhead, 2150; and the sagebrush flat and trailhead to upper Lavender Canyon, 2063 meters.

Hike Length and Time Needed If going into Salt Creek from the bottom end, you'll need to stay at least one night, maybe two, because of the length of the hike. One can reach most of upper Horse Canyon on foot in one long day by walking from Cave Spring, but a 4WD will get you there faster. This one is very sandy, and mtn. bikes don't help much. By coming down from the top, you can see most of the good stuff in Upper Salt Creek in one very long day; or spend one night there and see lots more. Upper Lavender can be seen on a day-hike from the Cathedral Butte area, but camping there one night will allow you to see more. If you get into Davis and Lavender with a 4WD, then you can see most of each canyon in one day. If memory is correct, you cannot camp inside the national park boundary in either Davis or Lavender. Instead, camp on BLM land below and walk or drive in during the day.

Water There's flowing water throughout Salt Creek, and intermittent springs and seeps in Davis and Lavender. There may be some in Horse Canyon, but that varies with the season. Always carry plenty of water in your car and in your pack.

Maps USGS or BLM map La Sal(1:100,000), or the plastic Trails Illustrated map Needles & Maze(1:62,500).

Main Attractions Anasazi ruins & pictographs, and the historic Kirk cabin, wagon and hay rake.

Ideal Time to Hike Spring or fall are best; summers are a bit warm.

Hiking Boots Any dry weather boots or shoes.

Author's Experience He has been into all of these canyons 2 or 3 times, or more, and from the top and bottom ends(except for the West Fork of Salt Creek).

Map 28, Salk Creek, & Horse, Davis & Lavender Canyons, Utah

BIG POCKET OVERLOOK &
UPPER SALT CREEK CANYON

CHINLE F.

MOENKOPI F.

ORGAN ROCK F.

CEDAR MESA
SANDSTONE

HORSE CANYON

TRAIL FORK

RUINS

EIGHTMILE SPRING
1585

HIGHWAY 13 KMS

2118

RUINS

DAVIS CANYON

TRAILS

RUINS

DEER

1908

UPPER JUMP

4 FACES

ALL AMERICAN MAN

RUINS

1955

SEEPS

2070

SALT

CREEK

RUINS

2064

SEEPS

1585

DEER

WEST

COW TRAILS

FORK

RUINS

ROUTE OVER

2130

HIGHWAY 15 KMS

CANYON

RUINS

TRAILS

BIG RUINS

BIG

RUINS

POCKET

RUINS

RUINS

2039

CLEFT ARCH

RUINS

RUINS

DRY

FORK

COW TRAILS

RUINS

KIRK'S CABIN
1817

LAVENDER

CANYON

KIRK'S HAY RAKE

2188

ARISTOCRATIC PASTURE

DEER TRAILS

BRIDGER JACK MESA

ROUTE OVER

SALT CREEK

EAST FORK

2240

4WD

P

2063

P

NPS BOUNDARY

TRAIL

2151

2150

P

CATHEDRAL BUTTE
2420

HORSE TRAIL

N

BRIGHT ANGEL

DUGOUT RANCH
NORTH COTTONWOOD CK.

SCALE 0 1 2 3 4 5 6 KMS

75

Looking down into the upper part of Horse Canyon and Trail Fork from the ridge-top above.

The old 1890's Kirk Cabin and wagon in the upper part of Salt Creek.

The Big Ruins, located in the upper end of Salt Creek Canyon.

Well-preserved matates and manos in Salt Creek Canyon.

The Maze(Under the Ledge), Utah

Location and Access This map shows the best hikes in The Maze(Under the Ledge) District of Canyonlands National Park. There are three ways to get into this very remote part of Utah. If you have a 4WD, drive along Highway 24 north of Hanksville to between mile posts 136 and 137(or between 133 and 134). Turn southeast and drive about 74 kms to the Maze(Hans Flat) Ranger Station. Get maps, camping permit and the latest information there, then head south to the Flint Trail. From the Flint Trail, drive down to the Maze Overlook, or the area called Land of Standing Rocks. Under normal conditions, about any vehicle can go down the Flint Trail, but you'll need a 4WD to get back up. If you have a car, then camp at the top of the Flint and use a mtn. bike to reach either trailhead area. If you have a 4WD or perhaps a HCV, you can drive to the vicinity of Hite on Lake Powell and to between mile posts 46 and 47. From the highway turn east onto the Hite Road heading toward The Cove and eventually to The Maze. Only a 4WD can make it to the Land of Standing Rocks. If you drive a car and don't have a mtn. bike, then you could go to The Needles District of Canyonlands and park at the Elephant Hill Trailhead. From there take several days of food and hike down Red Lake Canyon to the Colorado, then put your pack on an air mattress and swim across the river to Spanish Bottom. It's a short swim in slow waters. *This would be for experienced and tough hikers, and in warm weather only!* From Spanish Bottom you can reach the eastern part of the Maze. For more detailed information about trails and local history, see the author's book, *Hiking, Biking and Exploring Canyonlands N.P. and Vicinity.*

Trail or Route Conditions All the trails you see on this map are either hiker-made or are the remnants of former livestock trails. The trails on Chocolate Drop and Jasper Ridges are marked with cairns, as are many parts of all other trails. In a number of places you can see where early ranchers built trails over steep slickrock to get down into South, Pictograph, Water and Shot Canyons.

From The Maze Overlook, walk down along a good but slightly challenging trail to the bottom of South Fork. A short rope will help some people get large packs through or over a tight spot or two. From the Maze Overlook trailhead you can explore South and Pictograph Forks to as far as Standing and Chimney Rocks and Petes Mesa on day-hikes. From Chimney Rock Trailhead, you can walk to Spanish Bottom, or through Shot, Water, Pictograph and South Fork Canyons. Chimney Rock Trailhead is the most centrally located.

Jasper Canyon used to be a nice hike, but about a year after the author's book on *Canyonlands* mentioned above came out, the National Park Service closed it to hiking. It seems this is the only canyon in the Maze that never saw a cattle trail, therefore it has never been grazed. They'd like to keep it as pristine as possible and perhaps use it as a study area. It seems logical that you could still walk along Jasper Ridge into the lower end of the canyon and to the Green River, but they'll probably tell you that's off limits too! If so, you can still reach the Green River via a good trail in lower Water Canyon or via an emerging trail at the end of the Colorado River Overlook Trail. Powell used this route in 1869.

Elevations Maze Overlook, 1550 meters; Chimney Rock Trailhead, 1675; the Doll House, 1525; and Spanish Bottom, 1200 meters.

Hike Length and Time Needed Actual distances are difficult to calculate, so it's easier to figure distance in time walking. From Chimney Rock, down to the river via Water Canyon and return via Shot, is an easy day-hike. From Chimney Rock to the river via Jasper Ridge, would be an all-day hike for the average, maybe a long day for some! From Chimney Rock to The Maze Overlook and back would be an easy day-hike, unless you prefer to linger awhile in the canyons at the pictograph panel called the Harvest Scene. From the Doll House to Spanish Bottom is 3 hours round-trip.

Water Always carry plenty of water in your car and in your pack, but there is good drinking water in all or most of the springs or seeps shown on the map. Water is found at several locations in South and Pictograph, and in lower Water Canyon. Take it from the spring source or the beginning of the seep and it's normally good to drink as is. There are no beaver or livestock in any of these canyons today. Treat or filter river water.

Maps USGS or BLM maps La Sal and Hanksville(1:100,000), or the plastic Trails Illustrated map Needles & Maze(1:62,500). This is the best one. You can buy it at the Maze Ranger Station.

Main Attractions Deep, rugged, remote and colorful canyons, perhaps the second best pictograph panel on the Colorado Plateau(the Harvest Scene), plus other rock art panels.

Ideal Time to Hike Spring or fall. It's very warm in summer. If you're going to swim the Colorado at Spanish Bottom, then from sometime in June through the end of September, and on a warm day. The water in the Colorado is chilly even in summer because it comes out of Flaming Gorge Reservoir upstream.

Hiking Boots Any dry weather boots or shoes.

Author's Experience The author made most of his hikes here by walking from his boat on the rivers. From the Green River, he once hiked up Water and Shot Canyons to Chimney Rock, then down Water Canyon. Round-trip, 5 2/3 hours. From Spanish Bottom he hiked to Chimney Rock, then along Jasper Ridge to the river, then returned via Jasper Canyon to Chimney Rock, and back to Spanish Bottom. Round-trip, just over 9 hours. Next day he hiked from Spanish Bottom to Chimney Rock, then to Petes Mesa, the Maze Overlook, then up Pictograph Fork and back to Spanish Bottom. Round-trip, just under 10 hours. Another time he biked from the top of the Flint Trail down to The Maze Overlook, camped one night, then hiked up South Fork and down Pictograph. Round-trip, 7 hours walk-time. He biked back out that same afternoon. Another time, he hiked from Elephant Hill in The Needles, to the river, swam to Spanish Bottom and spent 3 days in Cataract Canyon. Most people will need much more time for these hikes, and others may have to choose shorter hikes

Map 29, The Maze(Under the Ledge), Utah

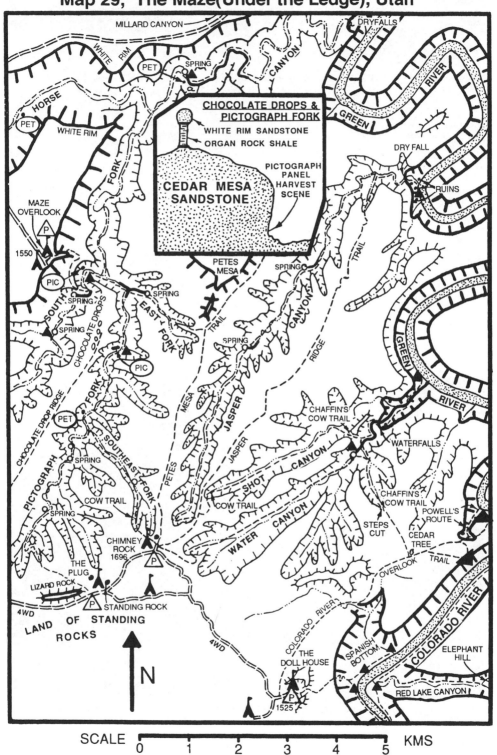

CHOCOLATE DROPS &
PICTOGRAPH FORK
WHITE RIM SANDSTONE
ORGAN ROCK SHALE
CEDAR MESA
SANDSTONE
PICTOGRAPH
PANEL
HARVEST
SCENE

MILLARD CANYON
DRYFALLS
WHITE RIM
RIVER
GREEN
HORSE
PET
PET
WHITE RIM
FORK
DRY FALL
RUINS
MAZE
OVERLOOK
P
1550
PIC
PETES
MESA
SPRING
TRAIL
SOUTH
SPRING
SPRING
FORK
EAST FORK
CHOCOLATE DROPS
SPRING
TRAIL
RIDGE
CANYON
PIC
SPRING
CHOCOLATE DROP RIDGE
PETES MESA TRAIL
JASPER
GREEN
RIVER
PET
CHAFFIN'S
COW TRAIL
WATERFALLS
FORK
SPRING
JASPER
PICTOGRAPH
SOUTHEAST FORK
COW TRAIL
SHOT CANYON
CHAFFIN'S
COW TRAIL
POWELL'S
ROUTE
SPRING
COW TRAIL
STEPS
CUT
CEDAR
TREE
CHIMNEY
ROCK
1696
P
WATER CANYON
OVERLOOK
TRAIL
THE
PLUG
LIZARD ROCK
P
4WD
STANDING ROCK
LAND OF STANDING
ROCKS
N
4WD
COLORADO RIVER
THE
DOLL HOUSE
SPANISH
BOTTOM
COLORADO RIVER
ELEPHANT
HILL
P
1525
RED LAKE CANYON

SCALE   0   1   2   3   4   5   KMS

Anasazi ruins at the bottom end of Jasper Canyon next to the Green River.

The Harvest Scene pictograph panel located in the lower end of Pictograph Fork.

Looking north into upper Jasper Canyon from the canyon rim.

Standing rocks in the Doll House area just west and above Spanish Bottom.

Lathrop Canyon and Neck Spring Trail, Utah

Location and Access These two trails are located in the Island in the Sky(formerly Grays Pasture) District of Canyonlands National Park. This is the northern third of the park which lies between the Green and Colorado Rivers. To get there, drive north out of Moab on Highway 191, or south from Crescent Junction and I-70, until you reach the highway running to Canyonlands and Dead Horse Point. This is just south of mile post 137. Drive west, then south into Canyonlands N.P. Upon entering, stop at the visitor center for the latest information on water, camping, etc., before proceeding. There are no facilities of any kind in this region, so go prepared with water, food, fuel, etc., for the duration of your trip. From the visitor center drive south less than a km to the first trailhead which is on the left or east side of the road.

Trail or Route Conditions Both of these trails are well-marked, and moderately well-used. The Neck Spring Trailhead is about 800 meters south of the visitor center. The Neck Spring Trail first descends into a shallow canyon and winds its way to the west toward Cabin Spring, then tops out on the rim and returns to the main road as shown. The Lathrop Canyon Trailhead is 3 kms south of the visitor center. This trail runs south across Grays Pasture, winds down through the Navajo Sandstone, contours southwest on the Kayenta bench, and finally drops down through the Wingate Sandstone on a rockslide. This last part is an old sheep trail. After you pass some old uranium mines, the trail heads down to the White Rim Road; then into and through lower Lathrop Canyon to the Colorado River along a 4WD road. For more information and history of trails, see the author's book, *Hiking, Biking and Exploring Canyonlands N.P. and Vicinity.*

Elevations Island in the Sky about 1850 meters; the Colorado River, 1220 meters.

Hike Length and Time Needed From the Lathrop Canyon Trailhead to the White Rim Road is about 8 kms. Add about 5 1/2 more if you walk the 4WD road all the way to the river, for a total one-way distance of 14 1/2 kms. Most people just walk down to the mines or the road and return the same day. The average hiker may want 6-8 hours, round-trip. To do the entire loop of the Neck Spring Trail is about 8 kms, and will take half a day.

Water Take plenty of water with you into this area, or you'll have to buy it at the visitor center. Cabin Spring has a good discharge, but the metal water tanks on the mesa above the spring are no longer in use. Neck Spring is maybe the best around. Treat Colorado River water.

Maps USGS or BLM map La Sal(1:100,000), or the plastic Trails Illustrated map Island in the Sky(1:62,500).

Main Attractions Upper Lathrop Canyon has several big alcoves, a steep drop, and a number of old uranium mine tunnels in the Mossback Member of the Chinle Formation. Some of these mines have been sealed because of cave-in danger. Also, a good chance to see desert big horn sheep. The author has seen big horns on two occasions where the trail begins to wind its way down through the Navajo ledges.

Ideal Time to Hike Spring or fall, but the higher elevations of this mesa top and cooler temperatures, makes summer hiking just barely tolerable.

Hiking Boots Any dry weather boots or shoes.

Author's Experience The author hiked the Neck Spring Trail loop in about 2 1/2 hours. Later he went down Lathrop Canyon to the White Rim Road and back in 4 1/2 hours.

Looking down into the upper end of Lathrop Canyon. Notice the trail running along the Kayenta Bench below.

Map 30, Lathrop Canyon and Neck Spring Trail, Utah

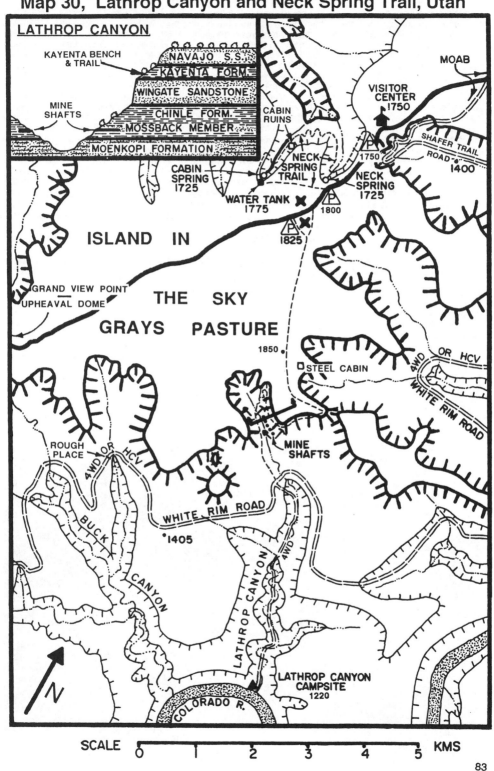

LATHROP CANYON

KAYENTA BENCH & TRAIL
NAVAJO S.S.
KAYENTA FORM.
WINGATE SANDSTONE
CHINLE FORM.
MOSSBACK MEMBER
MOENKOPI FORMATION
MINE SHAFTS

MOAB

VISITOR CENTER 1750

SHAFER TRAIL ROAD 1400

CABIN RUINS

P 1750

NECK SPRING TRAIL

CABIN SPRING 1725

NECK SPRING 1725

WATER TANK 1775

P 1800

P 1825

ISLAND   IN

THE   SKY

GRAND VIEW POINT
UPHEAVAL DOME

GRAYS   PASTURE

1850

STEEL CABIN

4WD OR HCV
WHITE RIM ROAD

ROUGH PLACE

4WD OR HCV

MINE SHAFTS

BUCK
CANYON

WHITE RIM ROAD

°1405

LATHROP CANYON

4WD

LATHROP CANYON CAMPSITE 1220

COLORADO R.

N

SCALE 0   1   2   3   4   5   KMS

Island in the Sky Trails, Utah

Location and Access All these trails are located in the Island in the Sky(formerly Grays Pasture) District of Canyonlands National Park. This is the northern third of the park which lies between the Green and Colorado Rivers. To get there, drive north out of Moab on Highway 191, or south from Crescent Junction and I-70, until you reach the highway running to Canyonlands & Dead Horse Point. This is just south of mile post 137. As you enter Canyonlands, stop at the visitor center for the latest information on trails, camping, etc., before proceeding. There are no facilities of any kind in this region, so go prepared with water, food, fuel, etc., for the duration of your trip. In the past, you could buy drinking water at the visitor center. Drive south on the paved road to the Grandview Point area.

Trail or Route Conditions There are short hikes on good trails to Mesa Arch, White Rim Overlook & Aztec Mesa, as shown. In the north, the Wilhite Trail(an old sheep trail) zig zags down a rock slide covering the Wingate Sandstone, then runs on to the White Rim Road. The Murphy Cattle Trail was built in 1918 by the Murphy Brothers. It runs down on top of another rock slide to the White Rim road via Murphy's Hogback or the Murphy Basin. The Government(or Gooseberry) Trail is an old sheep route, built in the 1930's, probably by the CCC's. It too zig zags down a steep rock slide covering the Wingate Sandstone. It ends at the Gooseberry Campsite. The Grandview Point(or Monument Basin) Trail has been built by hikers over the years and is steeper, less-used, usually unmarked and more exciting & challenging than the others. It drops down through a series of ledges in the Wingate Sandstone, then meets an old mining exploration track from the 1950's. The Grandview Point(Monument Basin) Trail is for experienced hikers only. In the early 1990's one hiker had to be rescued from a ledge there. For much greater detail & history of these trails, see the author's other book, *Hiking, Biking and Exploring Canyonlands National Park & Vicinity*.

Elevations The Island in the Sky Mesa, about 1900 meters; the White Rim Road around the White Rim, about 1500 meters

Hike Length and Time Needed From the trailhead down the Wilhite to the bottom of the steep part and back, about half a day; or a full day to reach the White Rim Road and return. For the Murphy, Government(Gooseberry) and the Grandview Point(Monument Basin) Trails, it would be about the same. That is, it'll take about half a day to get down the steep part and return; or a full day's hike to reach the White Rim Road and return. The best part of all these hikes is from the mesa top to the bottom of the steep part, which is always the Wingate Sandstone. From the bottom of the Grandview Point Trail is another route to the top of Junction Butte, as shown.

Water Have water in your car and in your pack, as there are no reliable sources on the mesa top. In summer take plenty of water on the longer hikes, maybe 2-3 liters per/person per/day.

Maps USGS or BLM map La Sal(1:100,000), or the plastic Trails Illustrated map Island in the Sky(1:62,500).

Main Attractions Great views of Canyonlands National Park and historic sheep and cattle trails.

Ideal Time to Hike Spring or fall, or some winter warm spells. This paved road is open year-round. Early summer brings insects and it's pretty warm(in mid-summer), even at these altitudes.

Hiking Boots Any dry weather boots or shoes.

Author's Experience The author has hiked down to the White Rim on all the longer trails on this map at least twice. The shorter trails have been done at least twice as well.

This is part of the constructed Murphy Trail as it zig zags down through the Wingate.

Map 31, Island in the Sky Trails, Utah

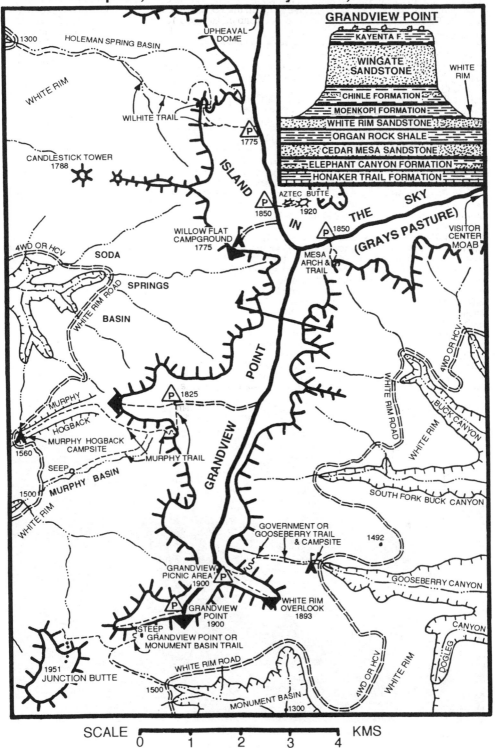

GRANDVIEW POINT

KAYENTA F.

WINGATE SANDSTONE

CHINLE FORMATION
MOENKOPI FORMATION
WHITE RIM SANDSTONE
ORGAN ROCK SHALE
CEDAR MESA SANDSTONE
ELEPHANT CANYON FORMATION
HONAKER TRAIL FORMATION

WHITE RIM

1300
HOLEMAN SPRING BASIN
UPHEAVAL DOME
WHITE RIM
WILHITE TRAIL
P 1775
ISLAND
IN
THE
SKY
(GRAYS PASTURE)
CANDLESTICK TOWER 1788
P 1850
AZTEC BUTTE 1920
P 1850
VISITOR CENTER MOAB
4WD OR HCV
SODA
SPRINGS
BASIN
WILLOW FLAT CAMPGROUND 1775
WHITE RIM ROAD
MESA ARCH & TRAIL
POINT
4WD OR HCV
WHITE RIM ROAD
WHITE RIM
BUCK CANYON
GRANDVIEW
P 1825
MURPHY
HOGBACK
MURPHY HOGBACK CAMPSITE
1560
MURPHY TRAIL
SEEP
MURPHY BASIN
1500
WHITE RIM
SOUTH FORK BUCK CANYON
GOVERNMENT OR GOOSEBERRY TRAIL & CAMPSITE
1492
GOOSEBERRY CANYON
GRANDVIEW PICNIC AREA 1900
P
P
GRANDVIEW POINT 1900
WHITE RIM OVERLOOK 1893
STEEP
GRANDVIEW POINT OR MONUMENT BASIN TRAIL
1951 JUNCTION BUTTE
WHITE RIM ROAD
1500
MONUMENT BASIN
1300
CANYON
DOGLEG
4WD OR HCV
WHITE RIM

SCALE   KMS
0   1   2   3   4

Upheaval Dome and Taylor Canyon, Utah

Location and Access The canyons on this map are located in the Island in the Sky(formerly Grays Pasture) District of Canyonlands National Park, which is between the Green and Colorado Rivers. To get there, drive north out of Moab on Highway 191, or south from Crescent Junction and I-70, to the highway heading into Canyonlands and Dead Horse Point. This junction is near mile post 137. Turn west and proceed to Canyonlands. Stop at the visitor center for the latest information on water, camping, trails, etc., then proceed south to Upheaval Dome and two trailheads nearby. Bring plenty of food, fuel and water with you. Years ago researchers thought Upheavals' circular features was a salt dome, caused by uplift of the Paradox Salt beds below. However, in the early 1990's after additional research, it's now believed to be the remains of a meteor impact crater.

Trail or Route Conditions From the Upheaval Dome car-park, there's one short trail to the rim of the Dome itself; another trail heads west and into the Upheaval Valley; and yet another runs northeast and into the Syncline Valley. These trails are well marked with stone cairns and have heavy use. One of the more popular treks would be to walk in either direction through the Upheaval and Syncline Valleys making a loop-hike around Upheaval Dome. On the western side of Upheaval, there's a trail-of-sorts into the center of the Dome from the crater's outlet. The last hike on this map involves walking from the Alcove Spring Trailhead, down what was once an old cattle trail--but which was modified by a bulldozer in the 1950's--into and through Trail and Taylor Canyons. It's possible to walk all the way to the Green River in Taylor Canyon, then return up Upheaval Canyon and back to the Dome Trailhead(or vise versa). For more hikes and history of this area see the author's other book, *Hiking, Biking and Exploring Canyonlands National Park and Vicinity*.

Elevations The Green River, 1240 meters; Upheaval Dome Trailhead, 1750; Alcove Spring Trailhead, 1725 meters.

Hike Length and Time Needed From the Dome Trailhead to the Green River via Upheaval Valley is 8 kms(a little longer through Syncline Valley). From Alcove Spring Trailhead, down Taylor Canyon to the Green River is about 18 kms. From either trailhead and a round-trip loop-hike through Upheaval and Taylor Canyons, is about 28 kms. Fast hikers can do the round-trip in one long day. An easy day hike would be to make a loop-hike around the Dome.

Water Have plenty of water in your car and for day-hikes take your own. Normally there's year-round water in the Syncline Valley and upper Upheaval Canyon. Alcove Spring has good water.

Maps USGS or BLM map La Sal(1:100,000), or the plastic Trails Illustrated map Island in the Sky(1:62,500).

Main Attractions The geology of Upheaval Dome. Note the geology cross section.

Ideal Time to Hike Spring or fall, or winter dry spells. Early summer brings lots of gnats.

Hiking Boots Any dry weather boots or shoes.

Author's Experience One day he walked down Syncline Valley to the river, then up Taylor and Trail Canyons back to his car in about 7 1/2 hours. Next day he walked into the Dome from the Dome Trailhead, then returned via upper Upheaval Valley Trail in 3 hours. You'll want more time than this.

Aerial view of Upheaval Dome, an ancient meteor impact crater. Looking to the northwest with the Green River in the far background(Fran Barnes foto).

Map 32, Upheaval Dome and Taylor Canyon, Utah

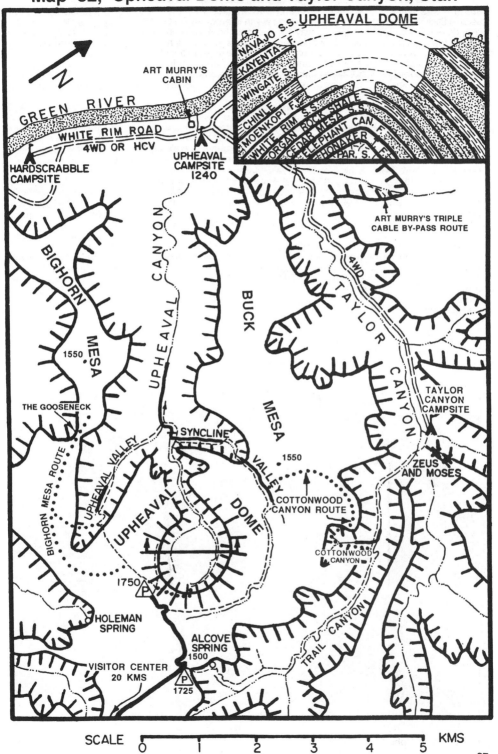

N

ART MURRY'S CABIN

GREEN RIVER

WHITE RIM ROAD
4WD OR HCV

HARDSCRABBLE CAMPSITE

UPHEAVAL CAMPSITE 1240

UPHEAVAL DOME

Navajo s.s.
KAYENTA F.
WINGATE S.S.
CHINLE F.
MOENKOPI F.
WHITE RIM S.S.
ORGAN ROCK SHALE
CEDAR MESA S.S.
ELEPHANT CAN.
HONAKER F.
PAR. S.

ART MURRY'S TRIPLE CABLE BY-PASS ROUTE

BIGHORN MESA

1550

UPHEAVAL CANYON

BUCK MESA

4WD

TAYLOR CANYON

THE GOOSENECK

BIGHORN MESA ROUTE

UPHEAVAL VALLEY

SYNCLINE

VALLEY

1550

TAYLOR CANYON CAMPSITE

ZEUS AND MOSES

COTTONWOOD CANYON ROUTE

UPHEAVAL DOME

COTTONWOOD CANYON

1750 P

HOLEMAN SPRING

ALCOVE SPRING 1500

TRAIL CANYON

VISITOR CENTER 20 KMS

P 1725

SCALE 0 1 2 3 4 5 KMS

Horseshoe Canyon, Utah

Location and Access The small bordered area on this map is the Horseshoe Canyon section of Canyonlands National Park. The remainder of Canyonlands is to the east about 15 kms. To reach this canyon, make your way to Highway 24, the road connecting I-70 and Green River, with Hanksville. Between mile posts 136 and 137(or 133 and 134), turn east onto a good dirt road signposted for the Maze District of Canyonlands. Drive about 40 kms to a junction. Turn right to reach the Maze(Hans Flat) Ranger Station; or turn left, and drive 11 kms to the normal west side hiker's trailhead for Horseshoe Canyon. 4WD vehicles may be allowed down into Horseshoe, but they must proceed to the ranger station, obtain a permit, then drive north from the ranger station. Along that east-side 4WD road is a side-track running west to the Deadman Trail. Always enter this country with good supplies of food, fuel and water.

Trail or Route Conditions On the east side of Horseshoe is the 4WD track to the bottom and the campsite. If you hike down the Deadman Trail you'll find a well-constructed cattle trail which enters the canyon less than a km above the Great Gallery. The trail in from the west side(actually the remains of an old road built by Phillips Petroleum) is the normal hiking route. From the trailhead simply walk down into the drainage and up-canyon to the pictographs. The canyon is mostly a dry creek bed. The adventurous and energetic hiker can walk all the way up-canyon(south) to the ranger station, or walk all the way to the Green River(north). Beyond this area it's a true wilderness experience. For a history of the Tidwell Ranch, etc., see the author's book, *Hiking, Biking and Exploring Canyonlands National Park and Vicinity*.

Elevations West side trailhead, 1600 meters; Deadman Trailhead, 1646; canyon campsite, 1425 meters.

Hike Length and Time Needed From the trailhead to the canyon bottom is about 2 kms, and another 4 or 5 kms to the best pictographs called the Great Gallery. This can be done in as little as half a day, but many people would rather spend most of a day in this canyon. You can also hike down-canyon to the Green River, a distance of about 25 kms one way(a 3 day round-trip), or up-stream as far and as long as you like. You can mtn. bike down to the Deadman Trail, then reach the Great Gallery in half an hour.

Water Always carry water in your car and in your pack. The author found 3 places in the creek bed where there was either flowing water or pools of water. Lower Horseshoe has good year-round running water.

Maps USGS or BLM map Hanksville(San Rafael Desert for the access route)(1:100,000), or The Spur, or the plastic Trails Illustrated map Needles & Maze(1:62,500).

Main Attractions Pictographs! This is the best group of pictographs on the entire Colorado Plateau. The style here is called "Barrier Creek", against which all other styles are judged.

Ideal Time to Hike Spring or fall, or some winter dry spells; or early mornings in summer.

Hiking Boots Any dry weather boots or shoes

Author's Experience The author hiked the normal route to the panels and returned in 3 hours. Another time he rode his mtn. bike down the road north from the Maze R.S. and visited the Tidwell Ranch; then rode to and hiked down the Deadman Trail and returned, all in 9 1/2 hours.

This panel of pictographs in Horseshoe Canyon is probably the best one in the world.

Map 33, Horseshoe Canyon, Utah

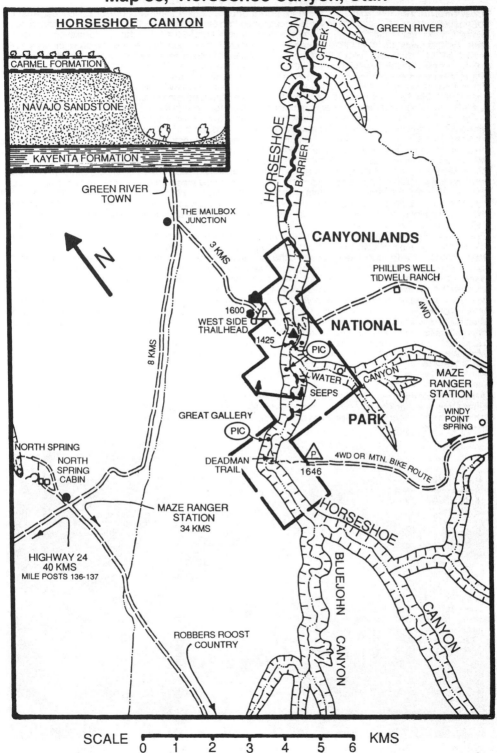

HORSESHOE CANYON

CARMEL FORMATION

NAVAJO SANDSTONE

KAYENTA FORMATION

GREEN RIVER

HORSESHOE CANYON

BARRIER CREEK

GREEN RIVER TOWN

THE MAILBOX JUNCTION

3 KMS

CANYONLANDS

PHILLIPS WELL
TIDWELL RANCH

4WD

N

8 KMS

1600
P

WEST SIDE TRAILHEAD

1425

NATIONAL

PIC

WATER
SEEPS

CANYON

MAZE
RANGER
STATION

WINDY
POINT
SPRING

GREAT GALLERY

PIC

PARK

NORTH SPRING

NORTH
SPRING
CABIN

DEADMAN
TRAIL

P

1646

4WD OR MTN. BIKE ROUTE

MAZE RANGER
STATION
34 KMS

HORSESHOE

HIGHWAY 24
40 KMS
MILE POSTS 136-137

BLUEJOHN

CANYON

ROBBERS ROOST
COUNTRY

CANYON

SCALE 0 1 2 3 4 5 6 KMS

89

Dirty Devil River, Utah

Location and Access The Dirty Devil River is found just to the southeast of Hanksville and east of Highway 95. This is the part where the river cuts deep into the mostly sandstone strata to a depth of about 425 meters. Before hiking here, stop at the BLM office in Hanksville for updated information on water, roads, etc., before going on. To do this hike, you should first proceed to mile post 10 on Highway 95, 16 kms south of Hanksville. Turn east and park in the vicinity of Pool Spring; there may be deep sand beyond! Another route into this canyon is to go down Poison Spring Canyon. Begin from mile post 17 on Highway 95; you'll need a 4WD or HCV to make it all the way to the Dirty Devil River.

Trail or Route Conditions Beaver Canyon is difficult to walk in because of the live beaver population and resultant brush, so you should walk(or mtn. bike or drive) the road running out to the west side trailhead to the Angel Trail at 1500 meters. From the main road, look for the side track running east 300 meters to the 4WD trailhead. Then head down a shallow drainage and low ridge to where Beaver Creek meets the river. This is the Angel Trail. Stone cairns mark the way. Plan to camp at Angel Cove Spring if you can; it has shade and good water, and is the best campsite on the whole river trip. Then it's down the Dirty Devil, crossing many, many times. Take a walking stick to probe the murky water for possible deep holes. Midway in the canyon you can walk along the faded remains of a mining road or track used for exploring the Moss Back Member of the Chinle Formation for uranium. Finally, walk up Poison Spring Canyon road, and hitch hike back to your car. A second car or a mtn. bike would eliminate a long road-walk or hitch hiking. One way to get into or out of the middle part of the Dirty Devil, is to drive to Burr Point. From there a route heads down over the ledges the turns north before dropping down into the gorge. See the Burr Point Hike in the author's other book, *Hiking and Exploring the Utah's Henry Mountains & Robbers Roost*. In that book, Robbers Roost Canyon is described in detail too.

Elevations Pool Spring, about 1425 meters; Angel Trail Trailhead, 1500; Angel Cove Spring, 1225; Dirty Devil Gauging Station, 1150; and the head of Poison Spring Canyon, about 1500 meters.

Hike Length and Time Needed Notice on the map some black dots and circled numbers. The numbers are distances in kms between the dots. If you walk the entire circle, it's 104 kms(65 miles!). For most people that's a long 4 day, or an easier 5 day hike. With a car shuttle(two cars or mtn. bike), you could shorten the hike considerably. Or you could exit(or enter?) at Burr Point, and shorten it even more.

Water Take plenty in your car, then you'll find good water sources at Angel Cove, Wall Spring, and in various other places in Poison Spring Canyon. At times you *may* find water in the water tap shown on the map. There is water year-round in the Dirty Devil. The author treated it with Clorox bleach, and it tasted fine except for the Clorox. However, Iodine tablets are better, or filter the water.

Maps USGS or BLM map Hanksville(1:100,000), or four maps at 1:62,500 scale.

Main Attractions Petrified wood in the Shinarump; a deep gorge; and complete solitude.

Ideal Time to Hike Spring or fall, but not during the short high water period in May. Summers are too hot, and wading in winter is out of the question--unless everything is iced-up.

Hiking Boots Wading boots or shoes.

Author's Experience He parked at Pool Spring, walked the road to Angel Trail, then went down the Dirty Devil and back to Highway 95 via Poison Spring Canyon where he got a ride for the last 6 kms. It took two full days and two half days, or about 27 hours total walk-time. This was during the first week of October, 1985.

The lower end of Larry Canyon before it enters the Dirty Devil River Gorge.

Map 34, Dirty Devil River, Utah

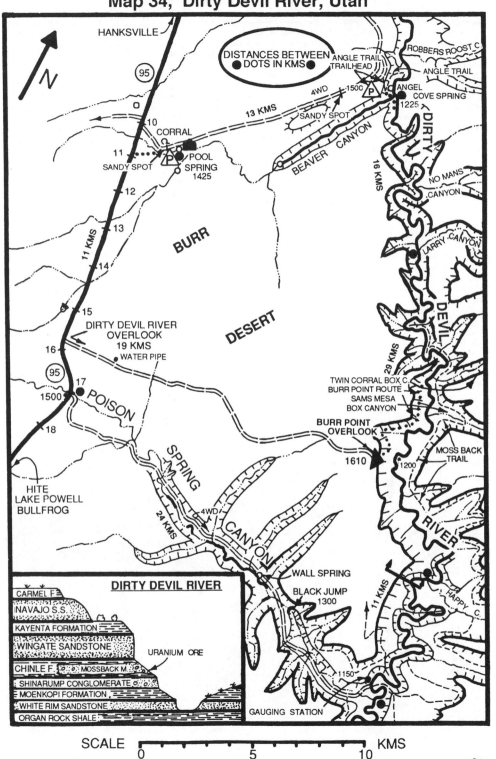

HANKSVILLE

N

95

DISTANCES BETWEEN
DOTS IN KMS

ROBBERS ROOST C.

ANGLE TRAIL
TRAILHEAD

ANGLE TRAIL

4WD   1500
SANDY SPOT   P

13 KMS

ANGEL
COVE SPRING
1225

10   CORRAL

11   P   POOL
SPRING
SANDY SPOT   1425

BEAVER CANYON

DIRTY DEVIL

NO MANS
CANYON

16 KMS

12

13

11 KMS

14

BURR

LARRY CANYON

15

DIRTY DEVIL RIVER
OVERLOOK
19 KMS

DESERT

16
WATER PIPE

29 KMS

95

17

TWIN CORRAL BOX C.
BURR POINT ROUTE
SAMS MESA
BOX CANYON

1500

POISON

18

BURR POINT
OVERLOOK

MOSS BACK
TRAIL

HITE
LAKE POWELL
BULLFROG

SPRING

1610   1200

DIRTY DEVIL RIVER

24 KMS

4WD

CANYON

WALL SPRING

BLACK JUMP
1300

11 KMS

HAPPY C.

DIRTY DEVIL RIVER

CARMEL F.
NAVAJO S.S.
KAYENTA FORMATION
WINGATE SANDSTONE   URANIUM ORE
CHINLE F.   MOSSBACK M.
SHINARUMP CONGLOMERATE
MOENKOPI FORMATION
WHITE RIM SANDSTONE
ORGAN ROCK SHALE

1150

GAUGING STATION

SCALE

0   5   10   KMS

Tributaries of Trachyte Creek, Utah

Location and Access Trachyte Creek and 4 of its main tributaries are located just south of the junction of Utah State Highways 95 (which runs between Hanksville and Blanding) and 276 (which runs to Lake Powell and Bullfrog Marina). They are situated east of Mt. Hillers and north of Mt. Holmes, both part of the Henry Mountains. There are four canyons you can enter to reach Trachyte Creek; Maidenwater, Trail, Woodruff and Swett Creek. The numbers along the highway are mile post markers. Use these to locate the right canyons and park accordingly.

Trail or Route Conditions Hiking in this area is simply walking down a canyon, sometimes in the dry creek bed, sometimes in water. The walking is generally easy, even in areas where there is abundant water, which makes for profuse plant growth. In Maidenwater Creek, pass to the north side of a dryfall to get down into the upper narrows. Halfway down Maidenwater, there used to be a chokestone, dryfall and deep pool. However, sometime in the early 1990's, this was gone and walking apparently easy--according to a report by one hiker. However, expect the unexpected, because these canyons change with every flood. In the lower part of Maidenwater there are some good narrows. Trail Canyon offers the easiest and fastest route into or out of Trachyte Creek. Going down Woodruff is an easy walk too, but near the bottom you'll have to skirt to the east side of one dropoff, then further down, wade or maybe even swim around one large chokestone to get through. Walking in Swett Creek down to Lake Powell is an easy hike with some nice Navajo and Wingate narrows. There are many other slot canyons in this area, some of which you can see as you drive along the highway and between mile posts 6 and 10. For more information and history of the area, see the author's book, *Hiking and Exploring Utah's Henry Mountains & Robbers Roost.*

Elevations Car-parks range from 1400 to 1450 meters. Lake Powell at high water is 1128 meters.

Hike Length and Time Needed To walk down Maidenwater and Trachyte, and up Swett is about 26 kms, or one very long all-day hike. Most would prefer to walk down Maidenwater and return up Trail Canyon, then drive to Swett Creek car-park and hike down from there. Down Swett and up Trachyte and Woodruff would be an easy day-hike. Walk on the Chinle Bench between Trachyte and Swett Creek. Woodruff is moderately interesting, while Trachyte and Trail Canyons are uninteresting. Maidenwater is the most interesting canyon in the area.

Water All of Trachyte, most of Maidenwater and Woodruff, and just a short stretch in Swett Creek have year-round running water. The upper parts of Maidenwater and in Swett Creek, water should be good to drink. There's also a spring with good water in lower Trail Canyon. There are cattle throughout the main Trachyte drainage, so purify or filter that water before drinking.

Maps USGS or BLM map Hite Crossing(1:100,000), or Mt. Hillers(1:62,500).

Main Attractions Deep pools, narrows, running water, and live beaver(Maidenwater Canyon).

Ideal Time to Hike Warmer weather, because of all the wading. Spring through fall.

Hiking Boots Wading boots or shoes; but in Swett and Trail Canyons use dry weather boots.

Author's Experience The author walked down Woodruff to the lake, then up Maidenwater(to the jumpoff, which is apparently gone now), then out via Trail Canyon. Round-trip was 9 1/2 hours. Next day it was 4 hours down and back out of Swett Ck.

This is one of several chokestones and resultant deep pools in Maidenwater Canyon.

Map 35, Tributaries of Trachyte Creek, Utah

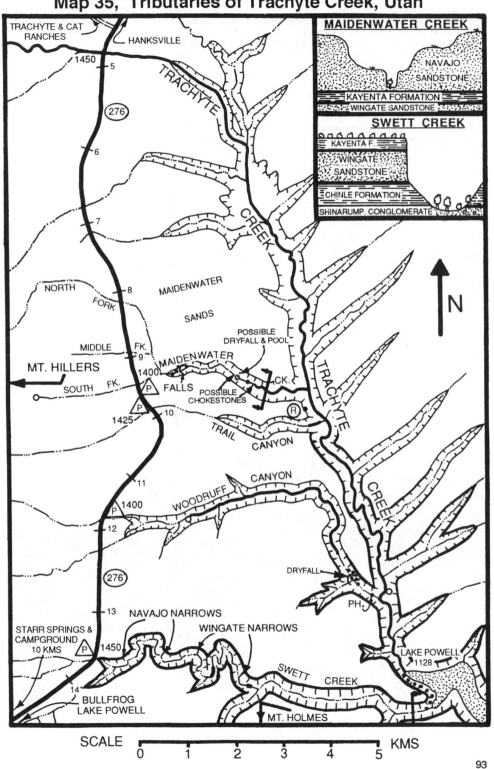

MAIDENWATER CREEK

NAVAJO SANDSTONE

KAYENTA FORMATION

WINGATE SANDSTONE

SWETT CREEK

KAYENTA F.

WINGATE SANDSTONE

CHINLE FORMATION

SHINARUMP. CONGLOMERATE

TRACHYTE & CAT RANCHES

HANKSVILLE

1450 — 5

276

— 6

— 7

TRACHYTE

CREEK

NORTH

FORK

— 8

MAIDENWATER

SANDS

MIDDLE FK.

— 9

MAIDENWATER

POSSIBLE DRYFALL & POOL

MT. HILLERS

1400

SOUTH FK.

P FALLS

CK.

POSSIBLE CHOKESTONES

R

1425

P

— 10

TRAIL CANYON

TRACHYTE

— 11

CANYON

P 1400

WOODRUFF CANYON

— 12

CREEK

276

DRYFALL

— 13

PH.

NAVAJO NARROWS

WINGATE NARROWS

STARR SPRINGS & CAMPGROUND 10 KMS

P 1450

LAKE POWELL

1128

SWETT CREEK

— 14

BULLFROG LAKE POWELL

MT. HOLMES

N

SCALE 0 1 2 3 4 5 KMS

93

Bowdie and Gypsum Canyons, Utah

Location and Access These two canyons form the northern half of the Dark Canyon Primitive Area, which is located due north of Natural Bridges National Monument and west of the Abajo Mountains. To the west lies the upper portion of Lake Powell. This is one of the more isolated hiking areas in this book, as one must drive at least 91 kms from Blanding, much of which is on dirt roads. You'll need other maps besides this one and should talk to BLM or Forest Service people in Monticello before setting out. Briefly, one can get to this area by driving north out of Blanding; or north and east from near Natural Bridges N.M.; or southwest up North Cottonwood Creek from near Newspaper Rock and the Dugout Ranch. Roads in this area are generally passable to all vehicles, but cars should be in good condition and you should go well prepared. These high altitude roads are basically for summer use only. Regardless of your approach route, follow signs to Dark Canyon Plateau or Beef Basin.

Trail or Route Conditions To get into Bowdie, start at the trailhead marked 2200 meters. There's a limestone ledge just beyond this car-park you'd have to work on with a shovel, even to get a 4WD over. So it's best to park and walk, or mtn. bike, in from there. Head northwest to a point between the north and south forks, then route-find down into either fork. There's a small stream for most of the way down, but eventually you'll come to a large waterfall. Going down, the author was stopped there once, but later he came up from the lake and found an easy route around this waterfall on either side. Just bench-walk a ways until you find a way back to the creek bed. Lower down there are other dryfalls and potholes, but you can get around them OK. For Gypsum, follow the route symbol northeast from the car-park to get down into Fable Valley. You can also get into the upper canyon from Beef Basin. The author hasn't been down Gypsum but he has come up from the lake to Gypsum Falls. In that area he once met Brian Blackstock who had just come down from Beef Basin. He said there were a number of waterfalls to skirt which made the going slow, but you can make it. Take a short rope to lower packs over ledges.

Elevations About 2200 meters at the car-park, and 1128 at Lake Powell's high water mark.

Hike Length and Time Needed It's about 24 to 26 kms from the car-park to Lake Powell via either canyon. If your goal is to reach Lake Powell, then it may take about 3-4 days round-trip in either canyon.

Water Carry plenty of water in your car. Bowdie has springs and running water as shown, while maps and other sources indicate the same in Gypsum.

Maps USGS or BLM maps Blanding and Hite Crossing(1:100,000), or Canyonlands National Park(1:62,500).

Main Attractions Very wild, remote, isolated, and unspoiled, with room to explore

Ideal Time to Hike Late spring to early fall--because of the high altitude approach roads all of which are made of dirt.

Hiking Boots Any dry weather boots or shoes, but there may be a little wading in lower Bowdie(?).

Author's Experience The author spent two days exploring and hiking upper Bowdie(9 and 11 hours each day). He has been up from the lake in Gypsum and climbed out where J. W. Powell did(east fork). It was about a half-day hike up to the waterfall in Bowdie and back to the lake. For more information and history of this region see the author's other book, *Boater's Guide to Lake Powell*.

There is year-round water in several parts of Bowdie Canyon. This is one pourover pool.

Map 36, Bowdie and Gypsum Canyons, Utah

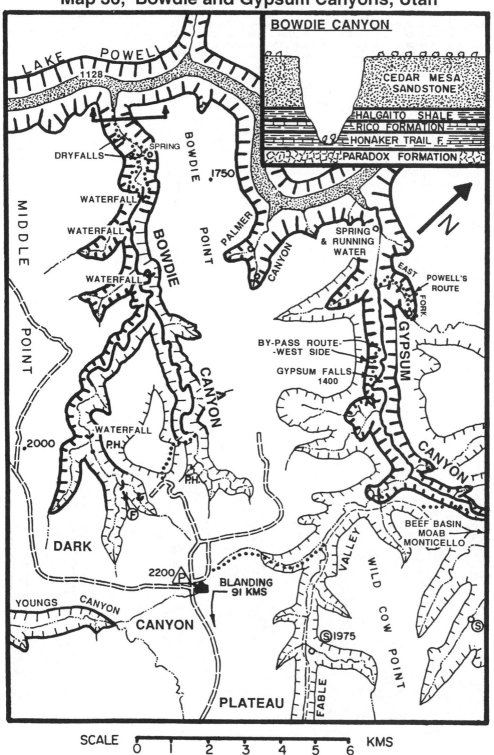

BOWDIE CANYON

CEDAR MESA SANDSTONE

HALGAITO SHALE
RICO FORMATION
HONAKER TRAIL F.
PARADOX FORMATION

LAKE POWELL

1128

DRYFALLS

SPRING

BOWDIE POINT

.1750

WATERFALL

WATERFALL

BOWDIE

WATERFALL

PALMER CANYON

SPRING & RUNNING WATER

EAST FORK

POWELL'S ROUTE

N

MIDDLE POINT

CANYON

BY-PASS ROUTE -WEST SIDE

GYPSUM FALLS 1400

GYPSUM

.2000

WATERFALL P.H.

P.H.

CANYON

F

DARK

BEEF BASIN
MOAB
MONTICELLO

VALLEY

WILD COW POINT

2200
P

BLANDING
91 KMS

YOUNGS CANYON

CANYON

S 1975

S

FABLE

PLATEAU

SCALE 0 1 2 3 4 5 6 KMS

95

Dark Canyon, Utah

Location and Access Dark Canyon and the primitive area by the same name is located west of the Abajo Mountains, north of Natural Bridges National Monument, and east of Lake Powell. To get there, drive to the junction of Highway 95 and the Natural Bridges road(between mile posts 91 and 92), and turn northwest. One km from the highway turn right, and drive through the Bears Ears, then to the corral at 2560 meters. Park there to hike down either Woodenshoe or Peavine Canyons. To reach the lower end of the canyon drive to the Hite Marina turnoff. About half a km up the highway from Hite Turnoff(right at mile post 49), turn east on a moderately good dirt road, and drive about 14 kms on Road #208a. Then turn northeast on the Squaw Rock Road(#209a) and drive about 4 kms to a point just north of Squaw Rock. Turn left for a short distance to a stock tank(pond) and park. This is the beginning of the Sundance Trail. If you have a 4WD, you can get even closer to the canyon. Serious hikers should get updated information from the rangers at Hite; or the Forest Service office in Monticello, before going into this country.

Trail or Route Conditions The Sundance Trail has stone cairns marking the route to and past a 4WD road and on into the lower end of Dark Canyon. Be observant of the route along this first part of the hike. When the trail reaches the canyon rim, it's easy to follow heading down. In the bottom of Dark Canyon, you walk in or beside the small stream and in several places use a trail around a dropoff or waterfall. If you enter the upper part of the canyon, there are cattle trails at the beginning of Woodenshoe and Peavine, and old roads in the upper sections of Peavine and Dark. Below the Confluence walk the easiest route in the canyon bottom.

Elevations Sundance Trailhead, 1700; Lake Powell at high water, 1128; the Confluence of Woodenshoe and Dark Canyons, 1775; corral at the head of Woodenshoe Canyon, 2560 meters.

Hike Length and Time Needed If you begin at the head of Woodenshoe and walk to Lake Powell, it's about 55 kms, or about 3 days--one way. Round-trip to Lake Powell and back would take most people about 6 days. Many people go down Woodenshoe, and up Peavine, in about 3 days. Others prefer to hike the bottom end. Most people would have little trouble going down the Sundance Trail and back in one day, but many do it in two.

Water There is year-round running water from just below Black Steer Canyon down to the lake. Some upper parts of the canyon have running water too, but it varies in amount from year to year.

Maps USGS or BLM maps Hite Crossing and Blanding(1:100,000), or Lower Dark Canyon, Fable Valley and Bear Ears(1:62,500).

Main Attractions Total solitude in a deep canyon; Anasazi ruins in the upper canyons.

Ideal Time to Hike Spring or fall in lower canyon; late spring to early fall in the upper end.

Hiking Boots Dry weather boots in the upper canyons; waders in the bottom end.

Author's Experience The author walked down Woodenshoe to Youngs Canyon, then back up Peavine in three long days, or 24 1/2 hours walk-time. Another time he ran down the Sundance Trail, had a skinny dip, and hurried back in 3 hours round-trip. He came up from the lake and into Lean-to Canyon on a third trip.

This is the lower end of Dark Canyon not far above Lake Powell and Cataract Canyon.

Map 37, Dark Canyon, Utah

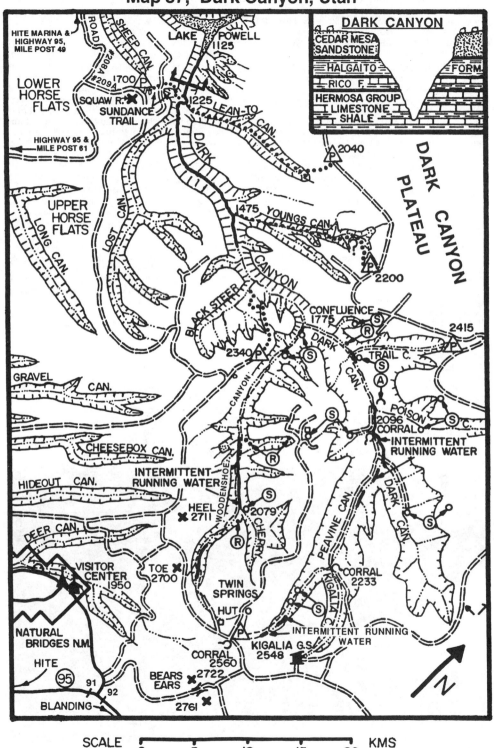

DARK CANYON

CEDAR MESA
SANDSTONE

HALGAITO

RICO F.

HERMOSA GROUP
LIMESTONE
SHALE

FORM.

HITE MARINA &
HIGHWAY 95,
MILE POST 49

SHEEP CAN.

ROAD #208A
#209A

LAKE POWELL
1125

1700 P

LOWER
HORSE
FLATS

SQUAW R.

SUNDANCE
TRAIL

1225 LEAN-TO CAN.

DARK

HIGHWAY 95 &
MILE POST 61

2040 P

DARK CANYON PLATEAU

UPPER
HORSE
FLATS

LONG CAN.

LOST CAN.

1475 YOUNGS CAN.

CANYON

2200 P

BLACK STEER CANYON

CONFLUENCE
1775

2415 P

2340 P

DARK CAN.

S

TRAIL C.

S

A

GRAVEL CAN.

S

POISON
2096
CORRAL

S C

CHEESEBOX CAN.

INTERMITTENT
RUNNING WATER

HIDEOUT CAN.

WOODENSHOE CANYON

R

INTERMITTENT
RUNNING WATER

DARK CAN.

DEER CAN.

HEEL
2711

S

CHERRY

2079

DARK CAN.

S

R

VISITOR
CENTER
1950

TOE
2700

TWIN
SPRINGS

PEAVINE CAN.

CORRAL
2233

KIGALIA

S

NATURAL
BRIDGES N.M.

HUT

P

INTERMITTENT RUNNING
WATER

HITE

95

91

92

CORRAL
2560

KIGALIA G.S.
2548

BEARS
EARS

2722

BLANDING

2761

N

SCALE 0   5   10   15   20   KMS

Lower White Canyon and The Black Hole, Utah

Location and Access The lower part of White Canyon is between Hite Marina on Lake Powell, and the Fry Canyon area. This entire hike is less than 2 kms from Highway 95 which makes for easy access. From Hite Marina or Natural Bridges N. M., drive along Highway 95 and park at mile post 57. Enter there and come out near mile post 55. *This canyon is for experienced hikers only.*

Trail or Route Conditions Before doing this hike, check the weather forecast; never go into this slot canyon if storm clouds are threatening. Also, check to see if there have been storms the day or night before up-canyon at Bridges N.M. In August, 1992, one family of 8 started hiking about 6 pm, then had to spend the night in the canyon. The next morning as they were starting out, they were caught in a flash flood. It had rained hard the night before at Bridges, and the flood took until 8 am to reach the Black Hole. They all luckily found a high place above the water, and were rescued that afternoon! From the car-park at mile post 57, walk north 250 meters on a trail to a little side canyon, then turn right and head down to the bottom. From there walk down-canyon. This is one of the best hikes around, but remember, only well-prepared hikers can make it all the way through the Black Hole. It's a dry canyon(except for the Black Hole), but you must have some kind of float device to take camera, clothes, maps, etc., through in a dry condition. A boater's life jacket works, but the best way is to line your pack with plastic bags and swim through with your pack on. To float better, include an empty water bottle or two in your pack. You must swim most of one 150 meter-long section, then swim or wade other potholes below. Best to go with several people, plus take 2 or 3 short ropes(5 meters) for any emergency and to help less-experienced members of your group down into the first part of the Black Hole. There are two routes out as shown on the map. Both are on the left about 200 meters apart. Look for hikers tracks and/or stone cairns. Or you can walk all the way to Lake Powell, but that makes it a very long hike.

Elevations M.p. 57, 1400 meters; Black Hole, 75 meters deep; m.p. 55, 1300; L. Powell, 1128 meters.

Hike Length and Time Needed In at m.p. 57, down-canyon, out at m.p. 55, and back to one's car is only about 12 kms. This can take 5-6 hours, but with the excitement of the Black Hole, plan to take all day. Lots of people are doing this hike now, mostly on weekends. Do it as a day-hike, because getting a big pack through the Black Hole is too difficult. For a short period of time in the late 1980's, there was a mass of floating logs in the Black Hole, which made getting through difficult. So be prepared for changes, which can occur with every flash flood.

Water It's a dry canyon except for a number of permanently filled potholes. There should be water in the potholes any time of year, but with an influx of hikers, it's best to take your own.

Maps USGS or BLM map Hite Crossing(1:100,000), or Browns Rim(1:62,500).

Main Attractions The Black Hole. *One of the best adventures in the hiking world.*

Ideal Time to Hike Hot summer weather with temps. 35°C or higher--preferably over 40°C. Black Hole water is frigid and hypothermia a real possibility. One hiker recorded the Black Hole water temp at 11°C(52°F) in summer! Keep clothes dry in plastic sacks and swim through in shorts. It was 41°C(106°F) the day the author went through, but he still got a touch of hypothermia(uncontrollable shivering). On a fourth trip, he returned to get better fotos and it was 34°C(94°F) outside. He thought he was going to freeze to death, and hardly got any fotos! That was on May 25, 1994.

Hiking Boots Wading boots or shoes.

Author's Experience On his 3rd try, the author was better prepared and made the entire hike in less than 5 hours. On his 4th trip, he also did it in 5 hours, then used a mtn. bike to get back to his car.

It's either wading or swimming the better part of one 150 meter section of the Black Hole.

Map 38, Lower White Canyon and The Black Hole, Utah

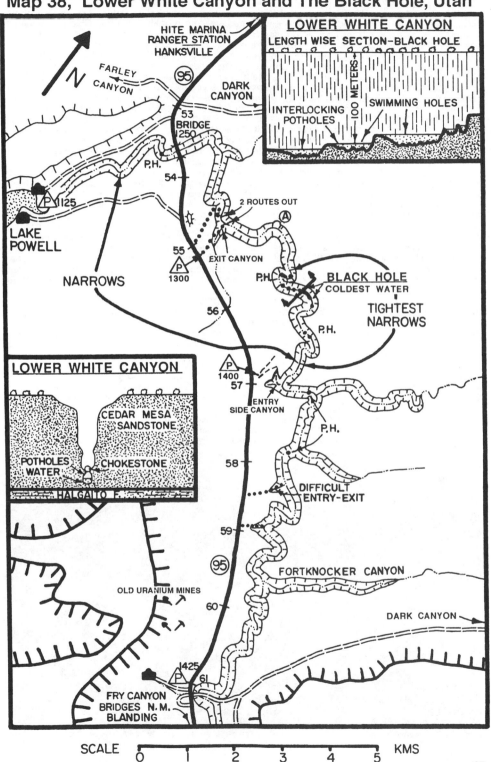

LOWER WHITE CANYON
LENGTH WISE SECTION-BLACK HOLE

100 METERS

INTERLOCKING POTHOLES

SWIMMING HOLES

HITE MARINA
RANGER STATION
HANKSVILLE

FARLEY CANYON

N

95

DARK CANYON

53
BRIDGE
1250

P.H.

54

2 ROUTES OUT

P 1125

LAKE POWELL

NARROWS

55
P 1300

EXIT CANYON

A

P.H.

BLACK HOLE
COLDEST WATER

TIGHTEST NARROWS

56

P.H.

LOWER WHITE CANYON

CEDAR MESA SANDSTONE

POTHOLES WATER

CHOKESTONE

HALGAITO F.

P 1400
57

ENTRY SIDE CANYON

P.H.

58

DIFFICULT ENTRY-EXIT

59

OLD URANIUM MINES

95

FORTKNOCKER CANYON

60

DARK CANYON

1425
P
61

FRY CANYON
BRIDGES N. M.
BLANDING

SCALE 0 1 2 3 4 5 KMS

Upper White Canyon and Tributaries, Utah

Location and Access The upper part of White Canyon and several tributaries are located about halfway between Blanding and Hanksville and immediately north of Highway 95. The part of White Canyon covered here is from Natural Bridges N. M. downstream to Fry Canyon Lodge. Access is very easy with no driving on dirt roads; except for one short side road, or if your intension is to reach upper Cheesebox Canyon. There are three main places you can park. If hiking down-canyon, proceed to mile post 81 and park on the north side of the road. Or at mile post 75, turn northeast and drive half a km on a dirt track and park opposite the mouth of Cheesebox Canyon. At Fry Canyon, park at the lodge or near the bridge and mile post 72. If you have a 4WD or HCV, then you might drive the Cheesebox Road to reach the upper end of Cheesebox or Gravel Canyons.

Trail or Route Conditions Walking is easy in White Canyon in the usually dry creek bed. Tributary canyons should offer something different. From the car-park at m.p. 81, there's an old and now blocked-off road(shown as a trail) running down to an overlook of Kachina Bridge. On the south side of the overlook, route-find down into Armstrong Canyon then into the main drainage. Opposite the mouth of Cheesebox Canyon near mile post 75, is an old faintly visible cattle trail running down into White. To enter at Fry Canyon, walk north and down-canyon on the east side of the drainage. Stay on the rim--don't get down in. About 200 meters up White Canyon from the mouth of Fry, look for a route into White. You may have to climb down the truck of one tree to get in(?). One of the best parts of this hike is the short walk up from White and into the bottom end of Fry Canyon(only from inside White Canyon can you reach these narrows). Just above the narrows look for Anasazi ruins. About one km into K & L is a dryfall and a spring. The author hasn't been into Hideout Canyon yet, but Scott Patterson states there are ruins, swimming pools, narrows and a dryfall, as shown on the map. The author went into Cheesebox and after swimming a ways, was stopped by narrows and a dryfall, with a mucky-looking pool beneath. Rumors say there's some great slots above near where the two forks meet. The author hasn't been to Gravel Canyon yet, but you enter it by parking on the highway at mile post 65, and walking across White and up Gravel from the bottom. The June, 1994 issue of Backpacker Magazine shows one of several good short narrows in Gravel that requires wading/swimming. Jim Ohlman, mentions there's another "Rescue Trail" in the upper end of the White Canyon.

Elevations Fry Canyon Lodge, 1620 meters; mile post 81, 1825 meters.

Hike Length and Time Needed From mile post 81 to Fry Canyon Lodge is about 26 kms along the bottom of White Canyon. This is one rather long day for most hikers. To do this you'd need two cars, or a car and a bike; or hitch hike back to your vehicle. You can also enter or leave the canyon at Cheesebox and mile post 75, cutting the distance considerably. Mile Post 75 is the best place to enter if you're going to Cheesebox, Hideout or K & L Canyons.

Water At Natural Bridges and Fry Canyon Lodge. Carry plenty in your car. Also at K & L Spring.

Maps USGS or BLM map Hite Crossing(1:100,000), or Natural Bridges(1:62,500).

Main Attractions Easy access, short narrows, ruins in lower Fry Canyon in south facing alcove.

Ideal Time to Hike Spring or fall, but if there's swimming in side canyons, then summer only.

Hiking Boots Any dry weather boots or shoes, but waders in Cheesebox, Hideout and Gravel Canyons.

Author's Experience The author hiked down Fry, up White Canyon to Kachina Bridge, then out at mile post 81. A 7 hour hike. He then hitched a ride back to his car. Years later he was stopped halfway up Cheesebox after swimming a ways, and has been up K & L Canyon to the dryfall, both on the same day.

Deep pool in lower Fry Canyon narrows just before it joins White Canyon.

Map 39, Upper White Canyon and Tributaries, Utah

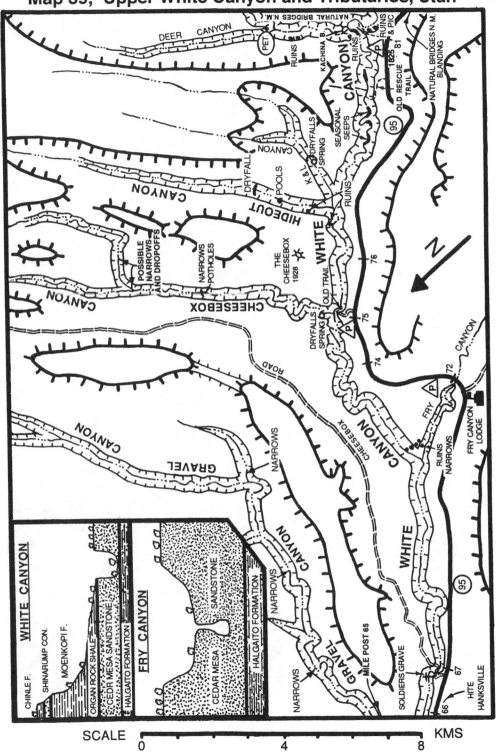

SCALE

0   4   8   KMS

101

Natural Bridges National Monument, Utah

Location and Access Natural Bridges National Monument is located about halfway between Blanding and Hite, and just north of Highway 95, which is the main access route. One can also reach Bridges from the south and Mexican Hat and the Navajo Tribal Lands via Highway 261. At a point 3 kms west of the junction of Highways 95 and 261, and between mile posts 91 and 92, turn northwest and drive 8 kms to the visitor center. There you can get the latest information on trail conditions, campsites, etc. Then drive to the 13 km loop-road, which takes you to the 3 main bridges the park is famous for.

Trail or Route Conditions The emphasis here is on the route in the canyon bottoms from Sipapu Bridge, down White Canyon to Kachina Bridge, then up Armstrong Canyon to Owachomo Bridge. About half of this distance is covered by walking in the dry creek bed, but in other places, a developing trail is appearing as this hike becomes more popular. As you drive along the park road, there are turnouts above each bridge. There you can park and walk down to each. If you don't have time for a long hike, you can visit each bridge from your car. There's also a trail on the mesa top connecting the three bridges for those who want to start walking at one, make a loop-hike along the canyon bottoms, then return to their car quickly. Trails near the bridges are well-used and signposted, so getting in and out of the canyon is no problem. The trail across the mesa is marked with cairns. Hikers are not allowed to camp in the canyon bottoms inside the park, but one can walk down White Canyon to the park boundary about 3 kms below Kachina Bridge and spend a night there. You can also drive outside the park boundaries and camp anywhere. Not far west of Sipapu Bridge are some reverse hand pictographs and some Anasazi ruins as shown on the map. There are other ruins and pictographs near where Armstrong and White Canyons meet near Kachina Bridge.

Elevations Viewpoints, 1800 to 1850 meters; the canyon bottoms 75 to 100 meters below.

Hike Length and Time Needed To do the loop-hike from Sipapu, Kachina and to Owachomo, then back to one's car is about 12 or 13 kms, or an all-day hike for many people. If two cars are available, the time can be cut in about half. If you have a mtn. bike, you could leave it at one bridge, drive to the other end, walk the canyon, then bike back to your car. To do the entire loop gives one a feeling of accomplishment, as well as a look at some good ruins and pictographs.

Water There are several big potholes or pourover pools, but no year-round springs in these canyons. Fill your water bottles at the visitor center or park campground.

Maps USGS or BLM maps Hite Crossing and Blanding(1:100,000), or Natural Bridges(1:62,500).

Main Attractions Three of the world's largest bridges, plus pictographs and ruins.

Ideal Time to Hike Spring or fall. Summers aren't so hot here because of the altitude.

Hiking Boots Any dry weather boots or shoes.

Author's Experience The author camped outside the park, then began in the morning. He made the loop-hike beginning and ending at Sipapu in about 4 1/2 hours.

Sipapu Bridge is the highest and widest of all bridges in this national monument.

Map 40, Natural Bridges National Monument, Utah

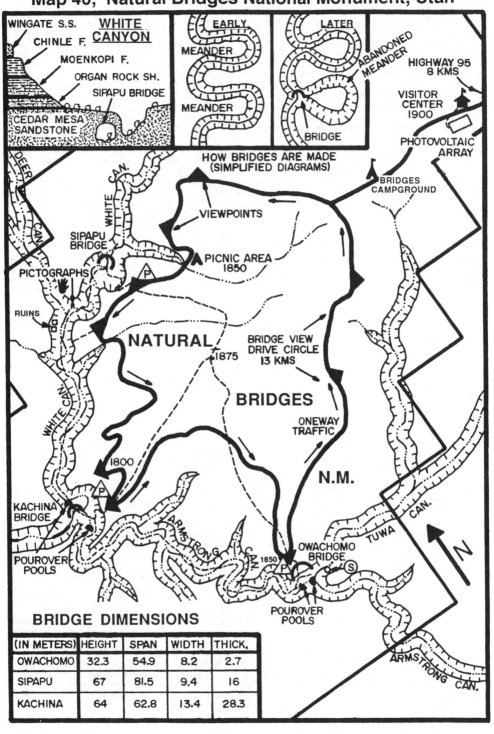

WINGATE S.S.
CHINLE F.
WHITE CANYON
MOENKOPI F.
ORGAN ROCK SH.
SIPAPU BRIDGE
CEDAR MESA SANDSTONE

EARLY
MEANDER
MEANDER

LATER
ABANDONED MEANDER
BRIDGE

HIGHWAY 95
8 KMS
VISITOR CENTER
1900
PHOTOVOLTAIC ARRAY

HOW BRIDGES ARE MADE
(SIMPLIFIED DIAGRAMS)

VIEWPOINTS

BRIDGES CAMPGROUND

DEER CAN.

WHITE CAN.

SIPAPU BRIDGE

PICTOGRAPHS

PICNIC AREA
1850

P

RUINS

WHITE CAN.

NATURAL

1875

BRIDGE VIEW DRIVE CIRCLE
13 KMS

BRIDGES

1800

ONEWAY TRAFFIC

N.M.

KACHINA BRIDGE

P

ARMSTRONG C.

TUWA CAN.

N

POUROVER POOLS

1850

P

OWACHOMO BRIDGE

S

POUROVER POOLS

ARMSTRONG CAN.

BRIDGE DIMENSIONS

(IN METERS)	HEIGHT	SPAN	WIDTH	THICK.
OWACHOMO	32.3	54.9	8.2	2.7
SIPAPU	67	81.5	9.4	16
KACHINA	64	62.8	13.4	28.3

SCALE 0   1   2   3 KMS

North and South Mule and Arch Canyons, Utah

Location and Access Mule and Arch Canyons are located about halfway between Blanding and Natural Bridges National Monument. They're also south of the Abajos and just north of Highway 95. To get there, first drive south, then west out of Blanding about 37 kms; or drive east from Bridges. At a point just west of the prominent Comb Ridge and between mile posts 107 and 108, turn north and drive about 4 kms to the mouth of Arch Canyon. One can park and camp there in a grove of cottonwood trees. To reach both upper forks of Mule Canyon, drive to within one km of the mesa top Mule Canyon Ruins(between mile post 102 and 103) and turn north onto the Texas Flat Road #263. Drive this good dirt road less than one km and park & camp near the bottom of South Fork of Mule; or a little further to enter North Fork. If you continue on, there's a good campsite at Dog Tank Spring. Continue still further, and you'll end up at the Arch Canyon Overlook at 2050 meters altitude. If you're looking for Anasazi ruins, Mule Canyon has many.

Trail or Route Conditions There are now hiker-made trails into both North and South Forks of Mule Canyon and it's easy walking right up the mostly dry creek beds. What water there is can be avoided and there's very little brush. As you walk up either canyon, the walls get higher and higher. Along the way are Anasazi ruins which are on the north side, facing south, and under overhangs. At the upper end of each there's a route over the ridge between the two, so with a little route-finding you can make a loop-hike without backtracking down the same canyon.

From the trailhead at the lower end of Arch, there's a very rough 4WD track up to the confluence with Texas. This canyon has been proposed for a Wilderness Study Area, but 4WD'ers are doing what they can to ruin that possibility. If you hike on weekends, you will have noisy competition. If 4WD's are allowed in the canyon, then mtn. bikes are also legal, so you could save a lot of walking by biking up to the confluence, then hike into the three tributaries. There are developing trails in each of these. One km from the trailhead there's a demolished ruin on the north wall and not far above that is a little side canyon to the north, which, according to one source, has a spring and an Anasazi ruin of some kind. Further up-canyon there are said to be other ruins as well, but the author didn't see any. At the confluence of Texas and Butts there's a big grove of Ponderosa pines, which makes a fine camping area. All vehicles stop there. If you have time, drive one km south from the highway in Comb Wash and turn west into Lower Mule Canyon. It's sandy, so watch out or you'll get stuck! Then hike a short distance to see several more Anasazi ruins on the right side going up-canyon.

Elevations Mule Canyon trailheads, 1800 meters; bottoms of the upper canyons, about 2000; the mouth of Arch Canyon, 1525; the confluence of Texas and Butts Canyons, about 1700; and the Arch Canyon Overlook, about 2050 meters.

Hike Length and Time Needed In either the North or South Forks of Mule Canyon, the distance from the car-park to the head of the canyon is about 10 kms. To walk up either and return the same way isn't a long hike, but with all the ruins to see, it'll take you a full day in each. If you decide you don't want to backtrack, then route-find over the divide between the two forks and enter the other. This way you could visit both canyons in one long day. The author found a route out of South Fork, but didn't take the time needed to find a route into or out of the North Fork. An informant has stated he did make the hike over the divide successfully, so it can be done. From the car-park at the mouth of Arch, up to Texas Canyon is about 13 kms, and another 7 or so to the head of Butts drainage. This hike can be done from the trailhead in one long day round-trip, but to explore the upper reaches of the canyons, 2 or 3 days are needed. A mtn. bike would shorten that hike.

Water The map shows running water at the time of the author's visit. In drier times there could be much less water in Mule Canyon, but the presence of so many Anasazi ruins indicates a permanent water supply. There's an excellent spring and campsite at the head of Dog Tank Draw. The author found running water in much of the Arch Canyon, as shown on the map by the heavy black line. There may be cattle in the area too, but the water seemed good to drink as is in the upper forks. If cattle are there when you arrive, better purify or filter it first to be on the safe side. There is also permanent running water in much of the drainage of Lower Mule Canyon below the highway.

Maps USGS or BLM map Blanding(1:100,000), or Bears Ears and Brushy Basin Wash(1:62,500), or perhaps the best one now is the plastic Trails Illustrated map Grand Gulch Plateau(1:62,500).

Main Attractions Mule Canyon is one of the best places on the Colorado Plateau to visit Anasazi ruins in a natural state, and without supervision. This means that each hiker must supervise him or herself as well as some companions in order to preserve these sites. The South Fork has more ruins than the North, but both canyons are worth a visit. One ruin in South Fork stands high against a wall and is uniquely different. Arch is a deep, rugged and scenic red rock canyon, with Ponderosa pines in the upper forks. It has several arches, some Anasazi ruins and a good water supply. The lower end of Mule Canyon, that part below the highway, has several well-preserved ruins as well.

Ideal Time to Hike Spring or fall, or winter warm spells(?). Also, the higher altitude of the upper canyons makes it just tolerable in summer. Early summer brings many large horse or deer flies which bite bare legs, so wear long pants then.

Hiking Boots Any dry weather boots or shoes.

Author's Experience The author spent two days in Mule Canyon, one in each fork. Each hike lasted about 6 hours. You'll surely want to spend more time there than he did. Later he walked from his camp at the mouth of Arch, up into Texas, and later into Butts Canyon. He spent a total of 7 1/2 hours in Arch Canyon. A mtn. bike would have made it faster and easier. Years later he drove to the mouth of Lower Mule and spent 5 hours seeing most of that drainage below the highway.

Map 41, North and South Mule and Arch Canyons, Utah

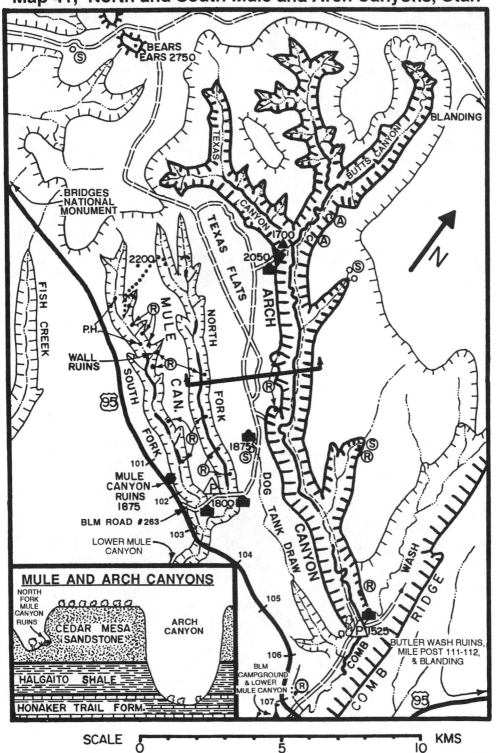

BEARS
EARS 2750

BLANDING

TEXAS

BUTTS CANYON

BRIDGES
NATIONAL
MONUMENT

TEXAS FLATS

CANYON 1700

2050

2200

ARCH

N

MULE CAN.

P.H.

WALL
RUINS

NORTH FORK

FISH CREEK

SOUTH FORK

95

1875

DOG TANK DRAW

101

MULE CANYON
RUINS
1875

102

1800

BLM ROAD #263

103

LOWER MULE
CANYON

104

105

WASH

RIDGE

COMB

1525

COMB

BUTLER WASH RUINS,
MILE POST 111-112,
& BLANDING

106

BLM
CAMPGROUND
& LOWER
MULE CANYON

107

95

MULE AND ARCH CANYONS

NORTH
FORK
MULE
CANYON
RUINS

CEDAR MESA
SANDSTONE

ARCH
CANYON

HALGAITO SHALE

HONAKER TRAIL FORM.

SCALE
0   5   10   KMS

Overlooking Arch Canyon and the extra thick Cedar Mesa Sandstone walls.

One of the better preserved Anasazi structures in the South Fork of Mule Canyon.

Well-preserved Anasazi ruins in the lower end of Mule Canyon.

Potsherds and corn cobs in Mule Canyon. Please leave these artifacts in place for others to enjoy.

Fish and Owl Creek Canyons, Utah

Location and Access Fish and Owl Creek Canyons are located to the south of the Abajo Mountains and Highway 95, to the east of Grand Gulch and Highway 261, and due north of Mexican Hat. To get there from the junction of Highways 95 and 261, drive south about 8 kms and look to the east or left side of the highway for Road #253(between mile posts 27 and 28). This minor junction is about 1 1/2 kms south of the Kane Gulch Ranger Station, the trailhead entry point to upper Grand Gulch. From the highway, drive southeast about 8 kms on Road #253 to an old drill hole site which is the trailhead. This is a good road for any vehicle, but it can be slick and muddy in wet weather. You can camp at the trailhead, but take your own water.

Trail or Route Conditions Because the trailhead is between these two canyons, people normally walk down one and up the other. The best way to do this loop-hike is to go in a clock-wise direction. The reason for this is, it's easier going down into Fish Creek, than to find, then climb up the route out of Fish Creek Canyon. To begin, walk north from the trailhead sign and register board on a good well-used trail until you reach a point on the canyon rim just up-canyon from where the two forks of Fish Creek meet. Then use a short rope to lower big packs down over the first ledge into the canyon. Below that ledge, and once in the drainage, it's easy walking down the creek bed or along hiker-made trails. There's water in places, but you can avoid wading. Later, you walk up Owl Creek until you reach a waterfall near the 1585 meter point. Use the trail on the north side, and follow it back to the trailhead. As you near the rim not far from the trailhead, you'll have a chance to visit an interesting cliff dwelling on the left under an overhang. There are a number of these in the canyon, but this one is right on the trail(see foto). The author has yet to see the upper part of Fish Creek or McCloyd Canyons; they could prove interesting.

Elevations Trailhead altitude is 1895 meters; confluence of the two canyons, 1460 meters.

Hike Length and Time Needed From the trailhead and down Fish Creek, then up Owl Creek and back to the trailhead is about 25 kms round-trip; a rather long day for most hikers. Many people do the hike in two days. If you have only half a day, you could enter Owl Creek Canyon, visit some ruins and one or two big pourover pools, and return.

Water There's running water in both canyons, but more in Fish Creek, which has minnows(small fish). Cattle apparently are no longer allowed to graze in the canyons, so the water should be unpolluted--except perhaps by careless campers! The running water shown on the map is as it was in 1986, after several wet years.

Maps USGS or BLM map Bluff(1:100,000), or Bluff and Cedar Mesa(1:62,500), or the best one now is the plastic Trails Illustrated map Grand Gulch Plateau(1:62,500).

Main Attractions This is one of the better places in Utah and the Colorado Plateau to visit Anasazi ruins in a wilderness setting. There are also ruins in McCloyd Canyon and the North Fork of Road Canyon.

Ideal Time to Hike Spring or fall, or perhaps winter dry spells, or morning-hike in summer.

Hiking Boots Any dry weather boots or shoes.

Author's Experience The author camped at the trailhead and did the round-trip in one day. The total walk-time 8 1/2 hours. In 1994, he returned for a quick look at the trail to the rim of Fish Creek, which took less than one hour round-trip.

These Anasazi ruins are very near the trailhead and drillhole in Owl Canyon.

Map 42, Fish and Owl Creek Canyons, Utah

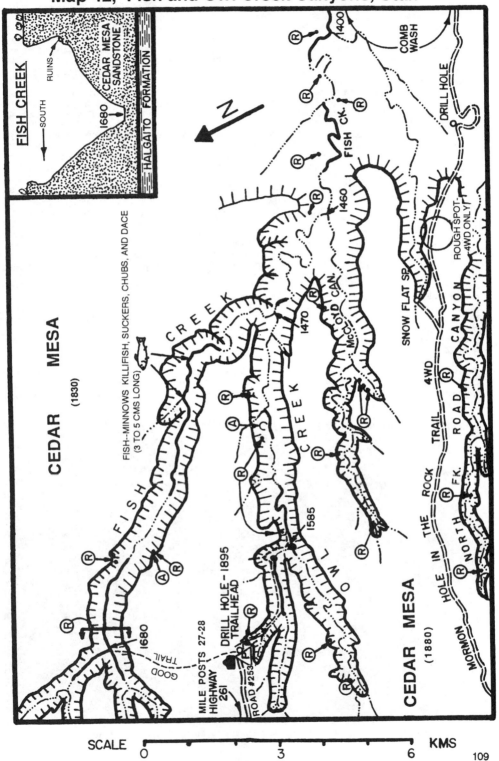

SCALE

0   3   6   KMS

109

Road and Lime Canyons, Utah

Location and Access Road and Lime Canyons are located due north of Mexican Hat and the Valley of the Gods, and east of State Highway 261. They are also just south of Fish and Owl Creek Canyons and east of Grand Gulch Primitive Area. To reach the trailheads, drive north from Mexican Hat, or south from near Natural Bridges N. M., to mile post 19 on Highway 261. At that point is a road heading east and a sign stating, Cigarette Spring--9 miles(14 kms). They may have changed the sign recently, but a sign is there. Drive east on this Cigarette Spring Road #239, to reach both Road and Lime Canyons. There is now a trail register at the gate about 1 1/2 kms from the highway. There are many places to camp, including Cigarette Spring, perhaps the best site around.

Trail or Route Conditions From the car-park & campsite at 1950 meters, which is just a short distance beyond a road junction and before you reach the gate shown, start walking northeast on a hiker-made trail. Soon you'll drop down into upper Road Canyon. Simply walk down the mostly dry creek bed, while keeping an eye on the north side for Anasazi ruins. At the bottom of the canyon you can explore the North Fork of Road Canyon before looking for a route out to the mesa top. Don't forget a compass; it helps when moving across the Cedar Mesa. Make your way to the end of the road at Cigarette Spring, then road-walk back to your car. You can also make Cigarette Spring your campsite and trailhead and hike from there. To get into Lime Canyon, locate the other car-park at 1950 meters. Get there from mile post 16 or Road #239. Walking in both canyons is easy.

Elevations Car-parks, 1950 meters; bottom end of drainages, 1500; Cigarette Spring, 1700 meters.

Hike Length and Time Needed From the head of Road, down-canyon and up Cigarette Spring Canyon to the spring is about 24 kms. To walk down Lime, then up on the mesa to Cigarette Spring, is about the same distance. From the spring back to one's car is from 8 to 10 kms. To do either of these loop-hikes would be a very long day for anybody. A mtn. bike placed at one end would save some walking. Also, shorter hikes can be made into either canyon and return the same way.

Water Always carry water in your car and in your pack. The author found short stretches of running water in Road, as shown by heavy black lines. He also hiked upper Lime Canyon in February and found lots of ice, indicating some water(?). Cigarette Spring has a good flow year-round.

Maps USGS or BLM map Bluff(1:100,000), or Cedar Mesa(1:62,500), but now the best one is probably the plastic Trails Illustrated map Grand Gulch Plateau(1:62,500).

Main Attractions Anasazi Indian ruins. One site in Road Canyon is called Seven Kivas Ruins(ceremonial sites). These are some of the best preserved kivas around. Good news, potsherds and other artifacts are still there, so let's all continue to leave them in place for others to enjoy.

Ideal Time to Hike Spring or fall, or perhaps winter warm spells. The elevations here are moderately high, so summer hiking isn't so bad.

Hiking Boots Any dry weather boots or shoes.

Author's Experience The author left his car-park camp, walked down Road, then took the short-cut to Cigarette Spring and walked back up Road #239 to his car in about 7 1/2 hours. He also day-hiked Lime Canyon in fresh snow from the road beginning at mile post 16. In 1994, he returned to have a quick look to as far as Cigarette Spring.

Some of the best preserved Anasazi ruins anywhere are found in Road Canyon.

Map 43, Road and Lime Canyons, Utah

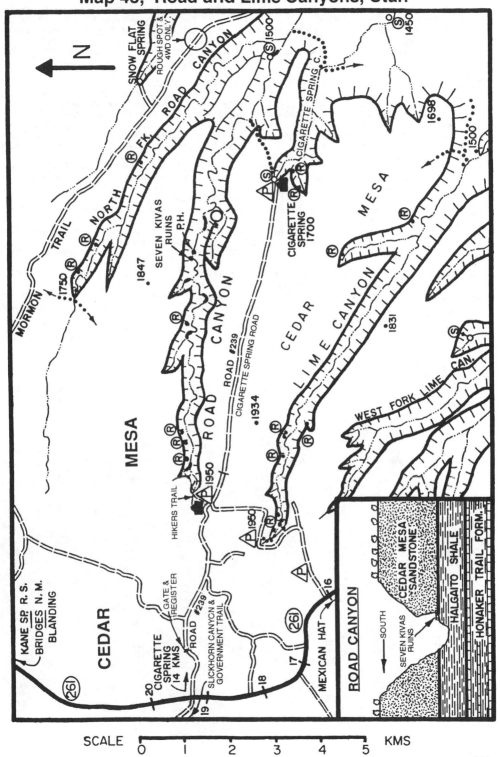

SCALE 0 1 2 3 4 5 KMS

111

Johns Canyon, Utah

Location and Access Johns Canyon is yet another of Utah's rugged canyons where Anasazi ruins can be found. It's located just west of State Highway 261, the paved road running between Mexican Hat and the area around Natural Bridges National Monument and the Kane Gulch Ranger Station. There are a number of ways to get into this canyon, but one simple way is via the dirt road running southwest from near mile post 17 on Highway 261. You can park about 4 kms from the highway and enter the upper part of the West Fork of Johns Canyon. The author has yet to try it, but you could probably park near mile post 16 and hike down in from there.

Trail or Route Conditions There are no trails into this canyon and it's rather rugged. There are very few routes into any of the forks because of the various layers of Cedar Mesa Sandstone. The sides of the upper canyons are like a wedding cake, with many cliffs and benches. Most of the ruins are near the rim and often cannot be seen from the floor of the canyon. If the main reason for going into Johns is to view ruins, it's best to stay on the rim when looking for these sites. One idea would be to rim-walk using a pair of binoculars. The author didn't make it all the way down the canyon, but there appears to be some dryfalls in each of the major forks, which may or may not be passable. It would be interesting to attempt to walk down either of the two upper forks and see if there is a route down. With all the ruins in the upper canyons there is surely a route down to the San Juan River. With a good map and perhaps a 4WD, you can also come up-canyon from near the San Juan River, but the author still isn't familiar with the lower end of the canyon.

Elevations The altitude of the rim of this canyon is about 1950 meters.

Hike Length and Time Needed Most people would prefer one or more day-hikes into Johns, but there appears to be some nice campsites down along the dry creek bed.

Water Always have water in your car and take some with you if day-hiking. There's one spring near the bottom end of the canyon, but it may be hard to reach. From the canyon rim and in February snows, the author could see cottonwood trees in the bottom of West Fork, so there's likely some water down there somewhere.

Maps USGS or BLM maps Bluff and Navajo Mtn.(1:100,000), or Grand Gulch and Cedar Mesa(1:62,500), or the plastic Trails Illustrated map Grand Gulch Plateau(1:62,500).

Main Attractions As mentioned, the main attractions here are the Anasazi ruins. USGS maps indicate about 7 sites in the Johns Canyon drainage, with several more at the head of Slickrock Canyon. Each site consists of two or three shelters and apparently were not occupied for very long.

Ideal Time to Hike The author made his hike in February, so some winter weather isn't so bad for hiking. Spring or fall are considered the ideal time however. With the moderate elevations, summers aren't so hot.

Hiking Boots Any dry weather boots or shoes.

Author's Experience The author spent half a snowy day in and along the rim of the West Fork and visited the first two ruins. There should be a route down into the canyon from each Anasazi site(?).

These ruins are found in the upper West Fork of Johns Canyon.

Map 44, Johns Canyon, Utah

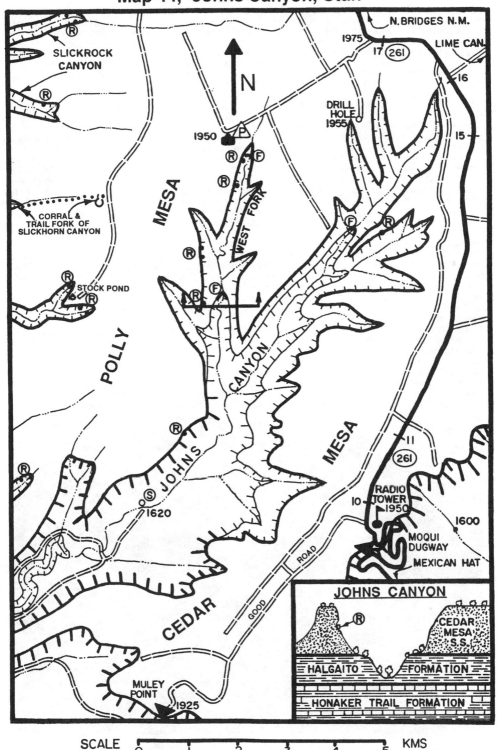

N. BRIDGES N.M.
1975
LIME CAN.
17 (261)
16
15

SLICKROCK
CANYON

DRILL
HOLE
1955

1950

CORRAL &
TRAIL FORK OF
SLICKHORN CANYON

WEST FORK

MESA

STOCK POND

POLLY

CANYON

MESA

JOHNS

11
(261)

RADIO
TOWER
1950

10

1600

MOQUI
DUGWAY

MEXICAN HAT

1620

ROAD

GOOD

CEDAR

MULEY
POINT
1925

JOHNS CANYON

CEDAR
MESA
S.S.

HALGAITO   FORMATION

HONAKER TRAIL FORMATION

SCALE   0   1   2   3   4   5   KMS

113

Slickhorn Canyon, Utah

Location and Access Slickhorn is one of many canyons on Cedar Mesa, a large uplifted area south of the Abajo Mountains and north of the San Juan River and Mexican Hat. To get there, drive along Highway 261, the main road linking Mexican Hat and Natural Bridges National Monument. Just north of mile post 19, and directly across the highway from Road #239(the one running out to Cigarette Spring and Road Canyon), turn west onto a graded dirt road. After a short distance there is a gate and trail register. Sign in please, and close the gate behind you. From there, continue on for another 3-4 kms and turn left at that major junction. If you turn right, you'll end up at Government Trail, a route down into Grand Gulch. Continue on to one of the 4 trailheads shown.

Trail or Route Conditions This must be a fairly popular canyon because there are now minor hiker-made trails down into each of the three main forks, plus into another un-named drainage which we'll call East Slickhorn. In First Fork walk right down the drainage, then when you come to the big dryfall, veer left and bench-walk along the little trail. After a ways, stone cairns mark a route down to the bottom. Not far below that, and to the right(north) above, are several structures, including a perfectly preserved kiva. Someone has placed a ladder there to help you in & out. There are several more ruins and running water at the bottom end of Second Fork. The bottom of Third Fork has good running water too, as does the bottom end of Trail Fork. Back on top now. At the head of Third Fork, are good campsites right where the road drops down into the slickrock drainage. Trail Canyon is easy to locate too. When you see the corral on the right, stop and park there, then head down the very flat shallow drainage. Down in, there's a constructed cattle trail bypassing a dryfall. Directly above that to the north are some good ruins on a ledge. East Slickhorn also has some ruins right under the top-most rim, and an old cattle trail down in. You should be able to hike all the way to the San Juan River in either of these two main canyons.

Elevations All trailheads are about 1875 meters; bottom of Trail Fork, 1585 meters..

Hike Length and Time Needed If one were to walk down First Fork and return up Trail Fork, that would be walking about 15 kms, trailhead to trailhead. That isn't very far, but it's slow walking in places, and you'll surely want to see some ruins, so plan on this being an all-day hike. Shorter hikes are possible too, as well as overnight camping.

Water On the last day of March, 1994, there was a fair amount of running water in the bottom, as shown. Some of it will dry up in summer, but surely some will be there year-round. If you're day-hiking, take water with you just to be sure.

Maps USGS or BLM maps Navajo Mtn. & Bluff(1:100,000), or Grand Gulch & Cedar Mesa(1:62,500), or the best one now is the plastic Trails Illustrated map Grand Gulch Plateau(1:62,500).

Main Attractions Well-preserved Anasazi ruins and solitude in a wild isolated canyon.

Ideal Time to Hike Spring or Fall. Possibly some winter warm spells, or summer cool spells.

Hiking Boots Any dry weather boots or shoes.

Author's Experience He walked from the trailhead in First Fork, had a peek at the kiva below, looked into Third Fork, then went out Trail Canyon, stopping enroute to see the Big Ledge Ruins. That took 7 3/4 hours, but then he had to road-walk back to his car. Total walk-time was 9 1/4 hours.

Well-preserved granary in Slickhorn Canyon.

Map 45, Slickhorn Canyon, Utah

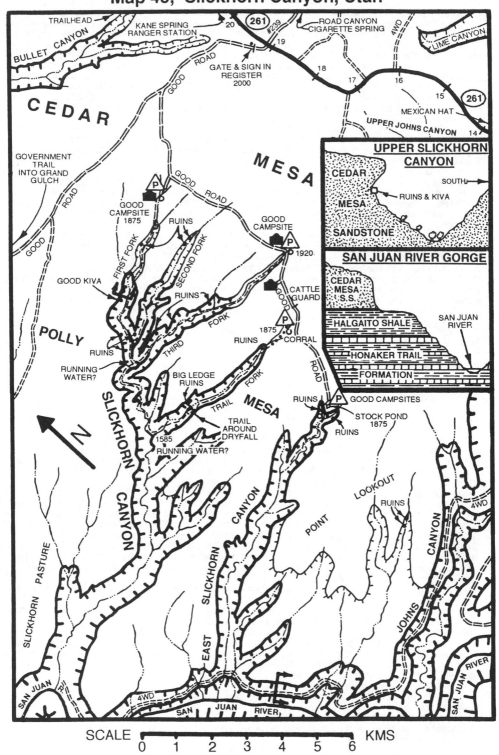

TRAILHEAD
KANE SPRING RANGER STATION
20
#239
19
261
ROAD CANYON CIGARETTE SPRING
4WD
LIME CANYON
BULLET CANYON

GOOD ROAD
GATE & SIGN IN REGISTER 2000
18
17
16
15
261

C E D A R
MEXICAN HAT
UPPER JOHNS CANYON
14

GOVERNMENT TRAIL INTO GRAND GULCH
GOOD ROAD
M E S A

P
GOOD CAMPSITE 1875
RUINS
GOOD CAMPSITE
P
1920.

UPPER SLICKHORN CANYON
CEDAR
MESA
SANDSTONE
SOUTH
RUINS & KIVA

GOOD KIVA
FIRST FORK
SECOND FORK
RUINS

CATTLE GUARD
GOOD ROAD

SAN JUAN RIVER GORGE
CEDAR MESA S.S.
HALGAITO SHALE
SAN JUAN RIVER
HONAKER TRAIL FORMATION

RUINS
THIRD FORK
RUINS
1875
P
CORRAL

POLLY
RUNNING WATER?

BIG LEDGE RUINS
FOURTH FORK
TRAIL
MESA

RUINS
P
GOOD CAMPSITES
STOCK POND 1875
RUINS

N

TRAIL AROUND DRYFALL
1585
RUNNING WATER?

SLICKHORN CANYON

SLICKHORN CANYON

LOOKOUT
POINT
RUINS

JOHNS CANYON
4WD

SLICKHORN PASTURE

EAST SLICKHORN CANYON

SAN JUAN R.
4WD
SAN JUAN RIVER

SAN JUAN RIVER

SCALE
0 1 2 3 4 5 6
KMS

115

Grand Gulch, Utah

Location and Access Grand Gulch is located just south of Natural Bridges N. M., southeast of Highway 276, west of Highway 261 and right in the middle of Cedar Mesa. The normal starting point is Kane Gulch Ranger Station, located about 7 kms south of the junction of Highways 95 and 261 and between mile posts 28 and 29. A second entry site is Todie Canyon. Drive along Highway 261 to near mile post 25, turn west and drive 1 1/2 kms to the trailhead. Bullet Canyon is the third entry point. Drive along Highway 261 to between mile posts 21 and 22, turn west and drive 3 kms to the trailhead. A fourth official entry point is now at Government Trail. Drive Highway 261 and just north of mile post 19, turn west. This is just across the highway from the road running to Cigarette Spring and Road Canyon. Drive 14 kms west following the signs to Government Trail. Collins Spring is a fifth entry/exit point. Drive along Highway 276 and near mile post 51, turn south on Road #218 and drive 10 kms to the trailhead. In dry weather, all these road are in good condition for any car.

This map is just an introduction to Grand Gulch, so be sure to have a good topo map in hand and talk to the rangers at Kane Springs before going in. If camping overnight, you now must pay $5 at each of the above trailheads. Put the money in an envelope and place in the safe box, then sign the register and go. Sign out at the same place when leaving. No fires allowed in the canyon; use stoves instead. Day hikers do it free, but are still requested to sign in & out.

Trail or Route Conditions The entire length of Grand Gulch is part of an official primitive area. The reason people go into this canyon is to see Anasazi ruins. Most of these are in the upper half of the drainage above Bullet Canyon. Over the years, hiker-made trails have slowly developed in the canyon, particularly downstream from Kane Gulch Ranger Station. To spread out the hiking, and especially camping pressure, it's recommended you enter some place besides Kane Gulch.

As you walk down-canyon, you will notice that all of these ruins are in protected alcoves facing south. Please observe the ruins, take fotos, leave a few footprints, but don't climb on or disturb the structures in any way. Also, please don't carry away potsherds or corn cobs. *It's against Federal law!* You can be arrested and fined for carrying away souvenirs. Leave them behind for others to enjoy.

The trail in from Kane Gulch is heavily traveled, but the one down through Todie Canyon is much less used. In fact it's just a crude hikers trail, which has never been improved. As you drop off the mesa, it's rather steep with lots of zig zagging back and fourth among large boulders making it slightly challenging for people carrying a large pack. One group of ruins in lower Todie totals 18 structures along one ledge. Bullet Canyon Trail is just hiker-made too, but Government Trail was apparently built in the 1930's by the CCC's. The author is just guessing on this, but it resembles lots of other CCC-made trails around southeastern Utah. It drops off the rim just south of Polly's Canyon. If memory is correct, the trail down to Collins Spring is also an old cattle trail.

Elevations Kane Gulch Ranger Station, Todie and Bullet Canyon Trailheads, about 1950 meters; Government Trailhead, 1730; Collins Spring is 1560 meters, and the San Juan River about 1128 meters, which is the high water mark of Lake Powell.

Hike Length and Time Needed One possible hike is to walk all the way down to the San Juan River and up Slickhorn Canyon to one of the car-parks there. This is about 110 kms. A strong hiker could do this in as little as 6 days, but he would need two cars, or a car and a mtn. bike. However, it would take most people a week or more for this marathon; maybe longer if you enjoy the ruins. Remember, most of the ruins are in the upper end of the canyon. A shorter trip would be to go down Kane Gulch and come back out at Bullet, a distance of about 39 kms, trailhead to trailhead. A good day-hike for a strong hiker might be to enter at Kane Gulch, and exit at Todie. Another car or mtn. bike at Todie, or Bullet Canyon Trailhead, would eliminate a road walk. Most people these days go in for one or two nights then come back out at the same trailhead. This eliminates the need for a shuttle of some kind.

Water Water is a problem in this canyon. The upper part of Grand Gulch has several minor springs and intermittent seeps, but always talk to the ranger and other hikers as to the whereabouts of water in the canyon. The amount changes from season to season. In late March, 1994, the author went down Government Trail to find a nice stream of water and lots of riparian vegetation, indicating it's there year-round. For him, getting across the stream with leather boots was a problem. It's seems that part of the canyon has lots more water than higher up. The main canyon down from Collins Spring is bone dry. Always have water in your car, and carry a couple of large jugs into the canyon so you can carry water from a spring to your camp. Please be considerate of other campers by not bathing in the potholes or seeps.

Maps USGS or BLM maps Blanding, Bluff, and Navajo Mountain(1:100,000), or Bears Ears, Cedar Mesa and Grand Gulch(1:62,500). Now perhaps the best one is the Trails Illustrated map Grand Gulch Plateau(1:62,500). This map shows all main road & trails, some ruins & pictographs, and more importantly springs! Buy this map at any area BLM office or national park visitor center.

Main Attractions The best place to see Anasazi ruins on the Colorado Plateau.

Ideal Time to Hike Spring or fall are by far the best, but it can be crowded, especially on Easter Weekend. Late March is a nice time, but you'll need a good sleeping bag, tent, stove, coat and gloves. Winter hiking is possible, but pretty cold for camping. Summers are a bit warm, but the upper canyon has moderate altitudes, so it isn't as hot as some places.

Hiking Boots Any dry weather boots or shoes.

Author's Experience The author started one afternoon, then walked all the next day, when he went down Kane and up Bullet. Total walk-time was 13 hours. He also day-hiked in from Collins Spring. In late March of 1994, he hiked down Government Trail, up-canyon to Big Man Pictograph Panel, and returned in just under 4 hours. Later that same afternoon, he hiked down Todie to the junction of Grand, then returned, all in less than 3 hours.

116

Map 46, Grand Gulch, Utah

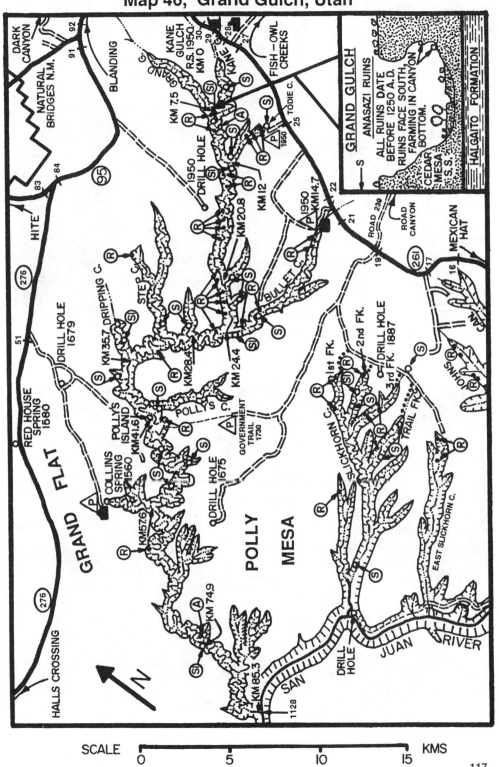

GRAND GULCH
ANASAZI RUINS

ALL RUINS DATE
BEFORE 1250 A.D.
RUINS FACE SOUTH,
FARMING IN CANYON
BOTTOM.

S →

CEDAR
MESA
S.S.

HALGAITO FORMATION

DARK CANYON
NATURAL BRIDGES N.M.
BLANDING
KANE GULCH
R.S. 1950
KM O 30
KANE
FISH—OWL CREEKS
TODIE C.
KM 7.5 GREEN
95
84
83
HITE
HALLS CROSSING
276
51
RED HOUSE SPRING 1580
DRILL HOLE 1679
276
KM 35.7 DRIPPING C.
STEP C.
1950
DRILL HOLE
1950
KM 20.8
KM 12
KM 14.7
25
22
21
ROAD 239
ROAD CANYON
19
261
17
16
MEXICAN HAT
BULLET C.
KM 28.4
KM 24.4
POLLYS C.
GOVERNMENT TRAIL 1730
POLLYS ISLAND
KM 41.6
COLLINS SPRING 1560
KM 57.6
DRILL HOLE 1675
POLLY MESA
SLICKHORN C.
1st FK.
2nd FK.
3rd FK.
DRILL HOLE 1887
TRAIL FK.
EAST SLICKHORN C.
KM 85.3
DRILL HOLE
SAN
JUAN
RIVER
1128
KM 749
GRAND FLAT
N

SCALE

0   5   10   15   KMS

Anasazi structures in Todie Canyon, a tributary of upper Grand Gulch.

Potsherds in Grand Gulch. Please leave in place for others to enjoy.

Underground Kiva with entry ladder in Slickhorn Canyon. Please don't walk on the roof.

Big Man pictograph panel in Grand Gulch near the Government Trail.

Introduction to the Escalante River Canyons, Utah

Location and Access In this edition, the author is including 9 different hiking maps in the Escalante River drainage. Because of this it seems necessary to do an introduction to this popular hiking region. The Escalante River begins high on the south slopes of the Boulder Mountains and flows into Lake Powell not far north of the Utah-Arizona state line. To get there, drive along Utah State Highway 12 which runs east from the Bryce Canyon area to the small towns of Escalante and Boulder. This same highway runs north over the Boulder Mtns. to Torrey just west of Capitol Reef National Park. You can also get there via the paved Burr Trail Road coming in from the Henry Mountains and Lake Powell to the east.

Canyons not Included The author believes he now has most of the better hikes in the Escalante included in this book. However, there are still many canyons left to explore. Some of them are Stevens Canyon, Moody Creek, Middle and East Moody Canyons, the upper Gulch, upper Deer Creek, Boulder Creek, Dry Hollow, Sand Creek, Pine Creek(The Box), Phipps Wash, Red Breaks, Fox and Scorpion Canyons, the Hole-in-the-Rock Trail and Llewellyn Gulch(directly south of Davis Gulch and the Hole-in-the-Rock Road). All of these except Llewellyn are discussed in Rudi Lambrechtse's book listed below.

Additional Information The very best sources of information about the Escalante is from the new Forest Service, BLM and National Park Service visitor center, located at the west end of Escalante. These offices are now all combined into one center and it's open from 8:00 am to 4:30 pm, Monday through Friday. Try to arrive in town so you can visit this place before going further. They will inform you of the latest road conditions and will give you some free handouts with pertinent facts and can sell you their latest maps.

Another good source is Rudi Lambrechtse's book, *Hiking the Escalante*. This is a good book; its only real flaw is it has no hiking maps! That author and publisher purposely wrote it that way to perhaps make it more challenging, and so you'd be required to buy and use the topographic maps of the area.

If the water of the lake drops any more than it was in early 1990's, then you can possibly hike into Cow, Calf and Fence Canyons in the very lower end of the drainage. If so, then this writters other book, *Boater's Guide to Lake Powell* may be of help.

Road Conditions The road called the Burr Trail is now paved from Boulder to the boundary of Capitol Reef National Park. From this main link, some pretty good roads run down to Little Death Hollow and Horse Canyon, and to Silver Falls Creek and the Moody Canyons. With more use, many of these roads are improving. Most of the roads shown on this map are graded occasionally. The Hole-in-the-Rock Road running south from near mile post 65 on Highway 12, is graveled for a ways and well-maintained. As you near the Hole-in-the-Rock Trail and Lake Powell, it deteriorates some. The last 5 kms or so is for 4WD's only. Side roads leading off from the Hole-in-the-Rock Road are often sandy, so discuss them with the NPS or BLM rangers in Escalante, and pay close attention while driving there.

Elevations Escalante, 1790 meters; Boulder, 2025; Lake Powell, 1128 meters(high water mark).

Water In some parts of this drainage getting good safe drinking water can be a problem. Before you leave either Boulder or Escalante, make sure you have several jugs full of water in your car. Always carry more than you think you'll need.

For the most part, the east-side canyons are mostly dry, at least a lot drier than the west-side drainages. If you plan to be out in the canyons along the Hole-in-the-Rock Road, about the only place to get a really safe drink is from the pipe filling Willow Tank. There are other watering troughs right on or near this main road, but the intake pipes seem to be under water in most, or it's mineral water. Other sites include Cat and Red Well. As for the east-side canyons, the only water near any road is at the mine and corral located where you begin to drop down into Moody Creek. Water in the two watering troughs comes from the mine above. It tastes good and seems safe to drink(?). Also, some of the small springs shown on the USGS maps dry up in times of drought, or are contaminated by cattle.

The Escalante River is of course a year-round flowing stream, but it has a large drainage basin with people, livestock and beaver throughout. If you drink it as is, most of the time perhaps nothing would happen. The old cowboys always drank it. But sooner or later you could pick up something. *Always treat, filter or boil Escalante River water.* Water from side canyons is usually better quality. To drink water as is, take it from a spring source. There used to be cattle in the lower main canyons during the winter months, but two of the grazing permits have been bought out by a conservation group. Since they left in 1991, willows are growing back and beaver dams appearing, so you still have to worry about giardia from beaver!

Maps Out of state hikers should have a Utah state highway map, and for everyone it's a good idea to also have a map put out by the Utah Travel Council called *Southeastern Utah*. This map covers the entire drainage. Buy it about anywhere. Next are the USGS or BLM maps *Escalante, Smoky Mountain and Navajo Mountain*. These three cover the entire drainage and are all 1:100,000 scale. Buy the BLM versions at the visitor center in Escalante. There are also the older and sometimes very much out-of-date maps by the USGS at *1:62,500 and 1:24,000* scale(15 and 7.5 minute). These are confusing especially if you need two maps, but at two different scales, to cover one hike. Another map that's very good for hiking, perhaps the best, is called *Canyons of the Escalante*. It's a two-sided plastic composite of a number of different USGS maps at different scales printed by Trails Illustrated. Buy this one at the visitor center in Escalante.

Ideal Time to Hike Spring or fall are the best times to hike, but if you're in a watery canyon and doing lots of wading, you may enjoy it in summer too. Otherwise it's pretty warm to be hiking. Also, some of the canyons in the upper drainage are much higher altitude than those closer to Lake Powell. Sometimes it may be best to visit these areas in summer weather. If the roads are dry, it's possible to hike the canyons near Lake Powell during some winter dry spells.

One problem you will encounter in some canyons during the summer season are deer and/or horse flies. They are prevalent along the Escalante River and The Gulch, but in the same time period, the author saw few if any along Calf Creek, Dry Fork of Coyote Gulch, Little Death Hollow or wading through Deer Creek. To be sure, take along a pair of light weight long pants. These flies like to bite bare legs.

Map 47, Introduction to the Escalanate River C., Utah

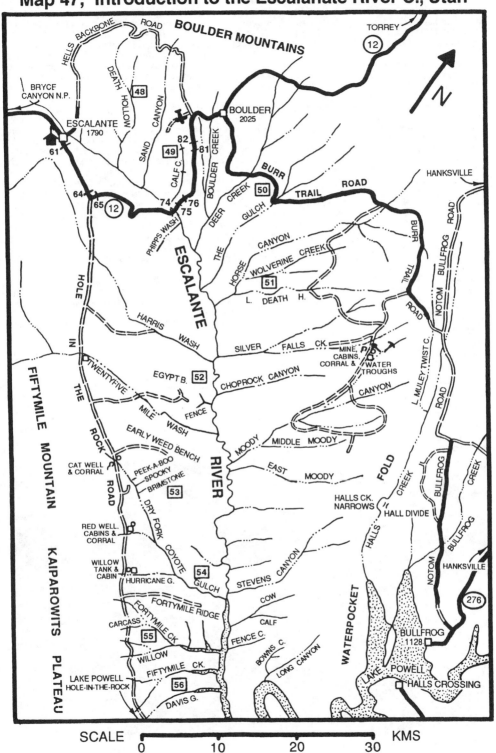

SCALE

0   10   20   30   KMS

Death Hollow, Utah

Location and Access Death Hollow is one of the upper tributaries of the Escalante River. The Hollow begins at the escarpment of the Boulder Mountains at over 3000 meters and flows south to the main Escalante River about 10 kms east of the town of Escalante. Get to the upper trailhead by driving along the Hells Backbone Road, which makes a northern loop between Escalante and Boulder. From town to town it's about 57 kms. This road is closed by snow from November to sometime in April, but it's in good condition for all vehicles during the summer season. You can also leave a vehicle in Escalante at the garbage dump. To get there, head southeast out of town about 1 km, then drive east past the cemetery and north to the garbage dump. You then walk north to the river. You can also park near the Boulder Airport which is near the head of Calf Creek.

Trail or Route Conditions From the upper trailhead, the first 18 kms or so is along a dry creek bed, then there's about 5 kms of good narrows. Normally there is running water or potholes in this section and you may have to float a pack across some pools. In some years it's easy, other times more difficult; it varies from year to year and season to season. Check the latest conditions at the visitor center in Escalante. In the area where the old Boulder Mail Trail crosses the canyon, it's wider with many pine trees along the year-round stream. Further down there are more moderate narrows and in the past one deep hole(which was only ankle deep when the author walked through it). You must take an air mattress to ferry packs in case a flood has scoured out that or other potholes. In the bottom end of Death Hollow and along the Escalante, there are many cow trails. For a good day-hike, park at the garbage dump east of town, and walk along the mail trail up the hillside just south of the "E" on the mountain, then follow the cairned trail into the middle of Death Hollow. From there walk downstream to the Escalante and back to your car. You could also walk the mail trail into the Hollow and camp, then explore upstream from there. You can also walk southwest from the trailhead near the Boulder Airport. This also takes you to the middle of Death Hollow, then you can walk up-canyon. This is a shorter way in than from Escalante.

Elevations Upper trailhead, 2700 meters; end of Death Hollow, 1625; Escalante, 1790; Boulder Airport, 2050 meters.

Hike Length and Time Needed The distance from the upper trailhead to Escalante is about 48 kms, one way. This is about a 3 to 4 day trip for most, as it's slow walking in the narrows section. If you have the luxury of two cars, fine; otherwise you'll have to hitch hike back. That could take another day. A mtn. bike could solve the shuttle problem, but it's a long peddle back to the trailhead.

Water There is pretty good water in the middle parts. There are no cattle, but some beaver--so take precautions. Running water begins about 18 kms from the trailhead. Treat, boil or filter the Escalante River water.

Maps USGS or BLM map Escalante(1:100,000), or Roger Peak and Escalante(1:62,500), or the plastic Trails Illustrated map, Canyons of the Escalante(1:62,500).

Main Attractions Good narrows and an entrenched canyon in the Navajo Sandstone.

Ideal Time to Hike Because of all the wading and possible swimming, you'll want to do this hike in warm weather; from about late May through the end of September.

Hiking Boots Wading boots or shoes.

Author's Experience The author parked just north of the cemetery and walked along the Boulder Mail Trail into Death Hollow. On that trip the water was no more than ankle deep(first part of June). He returned via the Escalante River in about 7 hours, round-trip. He later hiked the mail trail from the Boulder Airport, then walked up-canyon to about one km above where the water begins, then returned, all in 10 hours.

One of several swimming holes in the middle part of Death Hollow.

Map 48, Death Hollow, Utah

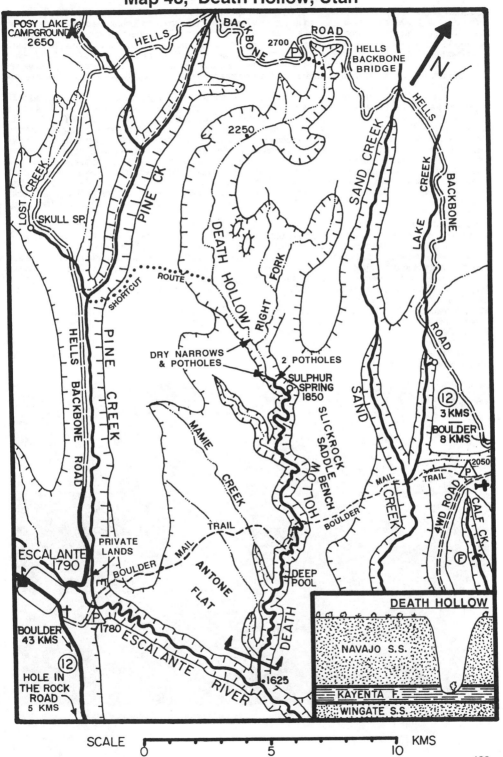

POSY LAKE
CAMPGROUND
2650

HELLS   BACKBONE   ROAD

2700

HELLS
BACKBONE
BRIDGE

N

2250

HELLS   CREEK

LOST CREEK

SKULL SP.

PINE CK

DEATH HOLLOW

RIGHT FORK

SAND CREEK

LAKE CREEK

HELLS BACKBONE ROAD

PINE CREEK

SHORTCUT ROUTE

DRY NARROWS
& POTHOLES

2 POTHOLES

SULPHUR
SPRING
1850

SLICKROCK SADDLE BENCH

SAND

DEATH HOLLOW

12
3 KMS
BOULDER
8 KMS

2050
P

TRAIL

MAIL CREEK

BOULDER

MAMIE CREEK

4WD ROAD

CALF CK.

F

ESCALANTE
1790

PRIVATE
LANDS

BOULDER   MAIL   TRAIL

ANTONE
FLAT

DEEP
POOL

DEATH

BOULDER
43 KMS

12

HOLE IN
THE ROCK
ROAD
5 KMS

P

1780

ESCALANTE   RIVER

·1625

DEATH HOLLOW

NAVAJO S.S.

KAYENTA F.

WINGATE S.S.

SCALE

0   5   10   KMS

Calf Creek, Utah

Location and Access Here is one of the best hikes around and one for the whole family. This is along both upper and lower Calf Creek, located about halfway between the towns of Escalante and Boulder. Calf Creek is one of the upper tributaries of the Escalante River system. To get there, drive along Highway 12 between Escalante and Boulder. To enter the bottom end, drive to between mile posts 75 and 76, and turn west into the BLM Calf Creek Campground. At the upper end of the campground is the trailhead for Lower Calf Creek Falls. To enter the upper part of the canyon, drive to mile post 81 on the narrow ridge between Calf and Dry Hollow. Drive north from there to the second dirt road on the left or west. Drive 200 meters and park under cedar trees.

Trail or Route Conditions From the Calf Creek Campground, simply walk up-canyon along a good and well-used trail. Along the way, you may see one Anasazi ruin high on the cliffs on the east side about one km up the trail. Not far above that is a small side canyon with some pictographs at the end. You'll have to wade the creek to get there. At the end of the trail is a large pool at the bottom of the 38 meter-high Lower Calf Creek Falls. This is one of the better waterfalls on the Colorado Plateau. Most people take a dip in the pool in summer. From the Upper Calf Creek Trailhead, head west over the rim and walk down a slickrock trail to the bottom of the drainage. Just before the bottom the trail splits in two; one goes west to the area above the falls, and the other goes down to just below the Upper Calf Creek Falls. There's another pool here too, below the 27 meter-high falls. There's an emerging trail going down along the stream to just above the lower falls. The area along the stream is an emerald green paradise in a sea of Navajo slickrock.

Elevations Calf Creek Campground, 1625 meters; lower falls, 1700; Upper Calf Creek trailhead, 1975, upper falls, 1800 meters.

Hike Length and Time Needed From the campground to the lower falls is about 4 kms and can be walked in about an hour, one way. It's only about 1 1/2 kms down to the upper falls from the upper trailhead. Some can do this round-trip in an hour, but most would like to make a half-day hike out of it.

Water Tap water is found at the campground, but with bathers and some beaver in the lower end of the canyon, best not to drink the creek water there. However, above the lower falls you should be able to drink the creek water as is. There's a good spring under the upper falls.

Maps USGS or BLM map Escalante(1:100,000), or Calf Creek(1:24,000), or the plastic Trails Illustrated map, Canyons of the Escalante(1:62,500).

Main Attractions Two waterfalls and swimming holes, riparian vegetation and pictographs.

Ideal Time to Hike Spring, summer or fall. Maybe even some warm dry spells in winter(?).

Hiking Boots Any dry weather boots or shoes.

Author's Experience Years ago he hiked from the campground to the pictographs, then in 1990 hiked to the lower falls and returned in 1 3/4 hours. Later that same afternoon, he walked down to the upper falls and back in less than an hour, round-trip.

Lower Calf Creek Falls, one of the best waterfalls on the Colorado Plateau.

Map 49, Calf Creek, Utah

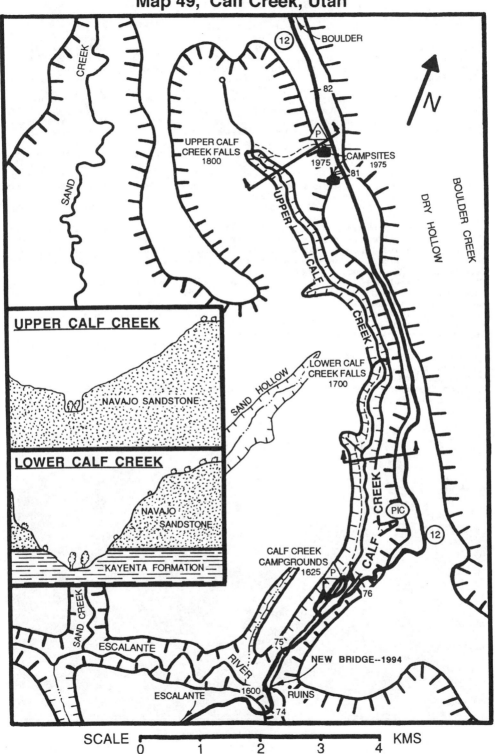

CREEK

SAND

(12) BOULDER

82

N

UPPER CALF
CREEK FALLS
1800

P

1975 CAMPSITES 1975

81

UPPER

BOULDER CREEK

DRY HOLLOW

CALF

UPPER CALF CREEK

NAVAJO SANDSTONE

CREEK

SAND HOLLOW

LOWER CALF
CREEK FALLS
1700

LOWER CALF CREEK

NAVAJO
SANDSTONE

KAYENTA FORMATION

CALF CREEK

PIC

(12)

CALF CREEK
CAMPGROUNDS
1625 P

76

SAND CREEK

ESCALANTE

RIVER

75

NEW BRIDGE--1994

ESCALANTE 1600 RUINS

74

SCALE |————|————|————|————| KMS
       0   1   2   3   4

Deer Creek and The Gulch, Utah

Location and Access The two canyons featured here are in the northeastern part of the Escalante River drainage and just southeast of Boulder. You can get there from Bullfrog by driving north on the Bullfrog-Notom Road and turning west on the paved Burr Trail. You will cross the upper parts of each of these canyons on the way. If you're coming from the north or west, first head to Boulder, then follow the Burr Trail east. From Boulder, Deer Ck. is about 12 kms; The Gulch 18-19 kms. There's a free BLM campground on Deer Ck., and good shady campsites on The Gulch(also sandy!).

Trail or Route Conditions On Deer Ck., walk south on a west-side cow trail for about 3 kms before entering the narrow stream channel. As the geology cross-section shows, it's a big canyon with a fairly narrow stream channel at the bottom. The bottom is a green paradise in a sea of slickrock. There's a hiker-made trail emerging, so it's not the bushwhack it used to be. The going is a little slow because you'll be in and out of the water all the way down to the Escalante. Deer Ck. is about the same size as Calf Ck., and it has fish in it. Walking down The Gulch is much easier and faster than along Deer Ck. The reason is, much of the upper end is dry and you can either walk in the dry creek bed or along cow trails which short-cut meanders. The upper part of The Gulch is broad and deep; the middle parts more shallow. The lower end is more entrenched. Rudy L. in his Escalante book talks about a waterfall in the lower end of The Gulch. Take a short rope to get a big pack around this one. The lower end of The Gulch and Halfway Hollow both have some good watery narrows which this author has yet to see. You can make a loop-hike of these two canyons connected by the Escalante River.

Elevations Deer Creek trailhead, 1730 meters; The Gulch trailhead, 1690; confluence of The Gulch and Escalante River, 1475 meters.

Hike Length and Time Needed It's 12-13 kms down Deer Ck. to the Escalante, and about 20 kms down The Gulch. Add another dozen river kms to this and it makes a 45 km round-trip hike. You'll need 3 or 4 days for this loop-hike. Otherwise, you can see about 2/3 of Deer Creek and return in one day; or walk down The Gulch nearly to Halfway Hollow and back on a very long day-hike. It's a 2, maybe 3 day backpack down to the river and return in either canyon.

Water Deer Creek has a good year-round flow and you could probably drink it as is, but it has a long drainage with a live beaver population. There are cattle in The Gulch, and likely some in upper Deer Creek. Drink from spring sources in either canyon if you can; otherwise better purify the water.

Maps USGS or BLM map Escalante(1:100,000), or King Bench and Red Breaks(1:24,000), or the plastic Trails Illustrated map, Canyons of the Escalante(1:62,500).

Main Attractions Challenging wilderness hiking for backpackers.

Ideal Time to Hike Because there's lots of wading, make it from late spring through early fall. The Gulch has many leg-biting deer/horse flies in summer, but few are found right along Deer Creek. Have a pair of light-weight long pants handy!

Hiking Boots Wading boots or shoes in both canyons.

Author's Experience In one long day the author hiked down about halfway to the Escalante in both canyons, then returned(half a day in each).

Typical scene along Deer Creek.

Map 50, Deer Creek and The Gulch, Utah

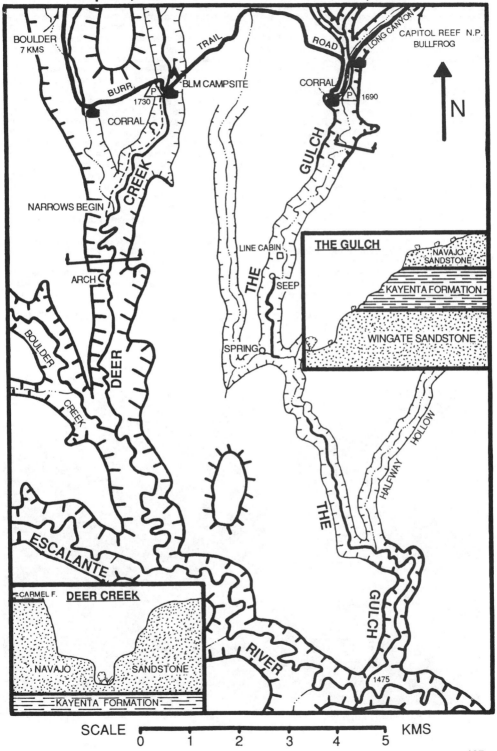

SCALE ⊢———0———1———2———3———4———5——⊣ KMS

Little Death Hollow, Utah

Location and Access The canyon featured here is called Death Hollow on USGS maps, but most add the "Little" to differentiate it from the other Death Hollow in the upper Escalante. To get there, drive along the now-paved Burr Trail Road between Boulder and Capitol Reef National Park. From this main road you can use one of two normally good roads to reach the trailhead. The most traveled one leaves the Burr Trail and heads south, first to the Petrified Forest site, then to Little Death Hollow Trailhead, where a corral is located. You can park and/or camp under some nearby cedars. The second one leaves the Burr Trail about 2 kms west of the Capitol Reef National Park boundary and heads south, then west. Parts of both roads forming this loop are made of clay, so they'll be impassable right after heavy rains. Be sure and have the Escalante map listed below when locating the roads to the trailhead.

Trail or Route Conditions From the corral, walk southwest along a dirt track. Cross a fence, then follow a cow trail into Little Death Hollow. It'll be wide open and uninteresting at first. Along the way look for petroglyphs on a large Wingate boulder to the right. Further down, the canyon constricts and around the first big bend is a spring. Cattle however are stepping right in it. From there you'll be in the Wingate part of the canyon. The further you go the tighter it gets. In the middle part of the narrows it turns to a slot and it's this way for a couple of kms. This one ranks high on the author's list of narrow canyons, and is surely the best of any made of Wingate Sandstone. You'll have about 7 or 8 chokestones to get around, but there are no major obstacles; however, things can change with every flood, so be prepared for anything! Just before you come to Horse Canyon, you'll see a water pipe running up to the southeast. It was an attempt by stockmen to pump water to the mesa top. To return to the trailhead, you can walk back the same way, or make a loop returning via Wolverine (Creek)Canyon. It too has narrows, but not as good as Little Death Hollow. In the upper end of Wolverine is a pile of petrified logs worth seeing. Stockmen bring 4WD's down Horse Canyon to the line cabin, so that drainage is uninteresting.

Elevations Trailhead, 1700 meters; junction of Horse Canyon and Little Death Hollow, 1490 meters.

Hike Length and Time Needed If you hike the loop suggested, it's about a 25 km walk, and an all-day hike. You can also use this route to enter the middle part of the Escalante River.

Water There's year-round running water in the lower part of Horse Canyon. There is also a good spring next to the line cabin, and two springs high above at the Navajo-Kayenta contact. There are two minor seeps in the middle of Wolverine. Because of water and sometimes cattle in the canyon, there are also lots of deer/horse flies in summer.

Maps USGS or BLM map Escalante(1:100,000), or Wagon Box Mesa and Moody Creek(1:62,500), or the plastic Trails Illustrated map, Canyons of the Escalante(1:62.500).

Main Attractions Very good narrows and petrified logs.

Ideal Time to Hike Spring or fall, but it's cool in the narrows in summer.

Hiking Boots Any dry weather boots or shoes, except right after rains, then maybe waders(?).

Author's Experience The author did the loop-hike suggested in just over 7 hours.

One of the best narrows anywhere is here in the Little Death Hollow. You must pass 7 or 8 chokestone like this along the way.

Map 51, Little Death Hollow, Utah

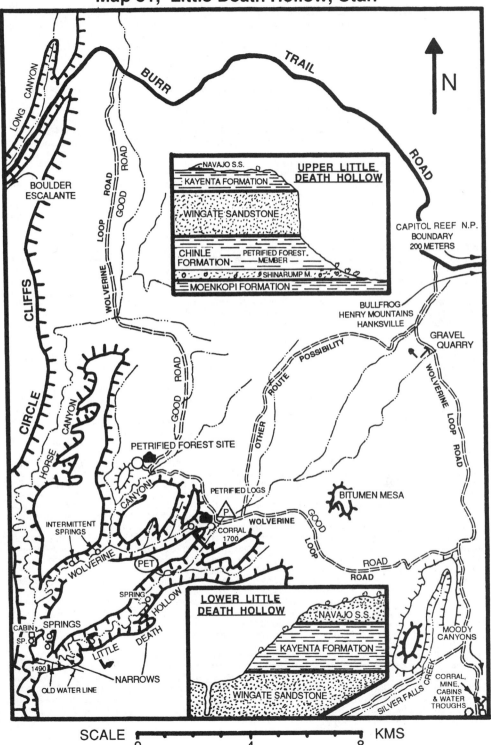

SCALE

0   4   8   KMS

Harris, Silver Falls, Choprock, Fence & Twentyfive Mile.

Location and Access This map shows some of the major canyons in the middle part of the Escalante River drainage. To reach the west side canyons, drive southeast out of Escalante on State Highway 12. Near mile post 65, turn south onto the Hole-in-the-Rock Road. About 22 kms from the pavement you'll see a sign on the left or east indicating the road to Harris Wash. From the Hole-in-the-Rock Road it's about 9-10 kms to the end of the road and car-park. There used to be a road going all the way down Harris Wash to the Escalante River, then up Silver Falls Creek, over the Circle Cliffs and down Halls Creek to the Colorado. Charles Hall first pioneered this road in 1881 after the settlers in Bluff on the San Juan River decided it was a better route than the one they took in 1880. Their original route was down the Hole-in-the-Rock. You can still make a hike along this historic route down Harris Wash but motor vehicles are no longer allowed.

To reach the trailhead on Twentyfive Mile Wash, continue southeast along the Hole-in-the-Rock Road from the Harris Wash turnoff another 10 kms. Turn left or east at the sign stating the way to Egypt, then drive about 5 kms to the bottom of Twentyfive Mile Wash. From there you can hike down to the Escalante, or continue up the road to the Egypt Bench Trailhead. From there you can descend into Fence Canyon or the middle part of Twentyfive Mile Wash.

To reach the east-side trailheads, drive east out of Boulder on the paved Burr Trail Road toward Capitol Reef National Park. About 2 kms west of the park boundary, turn south onto a road signposted as the *Wolverine Loop Road.* See the map on Little Death Hollow for a look at the upper part of that road. You can drive down into Silver Falls Creek almost to the boundary of the Glen Canyon National Recreation Area. If you continue south on this same east-side road, instead of turning west into Silver Falls Creek, and with the good BLM map in hand, you will drop down into the upper end of Choprock Canyon, or the Moody Canyons. Remember, all side roads in this area have some clay sections and can be very slick after heavy rains; otherwise, they're pretty good roads, which are graded about once a year.

Trail or Route Conditions In Harris Wash, you will begin walking along the old road, but much of the way down you'll be in the dry creek bed. After a ways water begins to flow and you'll be wading. Along the way are big undercuts, similar, but not as spectacular, as those in Coyote Gulch. Don't miss the historic Hobbs inscription along Silver Falls Creek. It's in one of two very deep undercuts. As you walk down Twentyfive Mile Wash, it'll be shallow and uninteresting at first, but then it gets deeper with flowing water in the lower end. The parts of Twentyfive Mile Wash with running water now(1994) has a number of beaver dams, but a pretty good hikers trail too. From the Egypt Bench Trailhead, walk over the edge of the bench on an old horse trail. You can follow it over the Navajo slickrock by observing the scratch marks made by horseshoes. At the bottom of the steep part, you can head due south to get into the middle part of Twentyfive Mile; or set your sights on northeast and walk along the northwest rim of Fence Canyon. You'll probably lose the trail on the flats, but at the junction of Fence and its North Fork, you'll find it again as it goes down the ridge to the east. Walk to the bottom and to the Escalante River. From along the Escalante, you can walk downstream to Twentyfive Mile Wash, then west, and return to Egypt Trailhead via one of about 3 trails or routes out to the north as shown on the map. There's at least one good panel of petroglyphs along Twentyfive Mile Wash. You can also walk into the lower end of Choprock where you'll see a 100 meter-long panel of petroglyphs about 200 meters from the Escalante on the north wall. Across from the petroglyphs is an old stock trail going up on top of a bench. There are some good Wingate narrows in the middle part of Choprock. Walking along the Escalante is easy as there are many cattle trails in the bottom which is covered with willows and cottonwood trees. In summer take long pants along the river because of deer/horse flies which bite bare legs.

Elevations Harris Wash Trailhead, 1550 meters; Silver Falls Creek Trailhead, 1600; Egypt Bench Trailhead, 1725; Twentyfive Mile Wash Trailhead, 1460 meters.

Hike Length and Time Needed To hike down Harris Wash to the Escalante is to walk about 17 kms. To do a round-trip to the river and back in one day is too much for most, but going down halfway and back would be easy. You could walk down Harris and up Silver Falls(about 30 kms) in two days easily--if you had a car on the other side. By starting on the east side, you can walk down Silver Falls Creek to the river and back in one day. Anyone should be able to walk from the Egypt Bench Trailhead, down Fence to the river and back in one day. Strong hikers can walk down Fence, then south along the Escalante and up Twentyfive Mile Wash and back to Egypt in one very long day; or two easy days. Have a good map in hand and count side canyons as you walk up Twentyfive Mile so you exit in the right place. In a couple of days you can walk down Fence and up into the narrows of Choprock and back.

Water There's water in all the major side canyons mentioned. Take it from spring sources. Purify or filter Escalante River water. Beware of water when cattle and beaver are in the area.

Maps USGS or BLM map Escalante(1:100,000), or Moody Creek(1:62,500), Red Breaks and Sunset Flat(1:24,000), or the plastic Trails Illustrated map, Canyons of the Escalante(1:62,500).

Main Attractions Deep canyons, wilderness solitude, petroglyphs and great scenery.

Ideal Time to Hike Spring or fall, but with all the streams around you can cool off in summer.

Hiking Boots Wading type boots or shoes.

Author's Experience Years ago he hiked down Harris Wash to near the Escalante and back in one day. In 1990, he camped at Egypt Bench Trailhead and hiked down Fence and into Choprock a ways before returning to Egypt. Round-trip was 4 3/4 hours. The old line cabin at the mouth of Fence Canyon was burned down by environmental fanatics in the winter of 1990. In 1994, he drove to near the end of the road in Silver Falls Creek and hiked down to the Escalante River, passing the Hobbs Inscription enroute. Also in 1994, he drove to Egypt again and saw the middle and best part of Twentyfive Mile Wash in one day, or about 6 hours walk-time.

Map 52, Harris, Sliver Falls, Choprock, Fence, and Twentyfive Mile Canyons, Utah

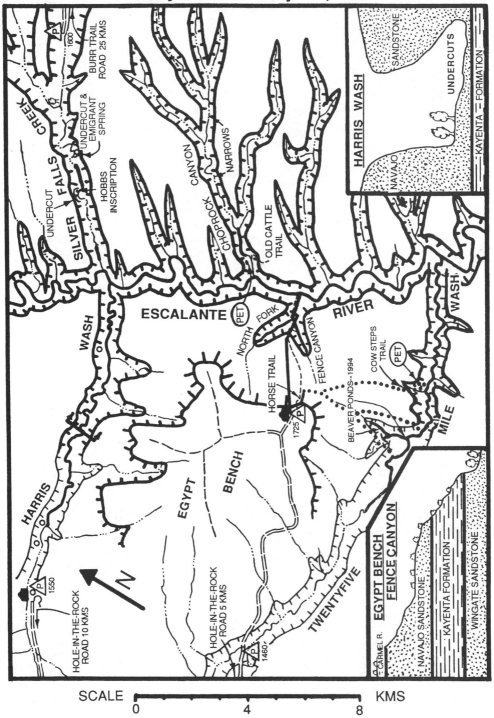

SCALE
0   4   8 KMS

131

Fall colors, desert varnish, and a cool sandy stream make Harris Wash a favorite.

The 1883 Hobbs inscription, and 1957 memorial plague in the middle part of Silver Falls Creek Canyon.

Petroglyph panel near the mouth of Choprock Canyon.

Petrified logs. These are found at the head of Wolverine Canyon.

Peek-a-boo, Spooky & Brimstone Gulches, and the Dry Fork of Coyote Gulch, Utah

Location and Access The short slot canyons featured here are in the upper part of Dry Fork of Coyote Gulch. To get there drive southeast out of Escalante on State Highway 12. When you come to mile post 65, turn southeast onto the Hole-in-the-Rock Road and drive about 39 kms or until you reach Cat Well, which is where you see a sign to the Early Weed Bench. You can camp under some cottonwood trees and begin hiking there if you like, but the normal trailhead is further along. Drive 4 more kms until you come to a lone cedar tree on the right with twin trunks. About 100 meters beyond that turn left and continue another 3 kms to the trailhead at 1494 meters altitude. You can camp at the trailhead, but there are no trees or shade.

Trail or Route Conditions From the normal trailhead, walk over the rim on a hiker-made trail and to the bottom of Dry Fork. About 150 meters to the right or east of where you reach the bottom is the beginning of Peek-a-boo Gulch. Near its bottom end is a natural bridge. There are now some steps cut which will get you into the lower end. About a km below the mouth of Peek-a-boo is Spooky Gulch. Not far after you enter this one, you'll have to remove your day pack and walk side-ways for most of its length. After a ways you'll come to a chokestone which is too much for most people to get around. You can however get into the upper end by climbing out of Dry Fork and onto the bench, then descend into the upper part. The lower end is exciting enough though. Watch out for that little *rattlesnake!* Continuing down Dry Fork, you'll pass a big chokestone in the narrows before arriving at Brimstone. You'll have to walk up Brimstone Gulch a ways before reaching the narrows. Remove your pack to get into this one too. The author was stopped by a log lodged in the crack, but with energy you can go further. That part was about the deepest, darkest and coolest hole the author has ever been in. You'll need high speed film to get any fotos here, or in Spooky. There are also some respectable narrows in Dry Fork which you can enter from either trailhead.

Elevations Cat Well, 1483 meters; normal trailhead, 1494; lower Brimstone, 1400 meters.

Hike Length and Time Needed It's less than a km to Peek-a-boo from the normal trailhead. From there you can see all these narrow slots in about half a day; or most of a day if you begin hiking at Cat Well.

Water Take plenty of water in your car and in your pack. The windmill well and the livestock watering trough at Cat Well are both open and the water suspect.

Maps USGS or BLM map Smoky Mountain(1:100,000), or Big Hollow Wash(1:24,000), or the plastic Trails Illustrated map, Canyons of the Escalante(1:62,500).

Main Attractions Extremely short, narrow, dark and fotogenic slot canyons. None better!

Ideal Time to Hike Spring or fall, but it's cool in these slots even in summer. Maybe winter too?

Hiking Boots Any dry weather boots or shoes(but waders after a storm).

Author's Experience The author walked in from the normal trailhead and explored all three of the gulches. The round-trip took 3 1/3 hours. At a later time he walked down to the bottom of the upper narrows in Dry Fork from Cat Well and returned in about an hour.

Spooky Gulch is so narrow you have to remove your pack and walk side-ways.

Map 53, Peek-a-boo, Spooky & Brimstone Gulches, and the Dry Fork of Coyote Gulch, Utah

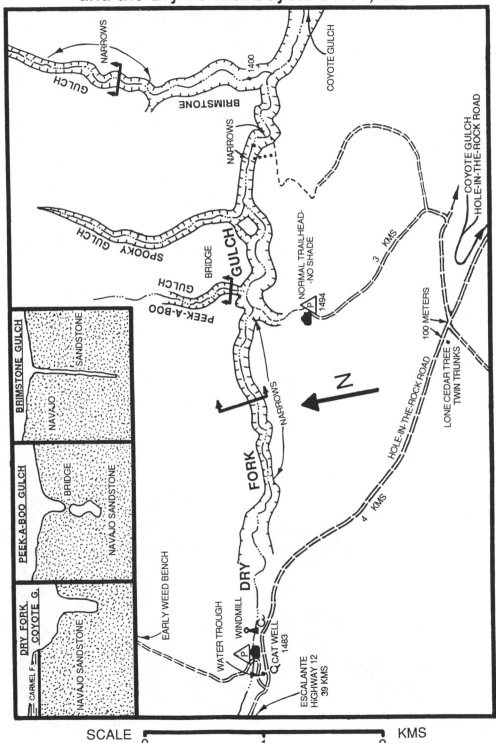

SCALE

0   1   2   KMS

Coyote Gulch, Utah

Location and Access To reach Coyote Gulch of the middle Escalante River, drive down the Hole-in-the-Rock Road about 50 kms which will put you at the Red Well. You can park and start hiking from there; or drive back up the road a ways, and drive northeast a couple of kms to get closer to Dry Fork. If you drive about 4 kms past Red Well you'll come to Willow Tank at the head of Hurricane Wash, another popular entry point. About 3 kms past Willow Tank is a signed road leading northeast along Fortymile Ridge. The road to the Jacob Hamblin Arch Trailhead and metal tank at 1450 meters is good for any car; but beyond that, it's sandy, so drive as fast as conditions allow and don't stop in deep sand! The Fortymile Ridge Trailhead is on a hill with firm ground.

Trail or Route Conditions Hiking from Red Well or Willow Tank is first along a shallow drainage, but gradually they deepen and become entrenched. Soon there is running water in each. As you near Jacob Hamblin Arch the canyon is very deep with huge undercuts up to 60 meters. In this part you'll be wading in a small stream much of the time. The canyon has many good campsites. Below J. Hamblin Arch there are many small sap springs coming out of the Navajo-Kayenta contact. Further down you walk under Coyote Bridge and there are several cascades and waterfalls; plus small Anasazi ruins on the north side. At the lower end, and half a km from the Escalante, look for an obvious trail heading south up a big sand slide. It's a good trail leading up to the top of the Navajo and a crack in the wall. From the top, walk southwest along a cairned slickrock & sandy track in the direction of Fortymile Ridge Trailhead. If you begin at the metal tank, walk due north until you reach the canyon overlooking J. Hamblin Arch. After you locate the arch, head for the slickrock ridge running down to the north just east of the arch. It's a little steep in places, but the author had no trouble going down with a big pack. From the bottom end of Coyote Gulch turn north and walk upstream to enter Stevens Canyon. Lower Stevens is an interesting canyon with some wading, climbing and maybe a little swimming.

Elevations Red Well & Willow Tank, 1390 meters; Jacob Hamblin Arch and Fortymile Ridge Trailheads, 1450; Lake Powell's high water mark, 1128 meters.

Hike Length and Time Needed From Willow Tank to the Escalante is about 20-22 kms(slightly longer from Red Well). This trip would take 2-3 days round-trip. You can shorten it by half if you start at Jacob Hamblin Arch Trailhead. Making a loop-hike from there can be done in one long day, but most would prefer to camp in the canyon one night. Using the Fortymile Ridge Trailhead would allow you to explore Stevens Canyon and others.

Water You can get a safe drink from a pipe at Willow Tank(maybe Red Well?), and at a number of springs in the canyon. Treat Coyote Creek and river water and always carry plenty in your car.

Maps USGS or BLM map Smoky Mountain(1:100,000), or Big Hollow Wash, King Mesa(1:24,000) and The Rincon(1:62,500), or the plastic Trails Illustrated map, Canyons of the Escalante(1:62,500).

Main Attractions Deep canyon, an arch & bridge, huge undercuts, outstanding scenery.

Ideal Time to Hike Spring or fall, but you can cool off in the creek in summer.

Hiking Boots Wading boots or shoes.

Author's experience The author once walked down Hurricane Wash nearly to Coyote Bridge and back in 7 hours. Several years later he walked from Jacob Hamblin Arch Trailhead, down to the river via Jacob Hamblin Arch, then back via the sand slide with a large pack. Two day total walk-time was only 7 1/2 hours. Still later he drove to Fortymile Ridge Trailhead and hiked into the lower end of Stevens Canyon. Round-trip was 5 1/2 hours.

From the bench above, one can see Coyote Gulch and Jacob Hamblin Arch below.

Map 54, Coyote Gulch, Utah

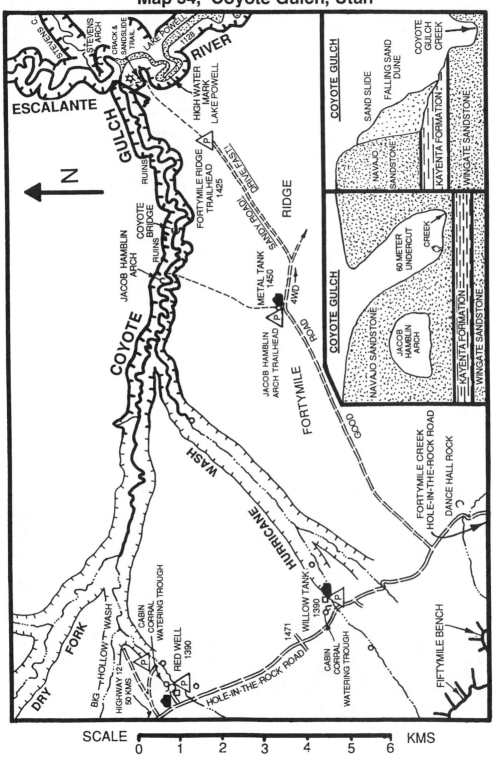

SCALE |———————————————————| KMS

0 1 2 3 4 5 6

Fortymile Gulch and Willow Creek, Utah

Location and Access These two interesting canyons are located in the lower end of the Escalante River drainage. Access is by boat from Lake Powell or by the Hole-in-the-Rock Road. To get there by car & road, drive southeast out of Escalante to mile post 65 and turn south onto the Hole-in-the-Rock Road. Drive between 70 and 75 kms from Escalante to any one of 6 trailheads along the main road as it crosses the upper tributaries of Fortymile Gulch and Willow Creek. Carcass Wash and Sooner Bench Trailheads are two of the most used. The best campsite in the area would be at Fortymile Spring, where you can wash up in a stock trough(the feeder pipe is under water though, so you may not be able to get a safe drink).

Trail or Route Conditions It's suggested you walk down one canyon and up the other making a loop-hike. For this description, we'll go down Fortymile and return via Willow. Starting at one of the three trailheads of Fortymile Gulch, walk down-canyon. You'll likely find a minor dropoff in each drainage, but you can surely find a way down into the bottom further on. The canyon slowly gets deeper and more entrenched in the Navajo Sandstone. About halfway down, water begins to flow year-round. After another two kms, you'll come to a waterfall. Nearby is a spring and a safe drink. From the falls on down, the canyon is very narrow at the bottom with large alcoves above. Two kms below the falls is a chokestone and a deep wading pool on the down-side. Not far below that is another chokestone and a swimming hole below it. However, the walls are close together and most people can straddle the pool by placing one foot on each wall and walking over it. Or you could swim or float the 3-4 meters across. At the junction of Willow Creek turn up-stream and after 1 1/2 kms you'll come to Broken Bow Arch. Nearby is another spring and a huge undercut for tent-less camping. From the arch up to the road there's not much to see along Willow Creek. When Lake Powell has low water levels, you may be able to walk into the North Fork of Willow Creek.

Elevations Trailhead altitudes, 1275 to 1350 meters; high water of Lake Powell, 1128 meters.

Hike Length and Time Needed To do the loop-hike suggested above is to walk about 18 kms. That's going down Carcass Wash, through Fortymile and up Willow Creek and Sooner Gulch to the main road. The average person will want all day for this hike. Having two cars or one mtn. bike will eliminate an hour or more of road-walking back to your vehicle.

Water Have plenty in your car. You can also get a safe drink at several springs in either canyon.

Maps USGS or BLM maps Smoky Mountain and Navajo Mountain(1:100,000), or The Rincon(1:62,500) and Sooner Bench(1:24,000), or the plastic Trails Illustrated map, Canyons of the Escalante(1:62,500).

Main Attractions Deep Canyon, good and slightly challenging narrows, and few people.

Ideal Time to Hike Spring, summer or fall, but with all the wading, not in a cold spell.

Hiking Boots Wading boots or shoes.

Author's Experience On his first trip, he went down from Sooner Bench through Willow to the lake and back. Another time he hiked up from the lake into each drainage. The last trip was down Carcass and Fortymile, and up Willow and Sooner. Road to road took less than 5 hours, but with the road-walk back to his car, it was 6 1/4 hours round-trip.

One way to get over this deep pool in Fortymile Gulch is to spread your legs and "walk the walls" above.

Map 55, Fortymile Gulch and Willow Creek, Utah

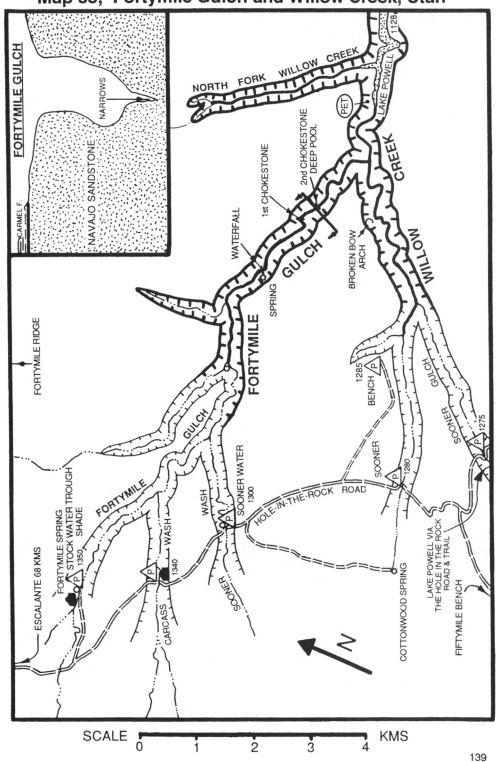

FORTYMILE GULCH

CARMEL F.

NAVAJO SANDSTONE

NARROWS

NORTH FORK WILLOW CREEK

1128

LAKE POWELL

PET

2nd CHOKESTONE
DEEP POOL

1st CHOKESTONE

WATERFALL

SPRING GULCH

FORTYMILE

FORTYMILE RIDGE

BROKEN BOW ARCH

CREEK

WILLOW

GULCH

1285

BENCH   P

SOONER

1275   P

FORTYMILE

GULCH

WASH

SOONER WATER
1300

P

1280   P

SOONER

HOLE-IN-THE-ROCK   ROAD

ESCALANTE 68 KMS

FORTYMILE SPRING
STOCK WATER TROUGH
SHADE

P   1350

WASH

P
1340

CARCASS

SOONER

COTTONWOOD SPRING

LAKE POWELL VIA
THE HOLE IN THE ROCK
ROAD & TRAIL

FIFTYMILE BENCH

N

SCALE   0   1   2   3   4   KMS

Fiftymile Creek and Davis Gulch, Utah

Location and Access These two canyons are in the lower end of the Escalante drainage and very near the end of the Hole-in-the-Rock Road. Drive south on this road about 75 to 80 kms from Highway 12. This will put you at one of the three trailheads shown on the map. This main road is good all the way to Davis Gulch, but then deteriorates rapidly below the "bad spot", as shown. There are almost no shade trees in the area, and the best campsite around is near the Griffin line cabin.

Trail or Route Conditions The best way into Fiftymile Creek is via its West Fork. Park at the eastern toe of Cave Point which is made of red Entrada Sandstone. Walk down the dry creek bed. After a ways you'll come to a dropoff. Get into the canyon below by walking along either side about 300 meters, then scramble down to the bottom. Near the end of West Fork is a cave where Fremont Indians once camped. In the main canyon you'll come to running water as shown. It flows off and on all the way to the lake. The map shows Lake Powell at its high water mark. There are a couple of exits as shown, and some pretty good narrows just above the HWM. The walk down South Fork of Fiftymile is longer. To hike into Davis Gulch, stop and park at the northern toe of Fiftymile Point. Park where there's a small hill made of talus, then walk north to the Carmel bench rim where you can observe the route before you. The only easy way into Davis is via the old stock trail. Start by walking to the left of a nipple-like rock, then veer a little to the northeast. Aim for the Carmel-capped hill marked 1346 meters, then head east and locate the trail while watching for the "green spot" on the opposite rim. In the bottom is a year-round stream, beaver, petroglyphs, and an arch and narrows upstream. This is the canyon where the now legendary Everett Ruess was supposed to have disappeared.

Elevations Trailheads, 1295, 1350 and 1375 meters; Lake Powell's HWM, 1128 meters.

Hike Length and Time Needed From the main trailhead on Fiftymile to the HWM of Lake Powell is about 10 kms, which makes the round-trip hike about 20 kms. Plan on an all-day trip. It's only 7-8 kms to the bottom of the Davis Gulch stock trail, but then you'll want to walk up-canyon and explore, so count on an all-day hike.

Water Have plenty in your car, because Soda Spring was dry upon the author's visit. Drink from the seep source in Fiftymile or other seeps down-canyon. There are beaver in Davis, so head upstream to a spring or treat the water near the lake.

Maps USGS or BLM maps Smoky Mountain and Navajo Mountain(1:100,000), or The Rincon (1:62,500) and Sooner Bench(1:24,000), or the plastic Trails Illustrated map, Canyons of the Escalante(1:62,500).

Main Attractions Fewer people than in canyons to the north, arches, petroglyphs, deep well-watered canyons.

Ideal Time to Hike Spring or fall. Summers are hot in these low altitude canyons.

Hiking Boots Dry weather boots except in the narrows of Fiftymile and in upper Davis.

Author's Experience The author has hiked up each canyon from the lake. Later he walked down from the road to the start of running water in Fiftymile in 2 1/3 hours round-trip. Also, he hiked down to the petroglyphs in Davis and back in 3 1/4 hours round-trip.

A line cabin and corral in the Escalante River drainage.

Map 56, Fiftymile Creek and Davis Gulch, Utah

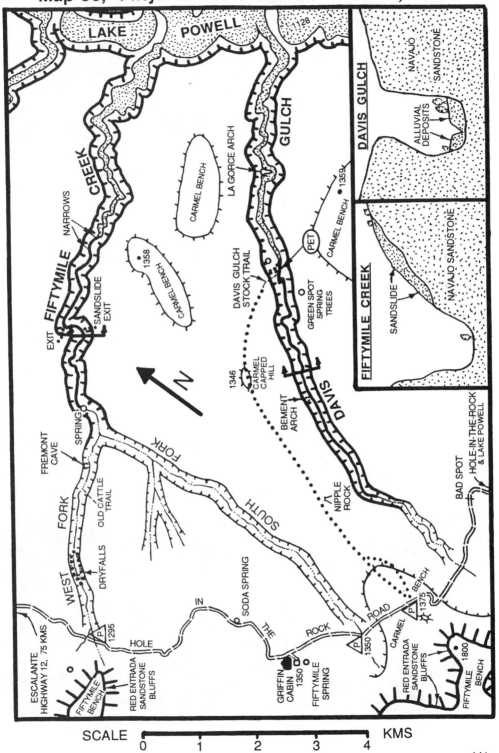

LAKE POWELL

128

DAVIS GULCH

NAVAJO SANDSTONE

ALLUVIAL DEPOSITS

FIFTYMILE CREEK

NAVAJO SANDSTONE

SANDSLIDE

FIFTYMILE CREEK

NARROWS

CARMEL BENCH

LA GORCE ARCH

GULCH

CARMEL BENCH • 1359

• 1358

CARMEL BENCH

EXIT

SANDSLIDE EXIT

SPRING

DAVIS GULCH STOCK TRAIL

PET

GREEN SPOT SPRING TREES

FREMONT CAVE

OLD CATTLE TRAIL

1346

CARMEL CAPPED HILL

DAVIS

BEMENT ARCH

WEST FORK

FORK

DRYFALLS

SOUTH FORK

NIPPLE ROCK

BAD SPOT

HOLE-IN-THE-ROCK & LAKE POWELL

SODA SPRING

• 1375

IN

THE

ROCK

ROAD

BENCH

ESCALANTE HIGHWAY 12, 75 KMS

P 1295

HOLE

P 1350

CARMEL

RED ENTRADA SANDSTONE BLUFFS

GRIFFIN CABIN 1350

FIFTYMILE SPRING

RED ENTRADA SANDSTONE BLUFFS

• 1800

FIFTYMILE BENCH

FIFTYMILE BENCH

N

SCALE 0 1 2 3 4 KMS

141

Bull Valley Gorge and Willis Creek, Utah

Location and Access The two interesting canyons featured here are Bull Valley Gorge and Willis Creek, both upper tributaries of the Paria River. They are located not far southwest of the small town of Cannonville, and directly south of Bryce Canyon National Park. To get there, drive south out of Cannonville on the road signposted for Kodachrome Basin, but turn to the right on the Skutumpah Road heading southwest toward Johnson Valley and Kanab. Or drive east from Kanab to Johnson Valley and turn north, then northeast on the Skutumpah Road. This road is sometimes closed in winter due to snow and/or slick clay beds, but is well-maintained and fairly heavily traveled in the warmer half of the year. Carry good topo maps to find the access routes, because there are no signs at these canyons. For more information and history of this region, see the author's book, *Hiking and Exploring the Paria River.*

Trail or Route Conditions To enter Bull Valley Gorge, park on the north side of the dirt bridge and walk west along the north side of the slot. After 300 or so meters, you can step into the dry creek bed, then walk down into the gorge. There's a log jam where the narrows begin, but you should have little trouble getting up or down this 5 meter-high dryfall. Take a short rope to raise or lower packs, especially a large one, and to help less-experienced hikers up or down. There's another steep alternate route into the upper part called the *Crack Route* on the map. This is easier than it first appears. There are a number of other possible entry or exits routes(E/E) in the lower gorge. To get into Willis Creek narrows, park on the Skutumpah Road where the small stream crosses the road, then walk into the slot. It has some fotogenic places in the upper end. You can walk down Bull Valley, up Sheep Creek and into and through Willis Creek back to the road. You can also enter or exit Willis Creek via Averett Canyon as shown on the map.

Elevations Bull Valley Gorge bridge, 1850 meters; Willis Creek, 1840; bottom end of Bull Valley Gorge, 1650 meters.

Hike Length and Time Needed Most people would be interested in visiting each canyon from the Skutumpah Road, then return the same way. It's the upper one or two kms in each which is most spectacular. But you can walk down Bull Valley, then up Sheep and Willis Creeks and to the road again. The round-trip from Bull Valley Gorge Bridge is about 25 kms. That's an all-day hike for most; a long all-day hike for some. A mtn. bike will eliminate a road-walk of 3 or 4 kms.

Water Willis Creek has running water year-round, but it may be suspect because of cattle in the area above during the summer season. Lower Sheep Creek has flowing water. Bull Valley Gorge is dry.

Maps USGS or BLM map Kanab(1:100,000), or Bull Valley Gorge(1:24,000).

Main Attractions Two good narrow canyons and the Bull Valley Gorge Bridge which was the scene of an accident on October 14, 1954, where 3 men were killed. The pickup is still wedged in the slot on the west side and at the bottom of the dirt bridge; and you can still see it from the canyon rim or bottom of the gorge. Also, the Averett Monument is where Elijah Averett was killed by Indians in August, 1866.

Ideal Time to Hike Spring, summer or fall. From April through October.

Hiking Boots Better take wading boots for both. Bull Valley is dry except for possible potholes below the bridge.

Author's Experience The author has been in each canyon on 2 or 3 occasions.

These narrows are above the dirt bridge in upper Bull Valley Gorge.

Map 57, Bull Valley Gorge and Willis Creek, Utah

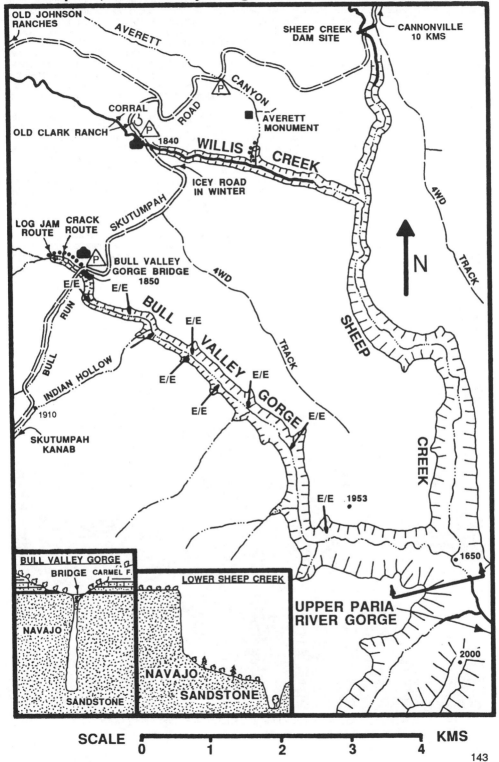

OLD JOHNSON RANCHES

AVERETT

SHEEP CREEK DAM SITE

CANNONVILLE 10 KMS

CANYON

ROAD

P

AVERETT MONUMENT

CORRAL

P

OLD CLARK RANCH

1840

WILLIS CREEK

ICEY ROAD IN WINTER

SKUTUMPAH

4WD TRACK

LOG JAM ROUTE

CRACK ROUTE

P

BULL VALLEY GORGE BRIDGE 1850

E/E

BULL

E/E

BULL RUN

4WD

VALLEY

E/E

INDIAN HOLLOW

E/E

GORGE

E/E

SHEEP

TRACK

1910

E/E

E/E

SKUTUMPAH KANAB

E/E

CREEK

N

E/E 1953

1650

UPPER PARIA RIVER GORGE

2000

BULL VALLEY GORGE

BRIDGE CARMEL F.

NAVAJO

SANDSTONE

LOWER SHEEP CREEK

NAVAJO

SANDSTONE

SCALE 0 1 2 3 4 KMS

143

Round Valley Draw, Utah

Location and Access Round Valley Draw is part of the upper Hackberry Canyon which is in the upper Paria River drainage. It's located about halfway between Kodachrome Basin State Park and Grosvenor Arch. Get there by driving along Highway 89 between Page and Kanab. Between mile posts 17 and 18, turn north onto the Cottonwood Wash Road and drive about 54 kms. There in the bottom of the upper valley you'll see a sign stating Round Valley Draw; and nearby another sign pointing to Rushbed Road. Turn south at that point and drive about 3 kms to the car-park as shown. Another way to get there would be to drive south out of Cannonville(which is just east of Bryce Canyon N. P.) on the paved road to Kodachrome Basin. At the Kodachrome Turnoff, continue east on a good graveled & graded road to the same area just described. The Cottonwood Wash Road will be slick right after heavy rains, but it's a well-used road in the warmer half of the year. This hike is also featured in the author's other book, *Hiking and Exploring the Paria River.*

Trail or Route Conditions From the Round Valley Draw Car-park, walk southwest in the open valley about 1 1/2 kms until you come to where the narrow slot begins. You can squeeze in at that point, but it's difficult. Best to rim-walk the slot on the northwest side for 200-300 meters, until you come to a stone cairn marking an easier way down in. This may be a little difficult for some, but just take it slow and it's not so bad. A short rope may help less-experienced members of your group down in. It's only 7-8 meters deep at that point. Further down there are 2 chokestones. The first is easy to pass on the left. The second you can slide or jump down over, or wiggle down on the left side by placing your feet on one wall, your back on the other(it's easier going down than up). The last problem is no more. There used to be a 2 1/2 meter dropoff in this section, but it's apparently been washed out, according to one hiker who wrote back. Canyons like this change with every flash flood, so expect the unexpected! In the lower end of the gorge you can exit up to the Slickrock Bench Car-park, then rim-walk back to your car.

Elevations Both car-parks, 1850 meters; bottom of gorge below Slickrock Bench, 1775 meters.

Hike Length and Time Needed It's about 5 kms from the Round Valley Draw Car-park to the junction of the Draw and the main Hackberry Canyon. If you were to hike down the narrows, exit at the Slickrock Bench route(a walk-up), and rim-walk back to your car, it would take about half a day, depending on how long you want to enjoy the narrows.

Water Take plenty in your car and in your pack. There are no reliable springs around. However, part of the time you may find water at the Round Valley spring and watering trough, as shown on the map.

Maps USGS or BLM map Smoky Mountain(1:100,000), or Slickrock Bench(1:24,000).

Main Attractions Another deep, dark narrow slot canyon in the Navajo Sandstone.

Ideal Time to Hike Spring or fall are best, but summers aren't so hot at this altitude and in the bottom of this slot canyon.

Hiking Boots Dry weather boots or shoes, except right after rains, then waders.

Author's Experience On one trip the author went down in late March and found half a meter of snow on the canyon floor. He went again in June, the driest month. Both were half-day hikes.

This is what it looks like above the slot known as Round Valley Draw.

Map 58, Round Valley Draw, Utah

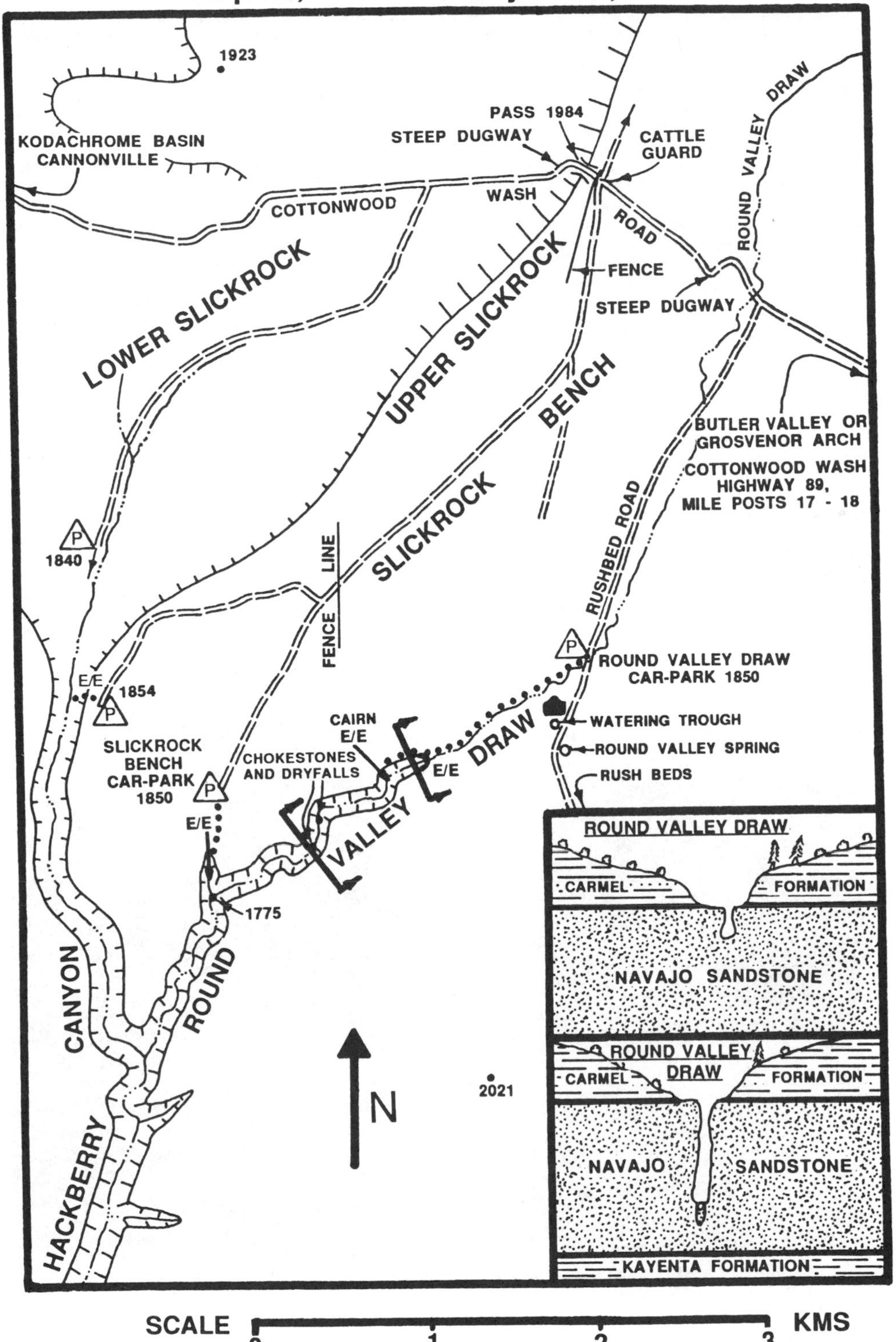

1923

KODACHROME BASIN
CANNONVILLE

PASS 1984
STEEP DUGWAY
CATTLE GUARD

ROUND VALLEY DRAW

COTTONWOOD WASH

ROAD

FENCE

STEEP DUGWAY

LOWER SLICKROCK

UPPER SLICKROCK

BENCH

SLICKROCK

BUTLER VALLEY OR
GROSVENOR ARCH
COTTONWOOD WASH
HIGHWAY 89,
MILE POSTS 17 - 18

P
1840

RUSHBED ROAD

FENCE LINE

E/E
1854
P

P
ROUND VALLEY DRAW
CAR-PARK 1850

CAIRN
E/E

SLICKROCK
BENCH
CAR-PARK
1850
P

CHOKESTONES
AND DRYFALLS

E/E

WATERING TROUGH

ROUND VALLEY SPRING

RUSH BEDS

VALLEY DRAW

E/E

1775

ROUND

CANYON

HACKBERRY

N

2021

ROUND VALLEY DRAW

CARMEL   FORMATION

NAVAJO SANDSTONE

ROUND VALLEY DRAW

CARMEL   FORMATION

NAVAJO   SANDSTONE

KAYENTA FORMATION

SCALE   KMS
0   1   2   3

145

Hackberry Canyon, Utah

Location and Access Hackberry Canyon lies between the Upper Paria River Gorge and the Cottonwood Wash or Canyon. It's also just southeast of Cannonville, and due north of the Paria River Ranger Station located on Highway 89. To get there, leave Highway 89(the link between Kanab and Page) between mile posts 17 and 18, and drive north for 20 kms to reach the bottom of the canyon; or about 50 kms to get to the top end of the drainage. Or from Cannonville, head south and follow the signs to Kodachrome Basin. When you come to the Kodachrome Junction, continue east on the Cottonwood Canyon(Wash) Road for another 10 kms, and turn south just before you climb up a steep dugway. This side road takes you into the main canyon. Drive as far as you can and park. Another entry possibility is to go to the top of the Slickrock Bench as shown on the map. You could also go down through Round Valley Draw, but that's another hike covered in this book. For a more detailed account of both, see the author's book, *Hiking and Exploring the Paria River,* which covers this canyon in better detail, including the history of the Watson Cabin.

Trail or Route Conditions From somewhere along the road route(see map), simply walk down Hackberry Canyon. It's easy walking because some ORV's still go down part way illegally, even though much of this canyon is part of a WSA. From the Slickrock Bench Trailhead, walk straight south and enter at the lower end of Round Valley Draw. Once in the canyon it's an easy walk all the way to the bottom. About halfway down the canyon, water begins to flow. Below that point, you'll get your feet wet on occasions. As you go downstream, you can exit to the west at two old cattle trails. You can also walk up a side canyon to see Sam Pollock Arch, or get onto an alluvial bench below Sam Pollock Canyon and inspect an old cabin built by Frank Watson. Just before you exit the canyon, you come to more narrows.

Elevations Trailheads, about 1850 meters; bottom end of canyon, 1440 meters.

Hike Length and Time Needed From either of the trailheads to the road at the bottom of the canyon is about 30 kms. This can be done in one long day, but most prefer to camp and do it in two. A mtn. bike would solve the car shuttle problem.

Water Throughout the lower half of Hackberry Canyon, and in several of the short side canyons. There are several good springs in the lower end, but cattle are in the canyon from November to May.

Maps USGS or BLM map Smoky Mountain(1:100,000), or Slickrock Bench & Calico Peak(1:24,000).

Main Attractions Good narrows, good water, a good arch, good campsites, and an interesting old cabin.

Ideal Time to Hike Spring or fall, but summers aren't so bad because of all the wading and shady campsites. Winter hiking might be possible(?).

Hiking Boots Wading boots or shoes

Author's Experience The author has been into this canyon 8 or 10 times now, while doing his book on the Paria River. On his first trip, he walked for a couple of hours the first day, camped, then all the next day. He then hitch hiked back to his car at Round Valley Draw. There might be 20 or more cars using this Cottonwood Canyon(Wash) Road daily in the warmer half of the year.

This narrow defile with giant cottonwood trees and running water is Stone Donkey Canyon.

Map 59, Hackberry Canyon, Utah

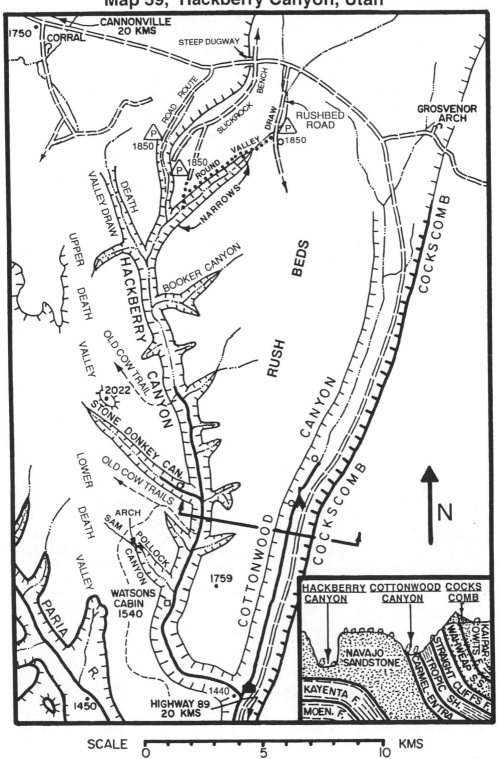

1750 CORRAL

CANNONVILLE 20 KMS

STEEP DUGWAY

GROSVENOR ARCH

ROAD ROUTE

SLICKROCK BENCH

VALLEY DRAW

RUSHBED ROAD

P 1850

P 1850

ROUND VALLEY

1850

P

NARROWS

COCKSCOMB

VALLEY DRAW

DEATH

UPPER DEATH VALLEY

HACKBERRY CANYON

BOOKER CANYON

BEDS

RUSH

CANYON

OLD COW TRAIL

• 2022

STONE DONKEY CAN.

COCKSCOMB

OLD COW TRAILS

LOWER DEATH VALLEY

ARCH

SAM POLLOCK CANYON

COTTONWOOD

CANYON

1759

PARIA R.

WATSONS CABIN 1540

N

• 1450

1440

HIGHWAY 89 20 KMS

HACKBERRY CANYON   COTTONWOOD CANYON   COCKS COMB

NAVAJO SANDSTONE

KAYENTA F.

MOEN. F.

STRAIGHT CLIFFS F.

TROPIC SH.

CARMEL-ENTRA

WINWEAP

KAIPAR

DOWITS. F.

SCALE   0   5   10   KMS

The Buckskin Gulch, Utah

Location and Access The Buckskin Gulch is located about halfway between Kanab and Page and south of Highway 89. Before hiking, stop at the Paria River Ranger Station information board between mile posts 20 and 21, for the latest information on camping, the whereabouts of good water in the canyon, and the latest weather forecast. A ranger no longer resides at Paria, so call the Kanab BLM office, 1-801-644-2672, for updated information. From the ranger station you have a choice of 4 entry points to the Buckskin. First, the White House Trailhead, discussed in the next hike. Second is the Middle Trail via Long Canyon. A third is called the Buckskin Trailhead, and the fourth and most popular is called Wire Pass. Get to Wire Pass Trailhead by driving south from Highway 89 from between mile posts 25 and 26. After 7 kms you'll come to the Buckskin Trailhead; then after another 6-7 kms you'll arrive at the Wire Pass. For more details and local history, see the author's book, *Hiking and Exploring the Paria River.*

Trail or Route Conditions From Wire Pass Trailhead walk east down the dry creek bed. In the past there used to be a chokestone and a 3 meter-high dropoff in this first section, but as of about 1993 that had disappeared--according to one report. However, this canyon changes with every flood, so expect the unexpected. At the confluence of the Buckskin and Wire Pass Canyons are some petroglyphs, then it's nearly 20 kms of narrows which average 4-5 meters in width for its length. In many places it's only one meter wide. Near the middle of the Buckskin is the Middle Trail, where you can exit on either side and camp, or take another look at the weather. *Have a good weather forecast for this hike!* Near the lower end of the Gulch, are several large boulders and a dryfall of about 3 meters. Steps have been cut in a couple of places, but be sure and take a *10 meter-long rope* just in case. A km up from the Confluence water begins to flow and there are several campsites. From there it's up the Paria, usually with running water, and to the White House Trailhead.

Elevations White House Trailhead, 1310 meters; Wire Pass, 1490; The Confluence, 1250 meters.

Hike Length and Time Needed It's about 33 kms from Wire Pass to the White House Trailhead. This can be done in one long day but with two cars(or a mtn. bike). Some people prefer to camp at the Middle Trail exit(but you have to carry lots of water!) or near the Confluence, and finish the hike the next day. Or continue down the Paria to Lee's Ferry.

Water The ranger station, White House Spring, lower Buckskin. Carry water through the Buckskin.

Maps USGS or BLM maps Kanab, Smoky Mountain and Glen Canyon Dam(1:100,000), or BLM map Hikers Guide to Paria(1:62,500), or Paria, Paria Plateau, and Lee's Ferry(1:62,500).

Main Attractions Longest and best all-around slot canyon hike in the world. Also petroglyphs.

Ideal Time to Hike May or June. In winter you often must wade in cold deep waterholes. Some deep wading at other times as well. From mid-July to mid-September is normally the monsoon season with flash floods. *Do this hike with a good weather forecast only!*

Hiking Boots Wading boots or shoes.

Author's Experience The author has been in the Buckskin and from all trailheads on 6 different trips. Once he walked from the Confluence to Wire Pass with large pack in 6 1/2 hours. Another time he walked from White House to the rockfall in the Buckskin and back in 7 hours. On his last trip, he left a mtn. bike at White House Trailhead, then drove to Wire Pass. He hiked down the Buckskin and up the Paria in 8 1/4 hours. The bike ride back to Wire Pass took nearly 2 hours, which made it a 10 hour day. After this long hike, the bike ride back to Wire Pass is extra tiring!

Most of Buckskin Gulch looks just like this.

Map 60, The Buckskin Gulch, Utah

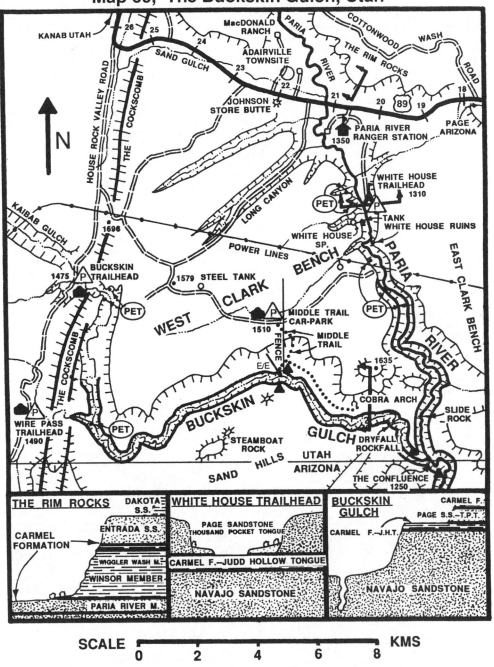

KANAB UTAH

MacDONALD RANCH

ADAIRVILLE TOWNSITE

SAND GULCH

JOHNSON STORE BUTTE

COTTONWOOD WASH

THE RIM ROCKS

PARIA RIVER

89

PAGE ARIZONA

PARIA RIVER RANGER STATION
1350

WHITE HOUSE TRAILHEAD
1310

PET

P

TANK
WHITE HOUSE RUINS

WHITE HOUSE SP.

BENCH

PARIA RIVER

EAST CLARK BENCH

HOUSE ROCK VALLEY ROAD

THE COCKSCOMB

LONG CANYON

1696

POWER LINES

KAIBAB GULCH

BUCKSKIN TRAILHEAD
1475
P

PET

1579 STEEL TANK

WEST CLARK

MIDDLE TRAIL CAR-PARK
P
1510

MIDDLE TRAIL

FENCE

E/E

PET

COBRA ARCH
1635

SLIDE ROCK

WIRE PASS TRAILHEAD
1490
P

THE COCKSCOMB

PET

BUCKSKIN

STEAMBOAT ROCK

GULCH

DRYFALL
ROCKFALL

SAND HILLS

UTAH
ARIZONA

THE CONFLUENCE
1250

N

THE RIM ROCKS

CARMEL FORMATION

DAKOTA S.S.

ENTRADA S.S.

WIGGLER WASH M.

WINSOR MEMBER

PARIA RIVER M.

WHITE HOUSE TRAILHEAD

PAGE SANDSTONE
THOUSAND POCKET TONGUE

CARMEL F.—JUDD HOLLOW TONGUE

NAVAJO SANDSTONE

BUCKSKIN GULCH

CARMEL F.

PAGE S.S.—T.P.T.

CARMEL F.—J.H.T.

NAVAJO SANDSTONE

SCALE

0   2   4   6   8   KMS

Paria River, Utah-Arizona

Location and Access One of the best known canyon hikes on the Colorado Plateau is the Paria River and its main tributary, the Buckskin Gulch. This information is meant only as an introduction to the Paria. Those who are serious about making this trip should obtain the author's other book, *Hiking and Exploring the Paria River,* for a more detailed route description. The previous map covers the Buckskin. To do this hike, drive along Highway 89 between Kanab and Page and to between mile posts 20 and 21. Turn south and first stop at the Paria River Ranger Station information board for the latest information on water and weather. A ranger no longer resides at Paria, so call the Kanab BLM office at 1-801-644-2672 for for information updates. From the highway drive 3 kms south to the White House Trailhead, the normal starting point for the hike to Lee's Ferry. *Never do this hike if bad weather is threatening.*

Trail or Route Conditions From the car-park, start walking along a trail, then in the creek bed. There is normally some water in the Paria and you'll be walking in it much of the time. The amount of flow in the river depends on upstream irrigation and the time of year, but it's seldom more than ankle deep. After about 7 kms the canyon narrows and it'll be that way for most of the trip. At the Confluence, most hikers walk up the Buckskin a ways before resuming the trip downstream. There are camping places just inside the Buckskin and good seep water. Three to four kms below the Confluence are several good springs and campsites. There are scattered campsites and springs all the way down to the Wrather Canyon area. From there on the canyon gradually opens up and becomes wider. Near Wrather and Bush Head Canyon you can climb out onto the rim. Generally speaking, there is no reliable drinking water below Bush Head.

Elevations White House Trailhead, 1310 meters; The Confluence, 1250; Lee's Ferry, 950 meters.

Hike Length and Time Needed From the White House Trailhead to Lee's Ferry is about 54 kms, or a 3 to 5 day hike, depending on side trips. If you go all the way to Lee's Ferry, you've got to have two cars, or hitch hike back to your car at White House Trailhead. There are people at Marble Canyon who run shuttle services. The BLM office in Kanab at 318 North, 100 East, telefone 1-801-644-2672, can give you a list of shuttle people you can contact. If you can make contact with the guy at the Paria Ranger Station, he could also give you a list. One solution to the car shuttle problem is to go down to about Wrather Canyon and return to White House. By doing this you'll have seen the best part of the canyon anyway. There are a number of alternatives.

Water Camp at or near one of many springs, or purify Paria River water. See the book mentioned above for more detailed maps and for locations of springs.

Maps USGS or BLM maps Smoky Mountain and Glen Canyon Dam(1:100,000), or Paria Plateau and Lee's Ferry(1:62,500), or the BLM map Hikers Guide to Paria(1:62,500). This is the best.

Main Attractions The best of the longer narrow canyon hikes on the Colorado Plateau.

Ideal Time to Hike Spring or fall. Summers are very warm, but it's always cool in the narrows.

Hiking Boots Wading boots or shoes. If going all the way to Lee's Ferry, wear shoes in good condition. It's a long hike!

Author's Experience The author has hiked into this canyon from both ends several times. On one day-trip, he walked down to the Confluence, up the Buckskin to the rock fall, then returned in 7 hours round-trip.

Hikers along the Paria River right at the Confluence of the Paria and Buckskin Gulch.

Map 61, Paria River, Utah-Arizona

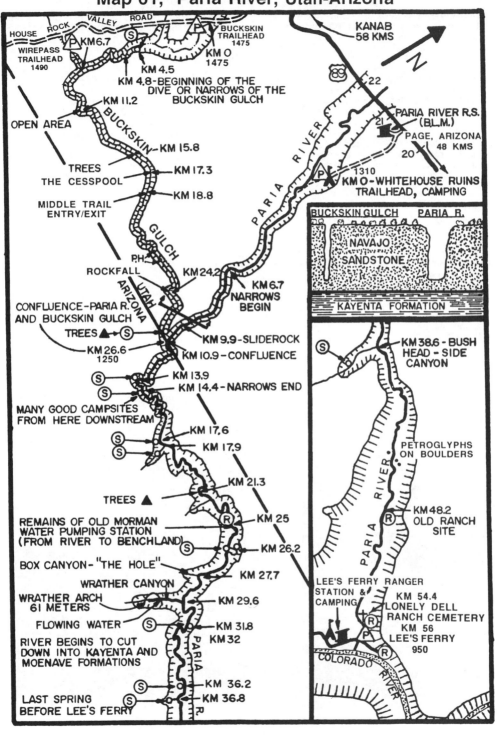

HOUSE ROCK VALLEY ROAD
WIREPASS TRAILHEAD 1490
KM 6.7
S
BUCKSKIN TRAILHEAD 1475
KM 0 1475
KM 4.5
KM 4.8-BEGINNING OF THE DIVE OR NARROWS OF THE BUCKSKIN GULCH

KANAB 58 KMS
N

89

22
21
PARIA RIVER R.S. (B.L.M.)
PAGE, ARIZONA 48 KMS
20

KM 11.2
OPEN AREA
BUCKSKIN

KM 15.8

TREES
THE CESSPOOL
KM 17.3

MIDDLE TRAIL ENTRY/EXIT
KM 18.8

PARIA RIVER

1310
P
KM 0-WHITEHOUSE RUINS TRAILHEAD, CAMPING

GULCH

P.H.
ROCKFALL
KM 24.2

UTAH
ARIZONA
KM 6.7 NARROWS BEGIN

CONFLUENCE-PARIA R. AND BUCKSKIN GULCH
TREES ▲→S
KM 26.6 1250

KM 9.9-SLIDEROCK
KM 10.9-CONFLUENCE

S
KM 13.9
S
KM 14.4-NARROWS END

MANY GOOD CAMPSITES FROM HERE DOWNSTREAM

S
S
KM 17.6
KM 17.9

KM 21.3

TREES ▲

REMAINS OF OLD MORMAN WATER PUMPING STATION (FROM RIVER TO BENCHLAND)
S
R
KM 25
KM 26.2

BOX CANYON-"THE HOLE"

KM 27.7

WRATHER CANYON
WRATHER ARCH 61 METERS

FLOWING WATER
S
KM 29.6

PARIA
KM 31.8
KM 32

RIVER BEGINS TO CUT DOWN INTO KAYENTA AND MOENAVE FORMATIONS

LAST SPRING BEFORE LEE'S FERRY
S
S
R.
KM 36.2
KM 36.8

Inset (upper right)

BUCKSKIN GULCH   PARIA R.
NAVAJO SANDSTONE
KAYENTA FORMATION

Inset (lower right)

S
KM 38.6 - BUSH HEAD - SIDE CANYON

PARIA RIVER

PETROGLYPHS ON BOULDERS

R
KM 48.2 OLD RANCH SITE

LEE'S FERRY RANGER STATION & CAMPING
KM 54.4 LONELY DELL RANCH CEMETERY
R
P
KM 56 LEE'S FERRY 950
R

COLORADO RIVER

SCALE |———————————| KMS
0   5   10

La Verkin Creek and Kolob Arch, Utah

Location and Access The canyons on this map are located about halfway between Cedar City and St. George, and just east of Interstate 15. This entire area is in the northern end of Zion National Park. Leave I-15 at Exit #40, near mile post 40, which is the road to Zion National Park, Kolob Section. About 200 meters off the highway is a visitor center, where you can get last minute hiking and trail information, maps, water and a free camping permit; then continue up the road to Lee Pass. Immediately over the pass is a car-park and the beginning of the trail into Kolob Arch and upper La Verkin Canyon. The road to the car-park is paved all the way and open year-round.

Trail or Route Conditions The trail from the Lee Pass Trailhead, down Timber Creek, up La Verkin Creek and into Hop Valley is a very good and well-used path that's marked all the way. This is a national park, so trails are always maintained. There is a less-used, but good trail up a side canyon to the north and to near the base of Kolob Arch, one of the largest natural arches in the world. Don't pass this one by, it's a good one. There's also a trail up La Verkin Creek, into Willis Creek and on to a series of springs called Birch Spring. From that point, there's a 4WD track running 3 kms to Kolob Reservoir, another possible entry point. In upper La Verkin Creek and in Bear Trap Canyon, the trail fades and you must do much of your walking in the stream itself. It's in this upper canyon area you'll find some good narrows. If you're heading that way, carry a walking stick to probe deep holes.

Elevations Lee Pass Trailhead, 1850 meters; the waterfalls in lower La Verkin Creek about 1525; and Birch Spring, 2200 meters.

Hike Length and Time Needed The hike from Lee Pass to the base of Kolob Arch is 11 kms, or an all-day hike round-trip for most people. From the pass to the junction of Bear Trap and La Verkin Canyons is about 13 kms, one-way. The author made it into both of these canyons from the pass in one day round-trip, but most hikers can't. It's best to get a camping permit at the visitor center, and do this long trip in two days. By doing so, you'll be able to see a lot more of the upper tributaries.

Water Year-round running water is shown with the heavy black lines. Timber, La Verkin, and Bear Trap all have good water. Get drinking water above summer swimming holes.

Maps USGS or BLM map St. George(1:100,000), or Zion National Park(1:31,680).

Main Attractions Kolob Arch is impressive, as are the sheer red cliffs of Navajo Sandstone. The upper sections of the canyons have some very good watery narrows.

Ideal Time to Hike For a hike to Kolob Arch, spring, fall, or possibly in winter warm spells. For the upper canyons where you'll be wading a lot, you'll want warmer summer-type weather.

Hiking Boots Dry weather boots or shoes on the trails, but waders in the upper canyons.

Author's Experience The author spent nearly 7 hours walking to Kolob Arch, into both upper La Verkin and Bear Trap Canyons, then back to the trailhead. That was on a cool April 2 day.

In the middle sections of La Verkin Creek is Kolob Arch, one of the biggest in the world.

Map 62, La Verkin Creek and Kolob Arch, Utah

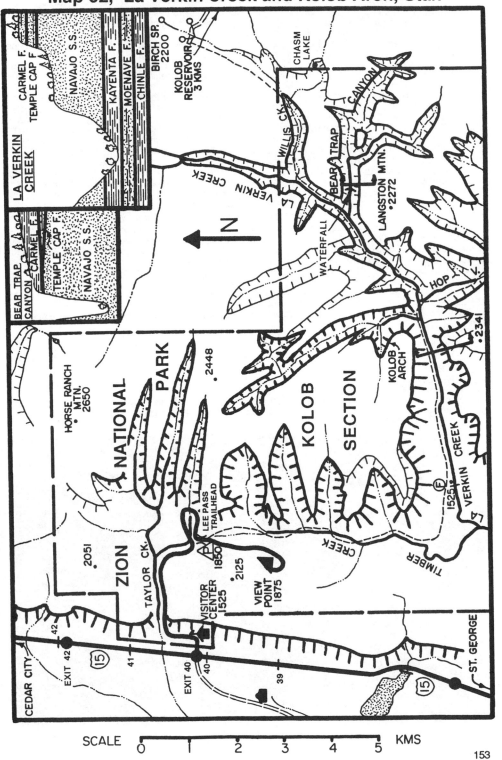

SCALE

0 1 2 3 4 5 KMS

153

Great West Canyon(Left Fork of North Creek), Utah

Location and Access Great West Canyon, which some prefer to call the Left Fork of North Creek, is located in the central part of Zion National Park. The upper end of the canyon is known as Wildcat Canyon, and the stream flowing through it is called North Creek. To get to the trailheads for a hike into the narrows or Subway part of Great West, drive to Virgin, a small town not far below the mouth of Zion Canyon. On the east side of town, look for the turnoff to the north between mile posts 18 & 19 and the sign indicating the road to Kolob Reservoir. There's a paved road all the way to the reservoir, but stop at a turnoff on the right just above the big "S" curve about 13 kms from Virgin. A second place to begin would be to drive further up the same highway and park at the Wildcat Canyon Trailhead.

Trail or Route Conditions At the lower trailhead, marked 1540 meters, look for the trail heading east. This path is used a lot and easy to follow. Walk east about one km to the rim, then head down to the creek below. Once in the canyon bottom, simply walk upstream, crossing the creek many times. There's a pretty good trail there today. The canyon gradually narrows and you'll eventually come to the feature called The Subway. Just a little ways into this keyhole-like-slot, there's a wall you must scale. There may be a tree trunk you can climb, but if that's gone, you may be stopped. There may also be a nylon rope hanging down which you can use to get up the waterfall. To do the entire canyon you really need to start at the top end. Drive to and park at the Wildcat Canyon Trailhead and walk eastward about 3 kms on a good trail, then turn southeast on a trail marked for Northgate Peaks. Soon after that, veer left or east, and walk along a developing trail down across Russell Gulch to the east side, then into the main canyon as shown. You can then walk down through The Subway. _You may need to swim one or more pools, and you'll need a rope 20-25 meters long_ to get down over one waterfall. There should be slings mounted to the wall you can use to rappel down and into a waist-deep pool. Just below that is The Subway, and another rappel(with a bolt and sling already mounted in the wall). The Subway is where the Kayenta Formation is exposed. Inquire at the Zion visitor center for more information about weather, routes and a hiking permit.

Elevations Lower trailhead, 1540 meters; Wildcat Canyon Trailhead, 2125 meters.

Hike Length and Time Needed Up to The Subway from the bottom is about 7 kms, or about a half-day hike round-trip. If you begin at the Wildcat Canyon Trail and walk down through The Subway, plan on a long all-day hike. You'll need a car shuttle for this one, or a mtn. bike.

Water This canyon has a year-round stream beginning below the mouth of Russell Gulch. There are good springs just above and below The Subway.

Maps USGS or BLM map St. George(1:100,000), or Zion National Park(1:31,680).

Main Attractions Narrows, potholes, and The Subway, in a deep forested canyon.

Ideal Time to Hike Because of all the wading and swimming, hike this in warm weather--from late spring to early fall. To get across the deep pools, line your pack with several plastic garbage can liners. _Stay out of this one if bad weather is threatening._

Hiking Boots Wading boots or shoes.

Author's Experience The author left his car at the lower trailhead, then mtn. biked up the paved road to the upper car-park. He then hiked down-canyon to his car. Bike-time was just over an hour; while hike-time was just over 5 hours. Total time, car to car, 6 1/4 hours.

This is one of two places where slings are mounted into the walls of Great West Canyon to help hikers down over ledges or waterfalls. This is the first rappel point.

Map 63, Great West Canyon(Left Fk. of North Ck.), Utah

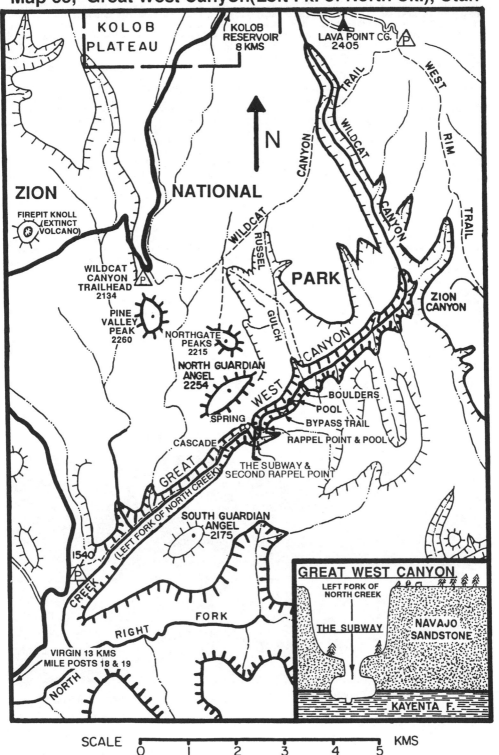

KOLOB PLATEAU

KOLOB RESERVOIR 8 KMS

LAVA POINT CG. 2405

P

N

ZION

NATIONAL

FIREPIT KNOLL (EXTINCT VOLCANO)

WILDCAT CANYON TRAILHEAD 2134

PINE VALLEY PEAK 2260

NORTHGATE PEAKS 2215

NORTH GUARDIAN ANGEL 2254

SPRING

CASCADE

THE SUBWAY & SECOND RAPPEL POINT

SOUTH GUARDIAN ANGEL 2175

CANYON

WILDCAT

TRAIL

WILDCAT

RUSSEL GULCH

PARK

WEST CANYON

BOULDERS POOL

BYPASS TRAIL

RAPPEL POINT & POOL

CANYON

WEST RIM TRAIL

ZION CANYON

GREAT

(LEFT FORK OF NORTH CREEK)

1540

P

CREEK

RIGHT FORK

NORTH

VIRGIN 13 KMS MILE POSTS 18 & 19

GREAT WEST CANYON

LEFT FORK OF NORTH CREEK

THE SUBWAY

NAVAJO SANDSTONE

KAYENTA F.

SCALE 0 1 2 3 4 5 KMS

155

Zion Canyon Trails, Utah

Location and Access The area shown on this map is in the heart of Zion Canyon and Zion National Park. To get there, drive east out of St. George, through Hurricane, Virgin and Rockville, and enter the canyon via Highway 9. Or you can drive west on Highway 9 from Mt. Carmel Junction, which is east of the park. When you reach the bottom of Zion Canyon, drive north and park at one of the trailhead parking lots. Be sure and stop at the visitor center at the bottom of the canyon before proceeding any further. They'll have the latest maps and information on trails and hiking conditions.

Trail or Route Conditions All hikes discussed here are along established and well-used national park trails. They are all signposted and you can't get lost. Because of this it's a good place to begin a hiking career. Some of the most popular hikes are along the Echo Canyon Trail. By parking at Weeping Rock, you can hike up the trail to overlooks on the canyon rim at either Observation Point, Cable Mountain, or to the top of Deer Trap Mountain. Observation Point probably has the best views of the lower canyon. For the cross-country hiker, there's a trail from Stave Spring running east to the trailhead on Clear Creek. Another trail begins about 2 kms north of the lodge(Grotto Picnic Grounds) and takes hikers to Angels Landing, the West Rim Viewpoint, and on to Kolob Arch and Lee Pass in the extreme northern end of the park. There's another trail running along the west side of the Virgin River to Sand Bench.

Elevations Zion Lodge is 1303 meters; while canyon rims average about 2000 meters.

Hike Length and Time Needed Here are some examples of hikes and estimated times. The East Rim Trail to Observation Point is 6 kms one-way, and about half a day round-trip. Weeping Rock to Cable Mountain, 13 kms one-way, and a long all-day hike. Weeping Rock to Deer Trap Mountain, 14 kms one-way--another very long all-day hike--or perhaps camp one night. Weeping Rock to Clear Creek(East Entrance) Trailhead 16 kms. This sounds like a half-day hike, but instead of back-tracking, you'd want a car or bike at the Clear Creek Trailhead for the return trip. Grotto Picnic Grounds to Lava Point Campground(on Great West Canyon Map), 21 kms one-way. This can be hiked in one day, but you'd need transportation back at the other end. The short but very scenic hike from the Grotto Picnic Grounds to Angels Landing is very popular and takes only about half a day or less round-trip. The view from on top is one of the best around.

Water Always carry plenty of your own water, but Stave and Cabin Springs offer drinks.

Maps USGS or BLM map Kanab(1:100,000), or Zion National Park(1:31,680).

Main Attractions Great views into one of the greatest canyons anywhere.

Ideal Time to Hike Late spring to early fall, with early morning hikes in summer. During some mild winter weather, you can hike some of the lower altitude trails.

Hiking Boots Any dry weather hiking boots or shoes.

Author's Experience The author camped outside the park, then hiked to Observation Point in the morning hours. That same afternoon was spent hiking to Angels Landing.

The Great White Throne stands tall and mighty, as seen from near Angels Landing.

Map 64, Zion Canyon Trails, Utah

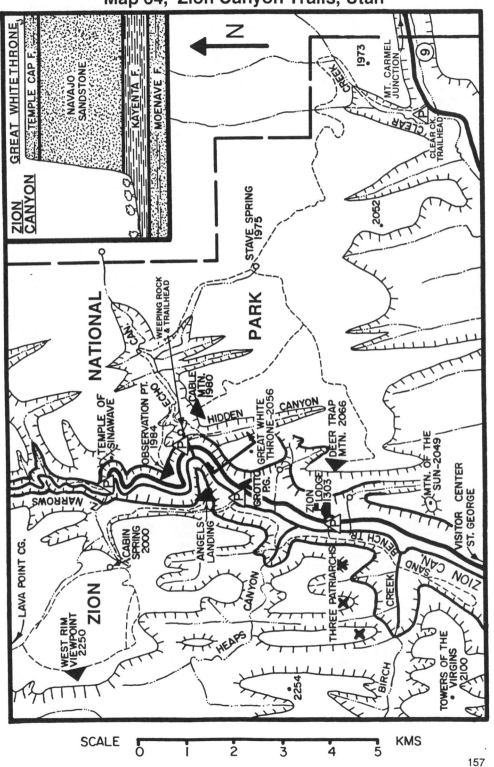

ZION CANYON

GREAT WHITETHRONE

TEMPLE CAP F.

NAVAJO SANDSTONE

KAYENTA F.

MOENAVE F.

N

1973

MT. CARMEL JUNCTION

CLEAR CK. TRAILHEAD

9

CLEAR CREEK

2052

STAVE SPRING 1975

WEEPING ROCK & TRAILHEAD

NATIONAL

PARK

ECHO CAN.

CABLE MTN. 1980

OBSERVATION PT. 1984

TEMPLE OF SINAWAVE

HIDDEN CANYON

GREAT WHITE THRONE-2056

DEER TRAP MTN. 2066

MTN. OF THE SUN-2049

NARROWS

GROTTO PG.

ZION LODGE 1303

CABIN SPRING 2000

ANGELS LANDING

VISITOR CENTER
ST. GEORGE

BENCH TR.

SAND CAN.

ZION CAN.

LAVA POINT CG.

ZION

WEST RIM VIEWPOINT 2250

HEAPS CANYON

THREE PATRIARCHS

BIRCH CREEK

TOWERS OF THE VIRGINS 2100

2254

SCALE

0   1   2   3   4   5

KMS

157

Zion Narrows, Utah

Location and Access The Zion Narrows is one of the most famous canyon hikes in the world. It has become famous because it has some of the best narrows around, and it's located in Zion National Park in southwestern Utah. This hike is found in the upper portions of Zion Canyon, the most-visited part of the park. There are two ways to enter this canyon. First, you can drive to the lower end of the narrows via the paved canyon highway, for short day-hikes up from the bottom. Or if you want to do the full length of Zion Canyon and all its narrows, you must drive east out of the park on Highway 9, the road running to Mt. Carmel Junction. About 3 kms east of the park's east entrance, and next to mile post 46, turn north onto a paved road. After about 3 kms, the pavement ends, but beyond that it's a reasonably good graded dirt road running the remaining 26 kms to the Chamberlain Ranch. The trailhead, which is about one km beyond the ranch house, has a place or two for camping, but you're supposed to get permission first if you do camp(this is a NPS suggestion). Chamberlain himself told this writer it's OK to camp, but please don't get out in the irrigated pasture next to the parking lot and tromp down the grass! BLM maps indicate the parking place is on public land, but it's not very flat. If you plan to camp somewhere in the vicinity of the trailhead, consider checking out some places along the road before arriving at the trailhead itself. Most of the land along the Chamberlain Ranch Road is controlled by the BLM.

Before heading for the upper end of the canyon, be sure you stop at the *visitor center* near the mouth of Zion Canyon just north or Rockville, and talk to the rangers about the weather and pick up a *hiking and/or a camping permit.* To get this permit, you must be at the visitor center before 5 pm the afternoon before you want to hike. Take a numbered ticket when you arrive and hang around. At 5 pm the rangers get the latest weather forecast. If the outlook is for clear skies and safe conditions, they will issue so many permits on a first come, first serve basis, for day-hikes; and so many other permits for those wanting to camp one night in the canyon. They will never issue a permit for two nights in the canyon, because they can't always count on the weather being good that far in advance.

Trail or Route Conditions From the car-park at Chamberlain Ranch, you first walk down-canyon on an old ranch road. This is private property, so stay on this main track. You'll pass an old tractor, a cabin, then you head downstream along the creek bed. Nowadays, there's a trail on one side or the other much of the way, but plan to be in the water wading about 1/4 of the time. It's slippery walking below the confluence of Deep Creek, so take a light-weight walking stick; perhaps a ski pole. It's also useful when probing the deeper holes. Along the way you'll encounter several rockfalls and deep pools, so have one air mattress per group to ferry larger packs across possible deep water. Because of these deep holes, small children should not attempt this hike. Use fast film, because of low-light conditions, and keep cameras in plastic bags or inside packs. Drowned cameras are common in this canyon because of the slippery rocks. You must camp in designated campsites, but camping is not allowed below Big Spring.

Consider seeing side canyons if you have time. The rangers will show you maps and give you more information regarding entry into upper Deep Creek, Kolob Canyon and others. Coming down Deep Creek involves a very long car-shuttle, and is really not worth it. To enter the main canyon via Kolob, you start at Lava Point and must have one very long rope and do at least two rappels. This is one of the best slots around and was featured on PBS. It's also where two boy scout leaders died in 1993. This is obviously for expert climbers and/or hikers only. Get more information about this hike at the visitor center. If you haven't the time for the full canyon hike, and don't want to drive the dirt road, there's a short paved trail up from the bottom car-park about 1 1/2 kms to where the canyon really narrows. Most people hike up-canyon several kms from there and return the same way. To do this day-hike from the bottom, you don't need a permit.

Elevations Chamberlain Ranch is about 1735 meters; bottom end of narrows, 1344 meters.

Hike Length and Time Needed From the ranch to the bottom car-park is about 20 kms. This can be done in one day, but it's a tiring hike, so many people do it in two. A car shuttle is a necessity, so look for other hikers in the visitor center, or hire help in Springdale to shuttle your car. If time is limited and/or the shuttle too difficult, just walk up-canyon from the bottom end. In about half a day, you can walk upstream to about Orderville Canyon, and see the lower end of the narrows.

Water Park rangers will tell you to treat all water, but that coming from side canyons(except Kolob) is generally safe to drink, except for possible human pollution! Water taken directly from springs should be OK. Treat river water. For day-hikes, most people just take their own.

Maps USGS or BLM map Kanab(1:100,000), or Zion National Park(1:31,680).

Main Attractions This canyon is at or near the top of the author's list of best hikes.

Ideal Time to Hike Because you'll be in the water and wet so much of the time, do this hike in warm weather, from early June through late September. If it's too early or late in the season, the water will be too cold and they won't give you a permit.

Hiking Boots Wading boots or shoes. Take a pair that are in reasonably good condition.

Author's Experience The author has been up from the bottom of the canyon three times trying to get better fotos. In 1992, he finally got a permit to stay one night in the canyon, so he camped near the trailhead, went down to the first campsite below Deep Creek, then explored up Kolob Canyon on that first day. The second day, he went up Deep Creek a ways, then returned to his car. From Deep Creek to his car at Chamberlain Ranch, with the large pack, took just under 4 hours. Returning to the trailhead saved a car shuttle. At another time he hiked all the way down Orderville Canyon to the main drainage and returned. That canyon and hike are on the next map.

Map 65, Zion Narrows, Utah

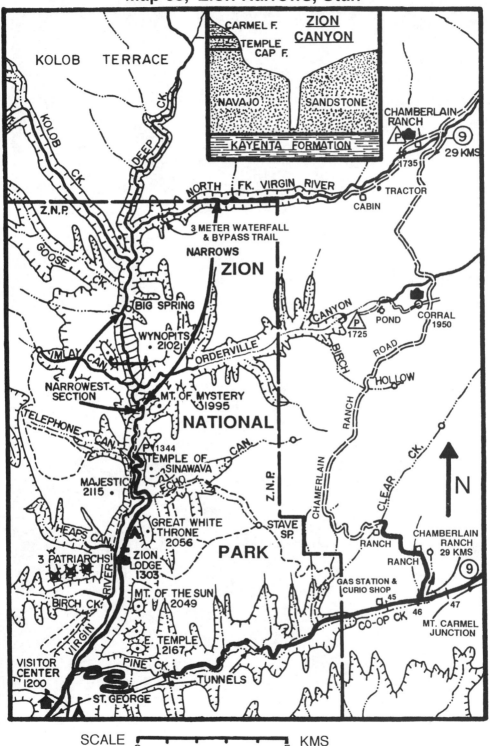

ZION CANYON

CARMEL F.

TEMPLE CAP F.

NAVAJO SANDSTONE

KAYENTA FORMATION

KOLOB TERRACE

KOLOB CK.

DEEP CK.

Z.N.P.

GOOSE CK.

NORTH FK. VIRGIN RIVER

3 METER WATERFALL & BYPASS TRAIL

CHAMBERLAIN RANCH

P 1735

9

29 KMS

TRACTOR

CABIN

NARROWS

ZION

BIG SPRING

WYNOPITS 2102

IMLAY CAN.

NARROWEST SECTION

TELEPHONE CAN.

ORDERVILLE

MT. OF MYSTERY 1995

CANYON

BIRCH

POND

CORRAL 1950

P 1725

ROAD

HOLLOW

NATIONAL

P 1344

TEMPLE OF SINAWAVA

CAN.

RANCH

CLEAR CK.

Z.N.P.

CHAMERLAIN

N

MAJESTIC 2115

ECHO

GREAT WHITE THRONE 2056

HEAPS CAN.

3 PATRIARCHS

ZION LODGE 1303

STAVE SP.

PARK

RANCH

CHAMBERLAIN RANCH 29 KMS

RANCH

9

BIRCH CK.

MT. OF THE SUN 2049

GAS STATION & CURIO SHOP

45

VIRGIN RIVER

E. TEMPLE 2167

PINE CK.

CO-OP CK

46

47

MT. CARMEL JUNCTION

VISITOR CENTER 1200

ST. GEORGE

TUNNELS

SCALE 0   5 KMS

The Narrows of Zion Canyon.

This is the waterfall in the upper end of the canyon known as the Zion Narrows.

One of 7 or 8 official campsites in the Zion Narrows.

The lower part of Orderville Canyon, just above Zion Narrows.

Orderville Canyon, Utah

Location and Access Orderville Canyon drains from the east into the North Fork of the Virgin River and Zion Narrows. The Lower half of this drainage is in Zion National Park. One of the nice things about this canyon is, it's easily accessible from both the top and bottom. The NPS requires hikers to get a permit for this hike(visitor center--Zion N.P.), mainly in order to give them the latest weather forecast. To get to the bottom end, drive up Zion Canyon to the end of the paved road. Most people seeing Orderville go up from there and see just the lower end. To get into the upper part, drive east out of Zion Park on Highway 9 in the direction of Mt. Carmel Junction. About 3 kms east of the east entrance of Zion, and next to mile post 46, turn left or north onto a paved road. After about 3 kms, the pavement ends(1992). After that continue on this same graded road in the direction of the Chamberlain Ranch. About 13 kms from Highway 9, look for a side road to the left or west. Drive down this about half a km and park under some big cedar trees next to a large corral. If you have a 4WD or HCV you can probably go down the road further, but there is some private irrigated pasture land below. It might be best to walk from the corral, or use a mtn. bike.

Trail or Route Conditions This description will be from the top down. From the corral, head down the moderately steep road to the drainage bottom, past a pond with stored irrigation water, and through the pastures to the end of the road. Walk down into the dry wash, and soon you'll come to a big dropoff of about 35 meters. Skirt it to the right or left. Further down near the park boundary, you'll come to a large chokestone with a 5-meter-high dropoff. Most people will want a rope to get down this, but the author made it down and back up without. Near Bulloch Gulch, you'll find running water for the first time, and not far below that two more chokestones which have created dropoffs of about 3 meters each. The author had no trouble with either of these, but some people might take a second and/or third rope along, just to make sure they get back up. Below that are several waterfalls and shallow pools. Finally you'll be at the Virgin River and Zion Narrows. The end of the Zion Canyon road is about 3 kms below the mouth of Orderville.

Elevations Trailhead, 1950 meters, mouth of Orderville Canyon, about 1375 meters.

Hike Length and Time Needed It's about 16 kms from the corral to the end of Orderville, but the walking is easy and fast. It can be done in one full day from top to bottom, and back. A mtn. bike or HCV would save time.

Water In the upper drainage, but don't drink it without purification. Good water in the lower end.

Maps USGS or BLM map Kanab(1:100,000), or Zion National Park(1:31,680).

Main Attractions Deep, cool narrows and easy access. One of the best narrows hikes around.

Ideal Time to Hike Late spring to early fall, but avoid doing this one if storms are threatening. If the weather outlook is bad, they won't issue a permit.

Hiking Boots Any dry weather boots or shoes for the upper part, but waders for the lower end.

Author's Experience The author camped near the corral, then rode a mtn. bike down to the end of the road. From there he walked all the way to the Virgin River, then returned, riding his bike all the way up the road(which means HC cars can make it in normal conditions). Total round-trip time from the corral and campsite was just over 6 hours.

This is the chokestone and 5 meter-high dropoff in Orderville Canyon.

Map 66, Orderville Canyon, Utah

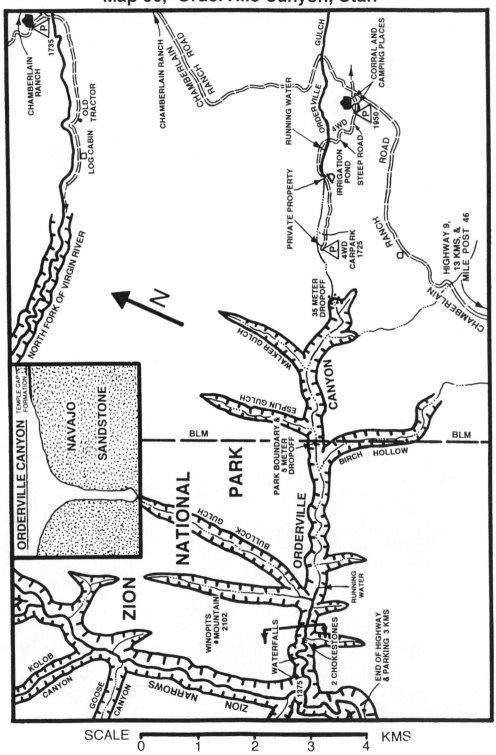

SCALE

0   1   2   3   4   KMS

Parunuweap Canyon, Utah

Location and Access Parunuweap Canyon lies between Mt. Carmel Junction and Springdale in southwestern Utah. The stream flowing through it is the East Fork of the Virgin River. The lower half of this canyon lies within Zion National Park. To reach the upper-most trailhead, drive south out of Mt. Carmel Junction to near mile post 81, and turn right onto a dirt road. Drive to a locked gate and park. One can also enter the canyon where Mineral Gulch and Meadow Ck. cross Highway 9 at mile posts 53 and 49. But the easiest and fastest way into the best part of this canyon is from near mile post 47. From there drive south on a pretty good road to either car-park at 1825 or 1750 meters. A little beyond the second car-park the road becomes very sandy and it's best to walk this last part to the river.

Trail or Route Conditions If you begin near Mt. Carmel Junction, you can walk(not drive) across private property with no problem while following a ranch road about 5 kms into the canyon. Beyond the end of the road, you merely walk in or along the creek. If you go down Meadow Ck. or Mineral Gulch, expect to route-find in the upper end, then pass through good narrows just before reaching the main canyon. If you use the route from mile post 47, you can walk all the way to the creek on a sandy 4WD-type road. Once in the canyon bottom, there are some deep holes and maybe a little quicksand in a narrows section called The Barracks. You never sink in too deep, but it can be scary the first time you try it. Take along a *walking stick* to help locate more solid walking. An *air mattress* would be handy in case you have to ferry packs across pools. *Also take a short rope*. There are several waterfalls in the lower Barracks, all of which are very near the park boundary. It's important you know where the boundary is, because in 1993, the NPS closed the canyon to hikers below the boundary. If it's closed when you arrive, you can still hike to about the waterfalls and return. Be sure to contact park rangers before you start--Tele. 1-801-772-3256.

Elevations Upper end of canyon, about 1550 meters; the waterfalls, roughly 1350 meters.

Hike Length and Time Needed From the trailhead near Mt. Carmel Junction to the waterfalls & park boundary is about 23-25 kms. You can shorten this by going down Meadow Ck. or Mineral Gulch. However, the shortest and easiest route in is via the road heading south from mile post 47. It's a 10 km mtn. bike ride & walk from the *steep place* to the river. With the lower end closed, this can be done as a day-hike by some people, but staying 2 days will allow you to really see the canyon. *Camp on higher ground only*. *This canyon is for more experienced hikers.*

Water There are many good springs along the way, but treat river(really just a creek) water.

Maps USGS or BLM map Kanab(1:100,000), or The Barracks & Springdale East(1:24,000).

Main Attractions Deep narrows near the park boundary, and in the lower end of Mineral Gulch.

Ideal Time to Hike Late spring to early fall. With all the wading, hike in warm weather only.

Hiking Boots Wading boots or shoes.

Author's Experience The author parked between mile posts 48 and 49 on Highway 9 and entered through Meadow Creek & Mineral Gulch. Near the park boundary, he had to lower his pack over one waterfall and ferry it across a deep pool. The trip took two days(14 hours total walk-time) in early June. He hitch hiked back to his car. This was all before the 1993 closure. In 1993, he biked & walked from the steep place to the river via the mile post 47 route, did lots of exploring, and returned, all in 5 1/2 hours.

The narrows of Mineral Gulch, a tributary of Parunuweap Canyon.

Map 67, Parunuweap Canyon, Utah

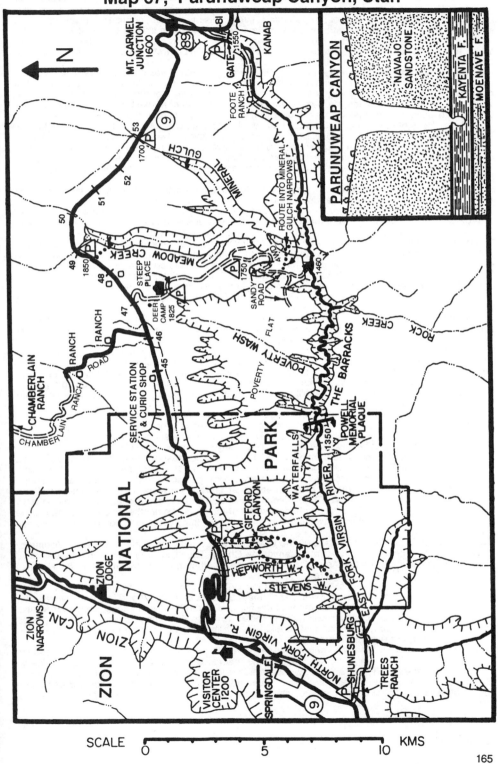

SCALE

0   5   10   KMS

Introduction to Hiking on the Navajo Nation, Utah and Arizona

Location The Navajo Nation occupies the entire northeastern corner of the state of Arizona and that part of southeastern Utah south of the San Juan River. This map covers about 1/4 of Navajoland. It shows all the hiking areas except for Canyon de Chelly, which is a little to the east near Chinle, Arizona. The good hiking areas on the Navajo Nation are in the area surrounding Navajo Mountain; south and southeast of Page, Arizona; along the Colorado River in Marble Canyon; and along the Little Colorado River Gorge northwest of Cameron.

Information about Hiking Permits Back in 1986, when the author was doing the first edition of this book, the tribal government at Window Rock, Arizona, was thinking of issuing hiking permits, but didn't get it started and going very fast. Over the years, he never got back to those people to ask what their policy was. Instead, he always went to the local *chapter house* for information to see if it was OK to enter a particularly canyon, or if there was any private land involved, or about road conditions. None of the people he talked to had ever heard of anyone having to have a permit to hike! By the way, a chapter house is a local seat of Navajo government, always located in or near a village or densely populated area. These seem similar to our town or county governments, plus it's a social gathering place with a large recreation hall and kitchen for town meetings and activities of various kinds. In some ways these facilities and functions are similar to some churches, but without the Christian end of it.

Over the years, the Window Rock government has created a policy regarding permits for hiking and camping on Navajo Nation lands, although almost no one knows what it is, not even those working at the various chapter houses! As of 1994, the only place you could get a permit was at the **Visitor Center at Cameron,** and at the **Navajo Tribal Parks & Recreation office at Window Rock, Arizona**. Here's the latest on permits taken from an information sheet in 1994. *A Camping Permit is $2 per person per night. A Backcountry Permit for hiking is $5 per person, $10 for 2-10 people, and $20 for more than 10 people. The Backcountry Permit is good for 1 to 14 days (if you will be camping, you also need a camping permit). For more information contact: Navajo Tribal Parks & Recreation, P.O. Box 308, Window Rock, Arizona 86515, Tele. 1-602-871-6645(or 6635, 6636).* As the author understands it, if you're just driving along a highway and pull off the road to camp or park for the night, a camping permit is not needed. Only if you're in the backcountry or perhaps at a trailhead(?), do you need a camping permit.

It must be noted that in all his hikes and traveling around the Navajo Nation, the author has never seen a Parks & Recreation officer on patrol, and doubts there are such personnel. It seems apparent that a permit does not guarantee help if you need it, or insure a rescue if you become lost. Only recently has the author gone to the trouble of getting permits, and that was because he was close to Cameron and it was convenient. If and when Tribal Parks & Recreation ever allow the chapter houses to sell permits, thus making it convenient for hikers, only then will they get full compliance and sell a lot of permits. For the serious hiker, it's recommended you go to the bother of getting a permit. If a rescue is somehow need, then you'll be on politically-safe ground. It may also save you from paying for the rescue!

For the popular Antelope Canyon near Page, you now(1994) just wait at the gate near mile post 299 and pay $5 to a member of the Begay family on the morning of your hike.

Places to Shop Probably the best place within Navajoland, or at least near the hiking areas, to get food, fuel and other goods, would be at **Page**. It has many gas stations, motels, shops and two large supermarkets. **Kayenta** might be the second best place with at least one large supermarket. **Tuba City** is also a sizable place with several small stores. **Cameron** has several motels, curio shops and gas stations(one has cheap Diesel), but not much else. Most people there drive to **Flagstaff** to shop. There are also gas stations, motels or small shops at **Marble Canyon, The Gap, Desert View, Kaibito, Gouldings Trading Post, Black Mesa, Inscription House**, and perhaps at **Navajo Mtn.**(In 1994, the mission store and gas station had closed, but there may be another one up the road to the north on the Utah side). It's best however to buy most of everything you'll need before getting into this region.

Maps USGS or BLM 1:100,000 maps on the Utah side are; Smoky Mountain and Navajo Mountain. In Arizona; Glen Canyon Dam, Kayenta, Canyon de Chelly, Tuba City, and Cameron. These maps cover the region included on this Area Map. There are also other maps at 1;62,500 or 1:24,000 scale, but they're listed under each hiking area.

Ideal Time to Hike For the most part, hiking on the Colorado Plateau and on Navajoland is usually done in the spring or fall months or March, April, May, September, October and into early November. Summers are pretty warm. However, many of the hikes featured here are in deep, dark, narrow and cool slot canyons, many of which will have potholes to wade or perhaps even swim through, so depending on the canyon, some can and should be done in warm weather. June, which is the driest month of the year, is a good time to do some of these hikes.

Road Conditions This Area Map shows the paved roads as solid black lines, while the important graded dirt roads are shown as dotted lines. All the main side roads are made of dirt; either clay or sand, and are graded. These are pretty good roads and one can often travel at or near highway speeds, or at least in your highest gear. Side roads branching off these graded roads are also pretty good, but never graded or maintained. In some areas roads are made out of clay, which can be very

Map 68, Introduction to Hiking on Lands of the Navajo Nation, Utah and Arizona

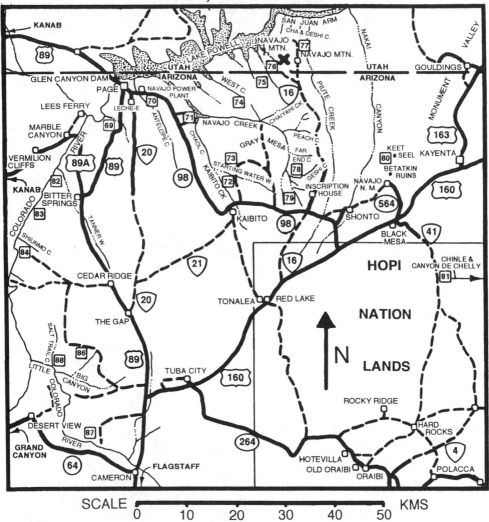

slick and muddy in wet weather; whereas others run over sandy regions. These roads are often good, but in some cases, deep sand can bog down 2WD's. In real sandy areas, roads are usually best in wet weather, because it makes the sand firm so you don't mire down. If a road seems well-used, you can bet there is one or more families living out there somewhere. Always take emergency-type gear and equipment in your car as recommended in the front of this book.

Water Holes Canyon, Navajo Nation, Arizona

Location and Access Water Holes Canyon is located about 10 kms south of the northern Arizona town of Page. The canyon is crossed by Highway 89 which runs between Flagstaff and Page. This drainage flows north and northwest into Glen Canyon and the Colorado River not far upstream from Lee's Ferry. This canyon has very easy access. To get there, simply drive along Highway 89 south of Page to mile post 542. This marker is next to the steel bridge spanning the Water Holes Canyon Gorge. Park on either side of the road north of the bridge. This canyon is yet another in the vicinity of Page which has some very narrow and fotogenic slots. It's also located on the Navajo Nation.

Trail or Route Conditions From where you park on the highway next to the bridge, simply climb over the fence and walk east over slickrock about 125-150 meters. There you'll be able to climb down into the upper part of the gorge, which is only about 15-20 meters deep at that point. From there walk up-canyon and into the best part of the drainage. From the highway bridge to the narrows it's only about one km. As the name implies, there will usually be some wading in water holes along the way. The best part is only about 150 meters in length. At the upper end of the slot is a dryfall, with normally a deep pool of water below. On the wall is carved the initials "TJH". You can't climb out at that point, but you can back-track a ways and get out of the bottom and onto a bench and continue walking up-canyon. The author doesn't know what's beyond the power lines, but the drainage is uninteresting at that point. Going down-canyon from the bridge are several steep entry points, most of which seem to be on the southwest side of the gorge. The canyon is deeper below the bridge, but doesn't seem to have many good tight slots.

Elevations Highway bridge and trailhead, 1400 meters; the Colorado River, about 950; and the bottom of the drainage in the far upper end, 1500 meters.

Hike Length and Time Needed It's only about one km from the car-park to the narrow slot going up-canyon. You can see this part in about an hour. Explorers may find something interesting further up-canyon. Going down-canyon to the Glen Canyon dropoff may take 1-3 hours, depending on what you do.

Water Have a good supply in your car, and take your own into the canyon.

Maps USGS or BLM map Glen Canyon Dam(1:100,000), or Leche-e Rock and Lees Ferry(1:62,500).

Main Attractions Easy access with fotogenic Navajo Sandstone narrows.

Ideal Time to Hike Spring or fall, but summers aren't so hot in the canyon narrows.

Hiking Boots Wading boots or shoes, except after a long dry spell. The author was there a day after a rainstorm and found the water holes knee-deep.

Author's Experience The author spent an hour in the upper narrows to as far as the power lines, then another hour rim-walking and exploring the lower end of the gorge.

Water Holes Canyon is one of the most fotogenic slots around.

Map 69, Water Holes Canyon, Navajo Nation, Arizona

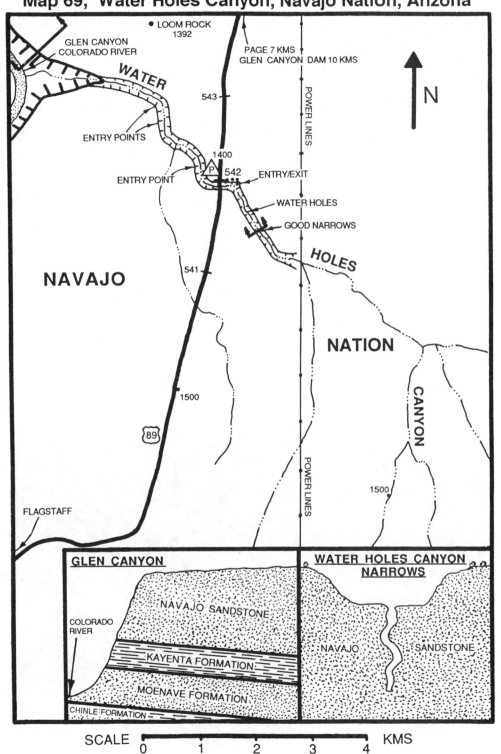

LOOM ROCK
1392

GLEN CANYON
COLORADO RIVER

WATER

PAGE 7 KMS
GLEN CANYON DAM 10 KMS

POWER LINES

N

543

ENTRY POINTS

1400

ENTRY POINT

P

542 ENTRY/EXIT

WATER HOLES

GOOD NARROWS

HOLES

NAVAJO

541

NATION

CANYON

1500

89

POWER LINES

1500

FLAGSTAFF

GLEN CANYON

WATER HOLES CANYON
NARROWS

NAVAJO SANDSTONE

COLORADO
RIVER

NAVAJO   SANDSTONE

KAYENTA FORMATION

MOENAVE FORMATION

CHINLE FORMATION

SCALE
0   1   2   3   4   KMS

169

Antelope Canyon, Navajo Nation, Arizona

Location and Access Antelope Canyon is located in northern Arizona just south of Lake Powell, west & south of the Navajo Power Plant, and immediately east of Page. Getting there is easy. Drive east out of Page 2 or 3 kms on State Highway 98, to just beyond mile post 299. You can park on the highway if visiting The Corkscrew part immediately north of the road; or turn right about 75 meters east of m.p. 299 and go through a gate. This gate is now locked with a member of the Begay family there to collect a $5 fee from each person going up-canyon to visit The Crack and other fine narrows above. There's a big sign at the gate stating it's open from 9 am to 4 pm. There should be someone there to collect the fee; if not, there will be a telefone number on the sign you can call. With a fone call, someone is supposed to come down, collect the fee, and open the gate.

Trail or Route Conditions To enter The Corkscrew part of the canyon, walk north down the wash from the highway bridge. This slot begins 250 meters from the highway and dives down fast. Several people have told the author there is a Navajo guide living in a trailer immediately east of this slot who charges money for taking people through. Also, as of March, 1994, the bolts and snap links in the walls had been taken out, apparently making it impossible to go down, even with ropes--unless you put your own hardware in, which would allow the last person to rappel down. Apparently that guide now uses ladders to take tourists through(?). You can also enter the bottom part from another short wash just to the west. If those steps are still good, you might get into the lower end a ways. To get to the upper parts of the canyon south of the highway, simply walk(or go in a 4WD) up the dry creek bed of Antelope to visit more narrows. Before arriving at The Crack, there are two other short, but very good slots along the way. These are not as deep and dark as The Crack, therefore more easily fotographed. The part called The Crack is only about 200 meters long, but it's about the darkest slot the author has seen. You walk right through it to reach more tight narrows further up-canyon(which this writer has not yet seen). Explorers can look for still more. Be sure and take higher speed film and a tripod when visiting either The Crack or Corkscrew.

Elevations Page, 1325 meters; the trailhead, 1260; The Crack, about 1350 meters.

Hike Length and Time Needed You'll need only an hour or two to see The Corkscrew. The walk from the trailhead to The Crack is about 5 kms along the sandy dry wash. If you have a 4WD, you can drive to The Crack. To visit all these known slots of Antelope Canyon will take you all day, but much of your time will be spent just taking pictures.

Water Take you're own and always have some in your car.

Maps USGS or BLM map Glen Canyon Dam(1:100,000), or Leche-e Rock(1:62,500).

Main Attractions Probably the most fotogenic, fotographed and famous slot canyon around.

Ideal Time to Hike Spring or fall would be best, but since you'll be in some deep, dark and cool narrows, it's fine in summer as well.

Hiking Boots Any comfortable walking shoes. Use wading shoes right after a storm.

Author's Experience He came up from the lake and visited The Corkscrew once, then later visited the narrows south of the highway. That trip lasted about 3 hours, round-trip, but he got a ride part way up.

This picture shows one of five narrow parts of Antelope Canyon.

Map 70, Antelope Canyon, Navajo Nation, Arizona

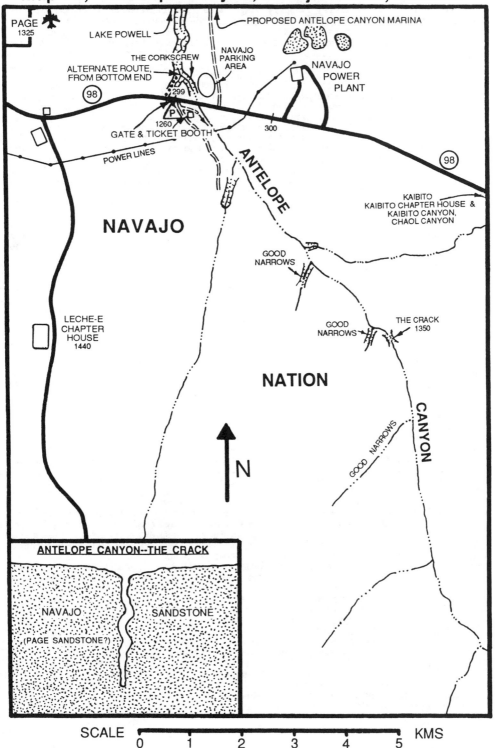

PAGE 1325

PROPOSED ANTELOPE CANYON MARINA

LAKE POWELL

THE CORKSCREW

NAVAJO PARKING AREA

NAVAJO POWER PLANT

ALTERNATE ROUTE, FROM BOTTOM END

98

299

1260

P

GATE & TICKET BOOTH

300

POWER LINES

98

ANTELOPE

KAIBITO
KAIBITO CHAPTER HOUSE &
KAIBITO CANYON,
CHAOL CANYON

NAVAJO

GOOD NARROWS

GOOD NARROWS

THE CRACK
1350

LECHE-E CHAPTER HOUSE
1440

NATION

CANYON

GOOD NARROWS

N

ANTELOPE CANYON--THE CRACK

NAVAJO

SANDSTONE

(PAGE SANDSTONE?)

SCALE   KMS
0   1   2   3   4   5

171

Lower Chaol Canyon & Middle Kaibito Creek, Peach Wash & Butterfly Canyon, Navajo Nation, Arizona

Location and Access This map covers most of Chaol Canyon from Lake Powell south to near where large power lines cross the canyon. It also shows Peach Wash and Butterfly Canyon. The upper part of Chaol, including the very good narrows of Kaibito Creek and Starting Water Wash, are covered on the next two maps. This mapped area is about a dozen to 20 kms southeast of Page on Navajo Nation lands, so don't forget your tribal hiking permit.

One way to enter the lower Chaol Gorge is to drive Highway 98 southeast out of Page to between mile posts 307 and 308, then turn northeast and drive a good dirt road to near the home of Owen Yazzie. From that area, turn southeast and drive a less-used road to the head of Butterfly Canyon. Cars should be parked on the rim. From there a 4WD road runs along the rim of Butterfly for about 3 kms, then it turns into a livestock trail. From that point, you can walk into lower Chaol or Butterfly Canyon.

A second way to get near this same area, but without passing anyones home, would be to drive along Highway 98 to a point near mile post 310, then turn northeast, pass under the power lines and continue to the end of the road, which is on the Carmel Rim just above the 4WD parking spot mentioned above. Any higher clearance car driven with care can make it to or near that point. From there you can route-find down off the rim and into the same two canyons. To reach the upper part of Peach Wash, continue to mile post 312 and turn northeast again. Follow that *very sandy 4WD road(leave cars at the highway!)* down to a garden, then parallel the power lines southeast until you come to a locked gate.

Trail or Route Conditions The main trail going into lower Chaol Canyon is a good one and used often by sheep & goats to get down to the year-round water of Kaibito Creek. From there you can walk downstream and up Navajo Canyon if you like; or head up Chaol Canyon. Not far from the trail is one of the most unusual waterfalls or cascades you'll ever see. It's called Lower Kaibito Falls, and has been created because of a limestone lens within the Navajo Sandstone. Just above and just below the falls are some good petroglyphs as shown on the map. Further up-canyon are a couple of good springs, and nearby, Anasazi ruins. The author has been up to the middle part of Peach Wash, which also has a small stream of water.

In the fall of 1993 and late winter of 1994, the author explored upper Starting Water Wash, plus a route down into the bottom end of Starting Water, then went up Kaibito Creek to another very fine waterfall, which we'll call Middle Kaibito Falls. It's about 5 kms below the power lines and blocks all traffic going upstream. About 1 1/2 kms above where Starting Water meets Kaibito, is a constructed sheep & goat trail leading out to the west rim. *Beware, the west-side roads leading to that trail are very sandy!* It seems there are no impediments in Chaol Canyon between Peach and Starting Water Canyons. Boaters interested in exploring by boat should see another book by the author titled, *Boater's Guide to Lake Powell.*

One of the best little slots around is down inside Butterfly Canyon. Walk from either of the nearby trailheads down to the bottom via the entry/exit route shown. Once down in, you can walk up or down-canyon. Be prepared for some potholes and wading--maybe even a little swimming--if you're there right after a good rain. Be prepared to crawl over and under a number of chokestones too. At one point you'll have to use a short rope to get down a 5-meter dropoff; or better still just chimney down as shown in the foto. It's easy in the one meter-wide slot. Down near the end of the canyon is a long wading or perhaps swimming pool.

The upper end of Peach Wash has some interesting but short narrows. There are several ways to get down off the top rim to a one km-long slot section below. One is via an old cattle trail as shown. Once down in, be prepared for wading. You'll need long ropes to get all the way down to the main canyon bottom.

Elevations The 4WD & car-park southeast of the Yazzie home, about 1550 and 1490 meters; Lake Powell's high water mark, 1128, Lower Kaibito Falls, 1135; Middle Kaibito Falls, 1320 meters.

Hike Length and Time Needed It's about 5 kms from either of the trailheads southeast of Yazzie's home, to the bottom of Chaol Canyon. From there it's another km up to the Lower Kaibito Falls. Getting to the falls will take less than 2 hours. Most people should be able to reach the mouth of Peach Wash and return in one easy day-hike. If you plan to go further up-canyon, take a pack and stay 2 or 3 days. For Butterfly Canyon, you could see parts of it in half a day, but if you plan to see it all, might as well take a lunch and make it a full day. Getting into the upper end of Peach Wash and back to your car shouldn't take more than about half a day, but it'll depend on where you park and if you've got a mtn. bike or not.

Water Lower Chaol Canyon, with a nice stream of water, appears to have some sheep & goats in it year-round, so take water from one of the springs or seeps shown on the map if you can. In Butterfly and upper Peach Wash there will be pothole water at times, but best to take your own.

Maps USGS or BLM map Glen Canyon Dam(1:100,000), or Leche-e Rock and Navajo Creek(1:62,500).

Main Attractions Waterfalls, petroglyphs, Anasazi ruins, great scenery, narrow canyons & solitude.

Ideal Time to Hike Spring or fall are best in Chaol, but in summer you can cool off in the creek. Butterfly can be hiked in summer, because you'll be down in a cool slot canyon.

Hiking Boots Kaibito Creek is a sizable stream which begins flowing 1 1/2 kms below the power lines, so take wading boots or shoes. Same in upper Peach Wash and Butterfly as well.

Author's Experience The author went up Navajo Canyon inlet twice in a boat to see the lower end of this canyon. He's also walked down into Chaol from the trailheads near the Yazzie home, as well as looked down into it from the power lines to the south. He has also been down most of the length of Butterfly, but swimming through big potholes on a cold November day wasn't that appealing. He has been in the narrows of upper Peach Wash, as well as lower Starting Water Wash, twice; once up to Middle Kaibito Falls.

Map 71, Lower Chaol Canyon & Middle Kaibito Creek, Peach Wash & Butterfly Canyon, Navajo Nation, Arizona

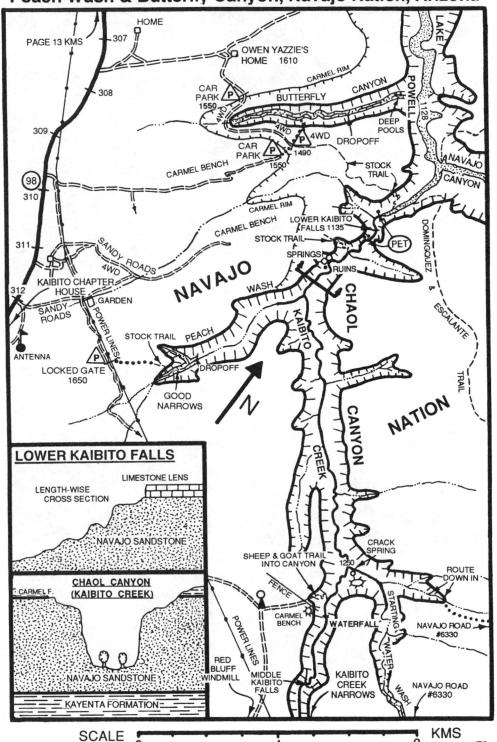

HOME

OWEN YAZZIE'S HOME 1610

PAGE 13 KMS

307

308

309

98

310

311

312

CAR PARK 1550

BUTTERFLY

CARMEL RIM

CANYON

LAKE POWELL

1128

DEEP POOLS

4WD

CAR PARK 1550

P

4WD 1490

DROPOFF

NAVAJO CANYON

CARMEL BENCH

STOCK TRAIL

CARMEL RIM

CARMEL BENCH

LOWER KAIBITO FALLS 1135

STOCK TRAIL

PET

SPRINGS

RUINS

DOMINGUEZ & ESCALANTE TRAIL

SANDY ROADS

4WD

KAIBITO CHAPTER HOUSE

GARDEN

SANDY ROADS

POWER LINES

ANTENNA

P

LOCKED GATE 1650

STOCK TRAIL

PEACH

DROPOFF

GOOD NARROWS

WASH

NAVAJO

KAIBITO

CHAOL

CANYON

NATION

N

KAIBITO CREEK

LOWER KAIBITO FALLS

LIMESTONE LENS

LENGTH-WISE CROSS SECTION

NAVAJO SANDSTONE

CRACK SPRING

SHEEP & GOAT TRAIL INTO CANYON

1250

ROUTE DOWN IN

NAVAJO ROAD #6330

CHAOL CANYON (KAIBITO CREEK)

CARMEL F.

FENCE

CARMEL BENCH

STARTING

WATERFALL

WATER

WASH

NAVAJO ROAD #6330

NAVAJO SANDSTONE

POWER LINES

RED BLUFF WINDMILL

MIDDLE KAIBITO FALLS

KAIBITO CREEK NARROWS

KAYENTA FORMATION

SCALE

0

4

8

KMS

One of the most unusual waterfalls around is in the lower end of Chaol Canyon. This is Lower Kaibito Falls, which has been created by a limestone lens within the Navajo Sandstone.

The lower end of Starting Water Wash, not far from where it enters Chaol Canyon.

Middle Kaibito Falls, one of the best around.

Getting into or out of one part of Butterfly Canyon requires some "chimney work". Take a short rope just in case.

Upper Kaibito Creek, Navajo Nation, Arizona

Location and Access This map features the narrows of upper Kaibito Creek located about 50 kms southeast of Page. The part of Kaibito Creek of most interest is about a dozen kms north of Kaibito Chapter House. To get there, drive along State Highway 98 between Page and Kayenta on Navajo Nation lands. Before proceeding to the canyon, first stop and talk to people at the Kaibito Chapter House for the latest on road conditions and any other information(1-602-673-3408). Ask for Philip Brown. From there, drive to mile post 334 and turn north on to Navajo Road #6330. This is on the east side of the canyon and is graded all the way to near Tse Esgizii Rock. A good place to park and/or camp is at the water trough shown. You can dip some water out to wash(and probably drink) and park under a tree. You can also continue north on a side road and park at a pile of rocks--marked 1740 on the map; or at a stock pond dam on the East Fork of Kaibito Creek marked 1680 meters. All roads on the west side of the canyon are very sandy.

Trail or Route Conditions From the water trough, walk west and down a little, then north along the shallow canyon rim to an entry point about 1 1/2 kms away. From the pile of rocks it's only about 500 meters to the same entry point. The canyon at that point starts to narrow, and for about one km it's one of the tightest, darkest and best all-around slots on the Plateau. It's as good as Antelope Canyon--maybe better. *Have a very good weather forecast before doing this hike. If a flash flood catches you there, you'll be a goner!* The lower end has two minor dropoffs, then it's so dark you'll need a flashlight, and ropes. *You'll need several long ropes to make it all the way through, because there are several big dropoffs south of the power lines, then another big dropoff or dryfall of maybe 50 meters just north of the power lines. North of that is the Middle Kaibito Falls of 20 meters.* To get out return the same way, as it seems practically impossible get all the way into Middle Kaibito. In East Fork, there's another very good twisting narrow slot in the bottom end beginning just below the 2nd stock pond dam. It too has a big dropoff going into Kaibito. You can park and/or camp there under some cedar trees. That side road is a little sandy and may be bad in *dry weather.* If so, lower the pressure in your tires to help get through(?).

Elevations Main trailheads, 1740 meters; Kaibito Chapter House, 1750 meters.

Hike Length and Time Needed From the car-park at the rocks, it's only about 2 kms to the dropoffs in the main narrows. You can likely do this in a couple of hours--more if it's your first look at a great slot canyon! A quick trip down East Fork to the dryfall is about one km, and can be done in an hour or two, round-trip.

Water At the one trailhead water trough, but always carry water in your car.

Maps USGS or BLM map Glen Canyon Dam(1:100,000). Ask if maps at 1:24,000 or 1:62,500 scale are completed yet? If so, they will be the best ones.

Main Attractions One of the best short slot canyon hikes on the Colorado Plateau.

Ideal Time to Hike Spring, summer or fall. It's cool in the bottom, even in the hottest weather.

Hiking Boots Use wading boots or shoes as there will surely be some water holes.

Author's Experience The author parked at the water trough and wandered a bit before entering the canyon at the southern entry point shown. He walked into the narrows halfway, but came back out because of the weather forecast--40% chance of rain around Page! Round-trip was 3 1/2 hours. That afternoon he spent 1 1/2 hours in the East Fork, then drove to the sand trap along the power lines road and parked. He hiked to the Chaol Canyon Overlook(twice).

One of the best slot canyons around is in the bottom of upper Kaibito Creek.

Map 72, Upper Kaibito Creek, Navajo Nation, Arizona

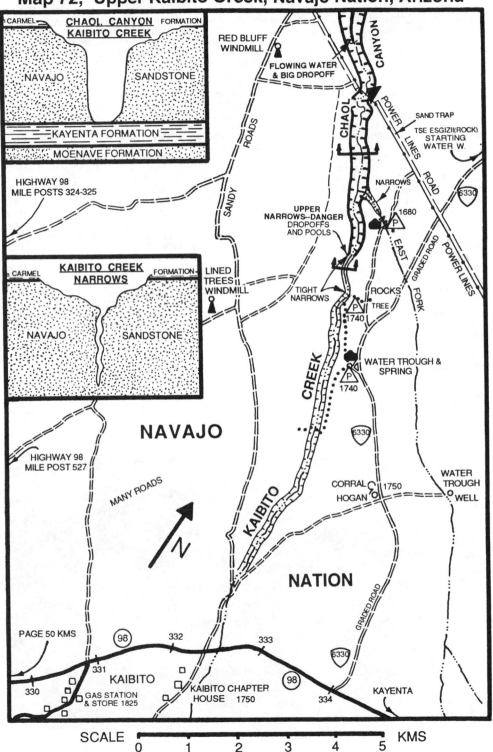

CHAOL CANYON KAIBITO CREEK

CARMEL FORMATION

NAVAJO SANDSTONE

KAYENTA FORMATION

MOENAVE FORMATION

HIGHWAY 98 MILE POSTS 324-325

KAIBITO CREEK NARROWS

CARMEL FORMATION

NAVAJO SANDSTONE

RED BLUFF WINDMILL

FLOWING WATER & BIG DROPOFF

CHAOL CANYON

SAND TRAP

TSE ESGIZII(ROCK) STARTING WATER W.

NARROWS

POWER LINES ROAD

6330

SANDY ROADS

UPPER NARROWS--DANGER DROPOFFS AND POOLS

1680
P

LINED TREES WINDMILL

TIGHT NARROWS

ROCKS

TREE

EAST FORK

GRADED ROAD

POWER LINES

P
1740

WATER TROUGH & SPRING

P
1740

KAIBITO CREEK

NAVAJO

HIGHWAY 98 MILE POST 527

MANY ROADS

N

6330

CORRAL

HOGAN

1750

WATER TROUGH

WELL

NATION

GRADED ROAD

PAGE 50 KMS

98

332

333

6330

331

KAIBITO

98

330

GAS STATION & STORE 1825

KAIBITO CHAPTER HOUSE 1750

334

KAYENTA

SCALE 0 1 2 3 4 5 KMS

177

Starting Water Wash, Navajo Nation, Arizona

Location and Access Starting Water Wash is a major tributary to the upper end of Kaibito Creek. It's carved out of the same rock as Antelope and Kaibito Canyons, therefore is one of the very best true slot canyons around. Drive east and southeast out of Page on Highway 98 until you come to Kaibito. Before leaving Kaibito, fill your fuel tank, water bottles and check in at the chapter house just downhill from the gas station & store(see the Upper Kaibito Creek map). Those people can update you on road conditions, etc. From Kaibito, continue east on Highway 98 until you reach mile post 334. At that point turn north on Navajo Road 6330. This is a well-maintained and graded road which heads north toward Tse Esgizii Rock, and is the same road you take if going to the narrows in Upper Kaibito. Drive along this main road for about 13 or 14 kms until you come to a sizable drainage which is the upper part of Starting Water Wash. Park near the bridge. Or you can make your way to one of the two car-parks marked 1690 meters, which puts you close to the route down into the lower end of the canyon. Side roads here are sandy, and *may be soft in dry weather!*

Trail or Route Conditions From the bridge at 1600 meters, simply walk down the dry creek bed. After about half a km, you'll enter the upper part of the narrows. At first the slot is rather shallow, but the further you go, the deeper it gets. It's narrow right from the beginning, often times barely shoulder width. Every so often there will be a place to exit the slot until you reach the point on the map called "last exit". After that it's really narrow and deep and perhaps with some wading in potholes. The canyon eventually opens up, but there will still be a slot at the bottom as shown on the geology cross-section. To get all the way down to Kaibito Creek via this main drainage of Starting Water, *you will need several ropes, a flashlight, and experienced climbers.* Hikers-Climbers have been all the way through this slot, so you may see belay anchors and maybe even a rope. If this isn't your cup of tea, you can walk cross-country from the other car-parks and enter the East Fork as shown. When going down the steep part, stay on the scree slopes to the right to avoid dryfalls. In the bottom there's running water and trees all the way to Kaibito Creek. Walk up Kaibito Ck. 5 kms to see Middle Kaibito Falls. There's also a sheep & goat trail out to the west side, as shown on the Map #71. *Have a good weather forecast before starting down this canyon!*

Elevations Bridge over Starting Water, 1600 meters; Stock Pond, 1690; Crack Spring, 1250 meters.

Hike Length and Time Needed It's about 6 kms from the bridge to the first dryfalls in the main drainage, and would be a short day-hike. To walk down the East Fork route into the bottom end and hike up Kaibito Ck. to the waterfall and back, will take all of one day.

Water Take water in your car and in your pack. There's good water in lower Starting Water Wash and Kaibito Creek.

Maps USGS or BLM map Glen Canyon Dam(1:100,000).

Main Attractions One of the best slot canyons anywhere.

Ideal Time to Hike Spring, summer or fall. It's cool in the bottom, even in summer.

Hiking Boots Wading boots or shoes.

Author's Experience He walked down the slot just past the "last exit" to find some pools and dropoffs, then got out and walked along the rim to the confluence of Kaibito. After that he rim-walked back to his car in a total time of 5 hours. Another time he explored the route down into East Fork as far as Kaibito Creek, and returned in less than 7 hours round-trip. On a third trip, he went up to the Middle Kaibito Falls, and to the west rim via the sheep & goat trail. That took 9 hours, round-trip.

The narrows of Starting Water Wash, above the big dropoffs.

Map 73, Starting Water Wash, Navajo Nation, Arizona

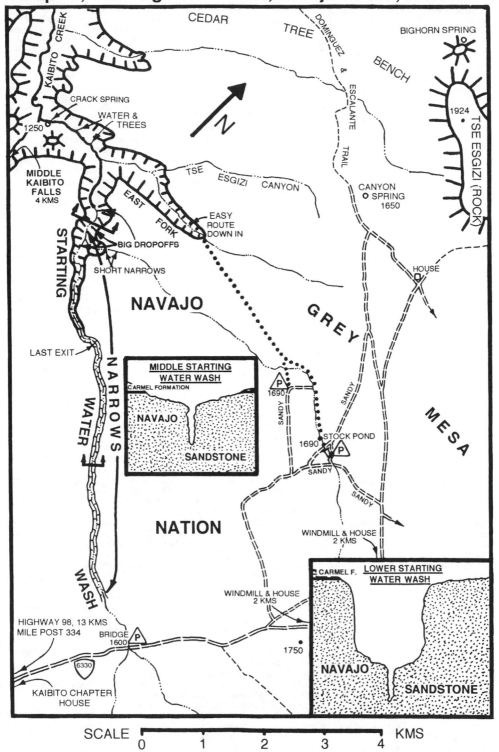

CEDAR TREE BENCH

BIGHORN SPRING

KAIBITO CREEK

CRACK SPRING

WATER & TREES

1250

MIDDLE KAIBITO FALLS 4 KMS

EAST FORK

BIG DROPOFFS

SHORT NARROWS

STARTING

NAVAJO

LAST EXIT

NARROWS

WATER

WASH

HIGHWAY 98, 13 KMS
MILE POST 334

BRIDGE
1600

6330

KAIBITO CHAPTER HOUSE

TSE ESGIZI CANYON

DOMINGUEZ & ESCALANTE TRAIL

1924

TSE ESGIZI (ROCK)

EASY ROUTE DOWN IN

CANYON SPRING 1650

HOUSE

GREY

MESA

P 1690

SANDY

Stock Pond
P 1690

SANDY

SANDY

WINDMILL & HOUSE 2 KMS

WINDMILL & HOUSE 2 KMS

1750

NATION

MIDDLE STARTING WATER WASH
CARMEL FORMATION

NAVAJO

SANDSTONE

LOWER STARTING WATER WASH
CARMEL F.

NAVAJO

SANDSTONE

SCALE
0   1   2   3   4   KMS

West Canyon, Navajo Nation, Arizona

Location and Access Featured on this map is one of the best slot canyon hikes on the Colorado Plateau. It's West Canyon, which is southwest of Navajo Mtn. and just south of Lake Powell. This hike is already covered in another book by the author called, *Boater's Guide to Lake Powell*, but there are ways to get in by land as well. Getting to this region by car and walking is a real adventure however, so it's recommend you get there by boat. *For those coming overland, the upper or southern end of this canyon is recommended for the experienced and adventurous hiker only.*

To start, drive along Highway 98 about halfway between Page and Kayenta. Near mile posts 337, 338 and 339 are two well-used roads heading north. They come together near the power lines, as shown on the map. From there continue north on the one main road. About halfway to Navajo Creek will be a fork--the main road runs left and toward Tse Esgizii Rock and Starting Water Wash, but don't take that one. Drive the one heading north past the two houses, as shown. There are many side tracks all along, but continue on the most-used road. Further on, you'll drop down off the clay beds of the Carmel Formation. Before reaching Navajo Creek there will be a couple of steep and rocky places, but ordinary 2WD vans and pickups make it down and back up all the time. Any HCV can make it to the bottom and back out under normal conditions, but some may want to use a mtn. bike for the last little ways. The author did it in his VW Rabbit Diesel after walking & clearing the way of rocks first. From Navajo Creek, make your way northwestward on the road running toward John Yazzilow's home, but stay along Jayi Creek until you reach the second HCV or 4WD parking place. *This is no place to get stuck or have a break-down, so be well-prepared.* Have a good running vehicle, because it's about 37 kms back to the highway. If you're doing this one via the overland route, be sure to have a Navajo hiking permit.

Trail or Route Conditions From your mtn. bike or vehicle, climb upon the bench to the left of the North Fork of Jayi Canyon, then head up-canyon to the northwest. Sometimes it's easiest to stay in the dry creek bed; other times it's easier to stay on the bench to the west. The Glen Canyon Dam map listed below shows a trail up this shallow drainage, but you won't see it until you reach the Carmel Bench near Honish Oosh Atiin. However, you won't go quite that far west. Instead, you'll want to veer toward the gap or divide between H. O. Atiin and Octagon Butte. Once there, drop down into the head of West Canyon. The author so far hasn't been down into this upper section, but he has gotten some of this information from Eberhard Schmilinsky. To bypass the really tough upper narrows, drop in at the first sheep & goat trail on the west side.

Once you get to the head of upper narrows, you'll have two choices of routes. You could go right down the main canyon but here's a list of things you'll need or do first: *Take several short ropes, no big backpacks, and at least 3 people. The reasons are: there are 7 or 8 dropoffs and potholes, it is ever changing, and you'll have to swim and may get a little hypothermia. If water levels aren't just right, one person alone couldn't get out! It's so dark in places you almost need a flashlight. This section is about 800 meters long. At the end of this is one big swimming pool, marked dropoffs on the map.* This upper part was done in July of 1992 and again in 1993 by a couple of Page people along with Eberhard Schmilinsky, who rented a helicopter to get there. Eberhard also thought you could get down into the canyon below the dropoffs, by going around on the east side and entering the canyon opposite the first sheep & goat trail. The author hasn't tried this route, but he has come up from the lake and exited the canyon on all three sheep & goat trails shown. Below the dropoffs, are some good narrows, then an area with cottonwood trees and springs(at the first trail out). Below that are more real good deep narrows with a couple of huge overhangs. Further down are more open spaces, a trail out to the east which you can use to reach the top of Cummings Mesa, a log hogan and a trail out to the west. Below that are more waterfalls, about half a dozen in all. In September of 1991, there was a nice little stream, as shown, but in July of 1992, it was mostly dry(?). There should always be some water around, so go prepared to wade or swim deep pools. Not far above the high water mark of Lake Powell, will be a spring and a swimming pool, then a sandy walk down the stream to the lake. You can bypass the swimming pool on the west if you like.

Elevations Navajo Creek, 1325 meters; divide between Jayi and West Canyons, 1525; and the high water mark of Lake Powell, 1128 meters.

Hike Length and Time Needed If you day-hike, you'll likely just make it to the upper end of the canyon, but wouldn't have time to see much. Strong hikers with a pack could get in and see most of the good parts and return in 3 days, maybe 4, if the upper 800 meter-long slot is taken. From a boat on the lake, you can walk all the way up to the dropoffs and return in one long day.

Water Have plenty in your car, but there will always be water in Navajo Creek and Jayi Canyon. Also, there will always be some water near the first sheep & goat trail, as well as in potholes. After rains, you'll find pothole water everywhere, but you'd better start with some with you, especially if it's hot weather!

Maps USGS or BLM maps Kayenta(for the road in) and Glen Canyon Dam for West Canyon(1:100,000), and/or Navajo Creek(1:62,500).

Main Attractions A top ten slot canyon, plus great scenery in the Navajo Creek country.

Ideal Time to Hike Spring or fall would be best for most of the hike, but if you take ropes and go down in from the upper end, then the dry and warm month of June would be best.

Hiking Boots Something good for a sandy hike and a wading boot or shoe as well.

Author's Experience The author has been there on 4 trips from the lake, then finally went in with his VW Rabbit. He parked halfway down the hill to Navajo Creek, then used a mtn. bike to reach the HCV or 4WD park. He walked to the place called Lone Rock, and returned to his car in a little over 8 hours. On a second trip by land, he parked right on Navajo Creek and explored nearby canyons.

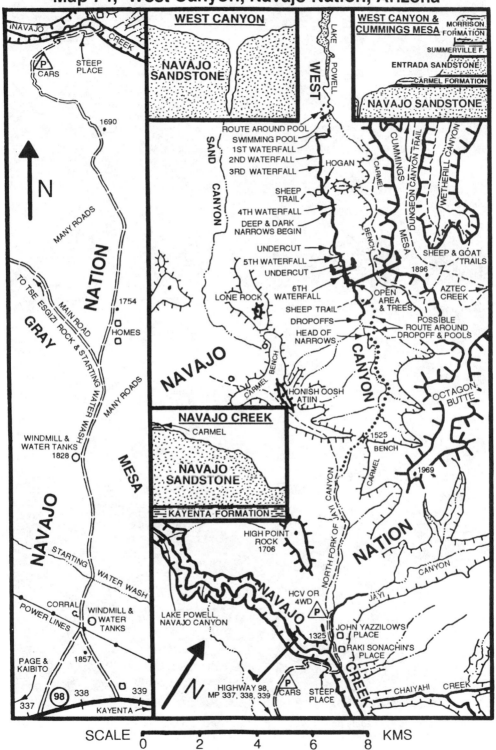

WEST CANYON

NAVAJO SANDSTONE

WEST CANYON & CUMMINGS MESA

MORRISON FORMATION
SUMMERVILLE F.
ENTRADA SANDSTONE
CARMEL FORMATION
NAVAJO SANDSTONE

NAVAJO CREEK

CARMEL
NAVAJO SANDSTONE
KAYENTA FORMATION

NAVAJO

CREEK

P CARS
STEEP PLACE
1690
MANY ROADS

N

TO TSE ESGIZI ROCK & STARTING WATER WASH
MAIN ROAD

GRAY

NAVAJO NATION

MANY ROADS

1754
HOMES

WINDMILL & WATER TANKS
1828

NAVAJO MESA

NAVAJO

STARTING WATER WASH

POWER LINES
CORRAL
WINDMILL & WATER TANKS
1857

PAGE & KAIBITO
98
338
339
337
KAYENTA

SAND CANYON

ROUTE AROUND POOL
SWIMMING POOL
1ST WATERFALL
2ND WATERFALL
3RD WATERFALL
HOGAN
SHEEP TRAIL
4TH WATERFALL
DEEP & DARK NARROWS BEGIN
UNDERCUT
5TH WATERFALL
UNDERCUT
6TH WATERFALL
SHEEP TRAIL
DROPOFFS
HEAD OF NARROWS
LONE ROCK

BENCH
CARMEL BENCH
HONISH OOSH ATIIN

LAKE POWELL

WEST

CUMMINGS MESA
CARMEL
DUNGEON CANYON TRAIL
WETHERILL CANYON

SHEEP & GOAT TRAILS
1896

OPEN AREA & TREES
AZTEC CREEK

POSSIBLE ROUTE AROUND DROPOFF & POOLS

CANYON

1525
BENCH
CARMEL BENCH
1969

OCTAGON BUTTE

NATION

NORTH FORK OF JAYI CANYON

HIGH POINT ROCK
1706

HCV OR 4WD
P

JAYI

CANYON

LAKE POWELL, NAVAJO CANYON

N

HIGHWAY 98, MP 337, 338, 339

P CARS

1325
STEEP PLACE
JOHN YAZZILOWS PLACE
RAKI SONACHIN'S PLACE

CHAIYAHI CREEK

SCALE 0 2 4 6 8 KMS

Typical scene in the middle part of West Canyon. It's best to approach this part from Lake Powell. Notice how small the hiker is at the bottom of the picture.

One of several waterfalls in West Canyon.

Another waterfall in the middle part of West Canyon.

Aztec Creek and Cummings Mesa, Navajo Nation, Arizona-Utah

Location and Access Aztec Creek and Cummings Mesa are located just west of Navajo Mountain. The creek begins in Arizona and flows north into Utah and Lake Powell. All the land here is part of the Navajo Nation. To reach this area, first drive along Highway 98, the road connecting Page and Kayenta. At the turnoff to Inscription House, between mile posts 349 and 350, drive north on the partly-paved Navajo Mtn. Road. After about 58 kms you'll come to a junction; to the right is Navajo Mtn. Mission, to the left the road to the Rainbow Lodge ruins and the trail to Rainbow Bridge. Take the road to the left or west, and drive about 8 or 9 kms to the dome rocks just south of a couple of homes which are near a large steel water tank. At that point turn south and drive another 3 or 4 kms to where you'll be on or near the Chaiyahi Rim and at the car-park at 1850 meters. To find the Cummings Mesa Trail, drive west from the dome rocks to the car park just west of Round Rock. The author walked the last part first, then drove there in his VW rabbit--but it's very sandy in places! *So beware of deep, dry sand!*

Trail or Route Conditions There are likely several trails or routes from the Chaiyahi Rim into the Aztec Creek drainage, one being from the 1850 meter car-park. Near the top, there's one big dropoff, but you can skirt around it to the left and get into a steep gully. Then it's easy walking in the canyon bottom. There's probably running water from there all the way to the lake. From Round Rock, a constructed trail runs down into the canyon, then it seems to disappear, but don't worry. Before reaching the bottom, look to the west and you can see a big slide area on the east face of Cummings. So although the trail disappears in the bottom you can walk toward the only possible route up to the mesa. Head that way, and you'll come to the trail on the other side. This sheep & goat trail was surely built by CCC's in the 1930's. See the author's book, *Boater's Guide to Lake Powell*, which shows 4 trails to the top of Cummings Mesa(page 245, 1991 edition). For those who want a long loop-hike, combine Aztec Creek, Cliff Canyon, and maybe Rainbow Bridge, with the trail coming up from the bridge to the Rainbow Lodge ruins(see the Rainbow Bridge hike for more information).

Elevations Trailheads are between 1750 and 1921 meters, while the lake is 1128 meters.

Hike Length and Time Needed From Round Rock to Cummings Mesa and back took the author 6 1/2 hours. From the 1850 meter car-park, down Aztec, up Cliff Canyon, and by trail and road back to one's car, is about 48 kms. Most people could do this in 3 days. To do this same route, but adding in the side trip to the bridge, it would be about 61 kms, or about 4 days.

Water Aztec Creek seems to flow all the way to the lake, but the author hasn't seen about 5 kms in the middle part which is near the Utah-Arizona state line.

Maps USGS or BLM maps Navajo Mountain and Kayenta(1:100,000), or Navajo Mtn.(1:62,500) and Chaiyahi Flat(1:24,000).

Main Attractions Wild solitude, interesting narrows, and a great hike to a tall mesa.

Ideal Time to Hike Spring or fall, or even warm dry spells in winter. Summers are hot.

Hiking Boots Dry weather boots or shoes for Cummings Mesa; waders for hikes along the creek.

Author's Experience The author has been into upper Aztec, has hiked up from the lake above(south) Cliff Canyon to the second waterfall, the entire trail to Rainbow Bridge from the lodge ruins, plus up all the trails to Cummings Mesa.

Looking north from the top of the east side of Cummings Mesa. The trail comes up the from the lower right-hand side of the rock slide shown.

Map 75, Aztec Creek and Cummings Mesa, Navajo Nation, Arizona-Utah

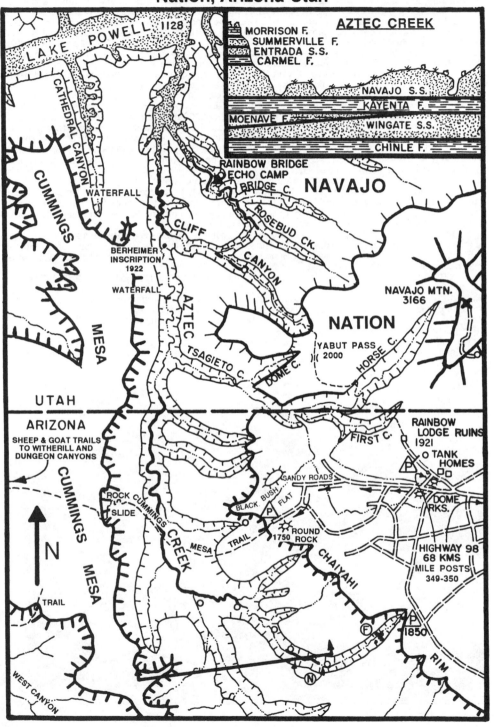

LAKE POWELL 1128

CATHEDRAL CANYON

CUMMINGS MESA

WATERFALL

BERHEIMER INSCRIPTION 1922

WATERFALL

AZTEC CANYON

CLIFF CANYON

TSAGIETO C.

DOME C.

AZTEC CREEK

MORRISON F.
SUMMERVILLE F.
ENTRADA S.S.
CARMEL F.

NAVAJO S.S.

KAYENTA F.

MOENAVE F.

WINGATE S.S.

CHINLE F.

RAINBOW BRIDGE
ECHO CAMP
BRIDGE C.

ROSEBUD CK.

NAVAJO

NAVAJO MTN. 3166

NATION

YABUT PASS 2000

HORSE C.

UTAH

ARIZONA

SHEEP & GOAT TRAILS TO WITHERILL AND DUNGEON CANYONS

CUMMINGS MESA

ROCK SLIDE

CUMMINGS CREEK

MESA TRAIL

BLACK BUSH FLAT

SANDY ROADS

ROUND ROCK 1750

FIRST C.

RAINBOW LODGE RUINS 1921

TANK HOMES

DOME RKS.

HIGHWAY 98 68 KMS
MILE POSTS 349-350

CHAIYAHI

RIM

WEST CANYON

TRAIL

N

SCALE 0 5 KMS

185

Cliff Canyon & Rainbow Bridge, Navajo N., Arizona-Utah

Location and Access The canyons featured on this map are found in extreme southern Utah and northern Arizona surrounding Navajo Mountain. To get there, one must drive in from Arizona through the Navajo Nation. Drive along Highway 98, the road connecting Page and Kayenta. Between mile posts 349 and 350, turn north on the Navajo Mtn. Road, and drive 68 kms on a good partly-paved all-weather road to the old Rainbow Lodge ruins. Follow the map carefully so you turn west at the right junction which is about 10 kms before arriving at the Navajo Mtn. Mission(which is now closed-1994). Before Lake Powell came to be, this was the normal route to Rainbow Bridge National Monument(a secondary route comes in from the north side of the mountain). Nowadays, many boaters visit this canyon, as the lake waters come up beneath the bridge. There may be a store north of the old Navajo Mtn. Mission near the school, but that's not been confirmed, so bring all supplies into this region.

Trail or Route Conditions This old trail from the lodge ruins isn't used much any more, but it's still good enough that you can't get lost. Simply follow it to the west and down into Cliff Canyon, over Redbud Pass and on down to the Lake. From Yabut Pass, you'll drop about 700 meters in 4 or 5 kms. This is the hard part when coming back out. At Echo Camp, near the lake, are the remains of the former tourist camp under a huge overhang. A more challenging route would be to park somewhere near the Navajo Mtn. Mission or the new school, and hire someone to drive you to the rim of Cha Canyon, then walk the North Rainbow Trail to the bridge, and return via the Rainbow Lodge route to get back to your car. A mtn. bike used as a shuttle could save time, money and road-walking.

Elevations From about 2000 meters, down to about 1128 at the bridge and lake.

Hike Length and Time Needed From the Rainbow Lodge ruins to the bridge is 21 kms. This can be done round-trip in one long day, but you'd better take two. Remember, the bottom end of this hike is at low altitude, and hotter than at the trailhead. To walk the entire route around Navajo Mtn. using the North Rainbow Trail will take 3-5 days, depending on how you arrange transport.

Water Always carry water in your car. About a km before the lodge ruins is a water tank and tap, as well as at several other locations along the road leading to the trailhead. Navajos use these wells daily. There's year-round live water in several places in Cliff and Bridge Canyons, as well as at Echo Camp. Water also flows past as the trail crosses Cha, Bald Rock, Nasja and Oak Canyons.

Maps USGS or BLM maps Navajo Mountain and Kayenta(1:100,000), or Navajo Mtn(1:62,500), Chaiyahi Flat and Chaiyahi Rim(1:24,000).

Main Attractions Deep and narrow Navajo Sandstone canyons and the largest natural bridge in the world. The lake provides great swimming, but it's sometimes too crowded with boaters.

Ideal Time to Hike Spring or fall are best but it can be hiked in summer and winter.

Hiking Boots Any dry weather boots or shoes.

Author's Experience The author camped, with the permission of a nearby resident, at the water tank near the lodge ruins. Next day he hiked to the bridge and back. Walk-time was 11 hours. He has also been up from the lake to the North Rainbow Trail in about all of the canyons shown.

The end of this hike is not a pot of gold, but the Rainbow Bridge itself.

Map 76, Cliff Canyon and Rainbow Bridge, Navajo Nation, Arizona-Utah

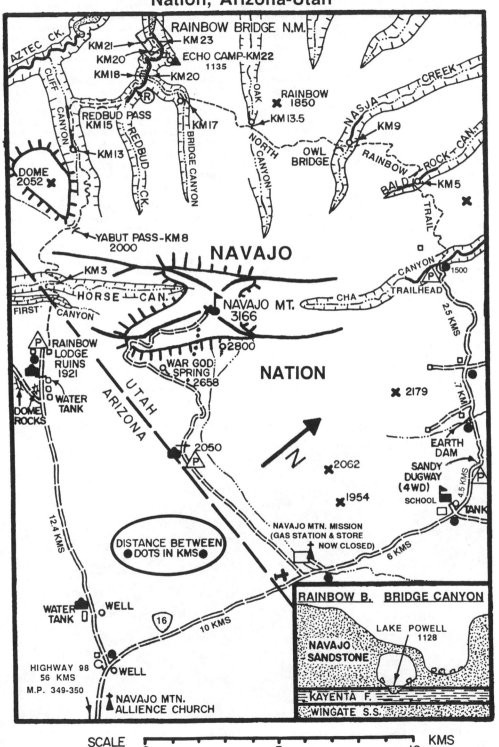

SCALE 0   5   10 KMS

Nasja, Bald Rock, Cha, Trail and Desha Canyons, Navajo Nation, Utah(Arizona)

Location and Access These canyons are all located just north of Navajo Mountain, which is just north of the Utah-Arizona state line. All this land is part of the Navajo Nation. There are two ways to reach the area; one is by boat and Lake Powell, the other by road from Arizona. Drive along Highway 98, the main link between Page and Kayenta. At the turnoff to Inscription House between mile posts 349 and 350, turn north onto the Navajo Mtn. Road and drive 66 kms to the now-closed Navajo Mtn. Mission. Continue north to an emerging settlement(perhaps a store, but no gas, so stock up before going!) at the water tap and crossroads. From there drive north about 2 kms to a viewpoint and park. If you have a 4WD, you can go down the steep sandy dugway and get back up. Navajos do it all the time with 2WD's, but they have to go like hell when coming back up!

Trail or Route Conditions For Desha Creek, walk down the dugway, and onto the road going to the old fruit farm, then head down along the stream to the lake. With a 4WD or mtn. bike you can use the road to lower Trail Canyon. From the end of that access road, a trail drops off the escarpment on the north side and ends at the lake. For Cha Canyon, drive or walk to the Cha Canyon Trailhead, then head down this watery canyon on a mostly west-side trail. The middle part of this deep canyon has large boulders with some wading. Walking is slow in Cha Canyon, but you can get to the lake with no major problems. This may be the most scenic and interesting canyon in the area. From along the North Rainbow Trail you can reach the lake easily along the nearly dry Bald Rock Canyon. For another interesting hike, walk west 9 kms from the end of the road on the North Rainbow Trail to Nasja Creek. The author hasn't seen the upper half of this canyon, but maps indicate it could have some good narrows in the Navajo Sandstone below the trail. In the lower end of the canyon are some impassable dryfalls, and a cattle trail which will get you to the lake(best described in the author's book *Boater's Guide to Lake Powell*).

Elevations Navajo Mtn. Mission, 1841 meters; viewpoint, 1700; Cha Canyon Trailhead, 1500 meters.

Hike Length and Time Needed From the viewpoint to Lake Powell via Desha Creek is about 11 kms one way, and an all-day hike round-trip. It should be an easy day-hike if you walk from the viewpoint to the lake via the Trail Canyon road and trail. From the viewpoint to the lake via Cha Canyon, is about 15 kms. This would take at least 3 days round-trip, but maybe one long day if you can drive to the Cha Trailhead. From Cha Trailhead to the lake via Bald Rock, one long day. From the viewpoint to the lake via Nasja Creek is about 26 kms, one way. This could take 3-4 days round-trip.

Water A well at the crossroads, Navajo Mtn. Mission, springs in Nasja(?) and running water in Upper Bald Rock, Cha and Desha Canyons.

Maps USGS or BLM map Navajo Mountain(1:100,000), or Navajo Mtn.(1:62,500).

Main Attractions Splendid scenery from the viewpoint, and solitude in little known canyons.

Ideal Time to Hike Spring or fall, or even some winter dry spells. Summers are very warm.

Hiking Boots Any dry weather boots or shoes, but you'd better take waders into Cha Canyon.

Author's Experience The author camped near the viewpoint, then searched for a route into Desha. He made it to the lake and returned in about 6 hours. He has hiked from the lake to the North Rainbow Trail in Bald Rock and Cha Canyons. He has also hiked out of Trail and halfway up Nasja Canyon from the lake.

A waterfall over a limestone lens of the Chinle Formation, in the lower end of Cha Canyon.

Map 77, Nasja, Bald Rock, Cha, Trail & Desha Canyons, Navajo Nation, Utah(Arizona)

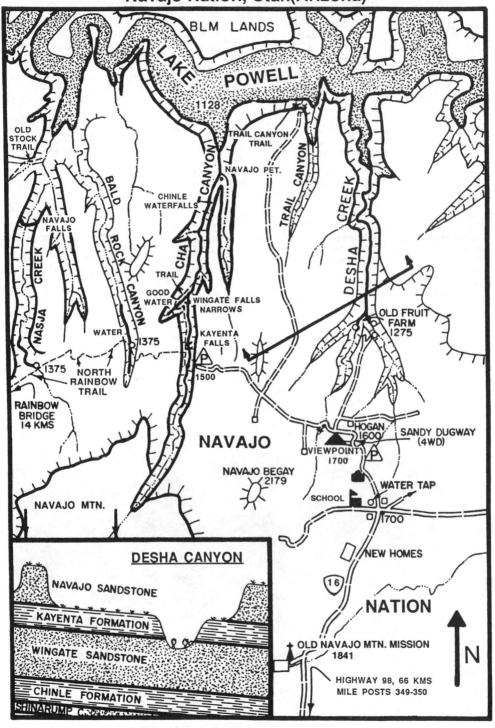

BLM LANDS

LAKE POWELL

1128

OLD STOCK TRAIL

TRAIL CANYON TRAIL

NAVAJO PET.

CHINLE WATERFALLS

BALD ROCK CANYON

CHA CANYON

TRAIL CANYON

DESHA CREEK

NAVAJO FALLS

NASJA CREEK

TRAIL

GOOD WATER

WINGATE FALLS NARROWS

KAYENTA FALLS

WATER

1375

OLD FRUIT FARM 1275

1375

NORTH RAINBOW TRAIL

P

1500

RAINBOW BRIDGE 14 KMS

NAVAJO

HOGAN 1600

SANDY DUGWAY (4WD)

VIEWPOINT 1700

P

NAVAJO BEGAY 2179

NAVAJO MTN.

WATER TAP

SCHOOL

1700

NEW HOMES

16

NATION

N

DESHA CANYON

NAVAJO SANDSTONE

KAYENTA FORMATION

WINGATE SANDSTONE

CHINLE FORMATION

SHINARUMP CONGLOMERATE

OLD NAVAJO MTN. MISSION 1841

HIGHWAY 98, 66 KMS MILE POSTS 349-350

SCALE 0 — 4 KMS

189

Far End and Geshi Canyons, Navajo Nation, Arizona

Location and Access The two canyons on this map are situated south of Navajo Mtn. They're also just north of Inscription Chapter House & Ruins. To get to Geshi, drive along Highway 98 between Page and Kayenta. Between mile posts 349 & 350, turn north on the Navajo Mtn. Road. Drive to between mile posts 7 & 8, which is where the Full Gospel Church is located. Turn left or west on one of several roads near that church, and locate the one main road leading out onto a point south of Geshi. Drive about 4 kms over a sometimes-rough road to where you'll be looking down a steep sandy dugway. Park under a large piñon tree. Most HC cars can make it there; or use a mtn. bike. A second trailhead is back up the road to the east. In the middle of a large flat, turn north, drive about 1 km, and park before you get stuck in sand. To reach the upper end of Geshi, drive along the Navajo Mtn. Road to a point just north of mile post 12, and park on the west side. To get into upper Far End, drive to the Hilltop Well, then turn southwest. At the middle of the fenced corn field, turn right and drive to the rim as shown.

Trail or Route Conditions Beyond the western car-park to Geshi is a very sandy old road, so walk from there. The road soon turns into a good trail and eventually drops off the top into a short side drainage to the north. From the second trailhead, walk north on remnants of a road, which is very sandy. It soon changes into a trail. Finally at the west side of a deep drainage, it drops off the rim. Near the top are steps cut in the very steep slickrock wall. From the trailhead in upper Geshi, climb over the fence and walk north into the head of a short side canyon. Climb down a steep route-trail to the main drainage. From there, livestock trails go down-canyon. From the end of the side road to Far End, walk down the slickrock to the west to a steep nook in the canyon wall. There's a steep trail leading down to first, a bench, then up-canyon(east) a ways, then down again to the bottom. Lots of trails in the bottom, but you'll have to route-find to get down there..

Elevations Trailheads, 1925, 2100 & 2150 meters; lower parts along Navajo Creek, 1550 meters.

Hike Length and Time Needed From one of the 1925 meter car-parks, down into lower Geshi and back up to the other 1925 m. car-park, will take most people all-day. With a mtn. bike left at the 2100 m. car-park, then driving to the north 1925 m. trailhead, a strong hiker could hike the length of Geshi in one long day. It should take only 2-3 hours to climb down into the upper end of Geshi to visit the ruins, and return. One could visit the upper end of Far End in as little as half a day.

Water There's running water throughout both canyons, but it may be polluted by livestock. So you'd better take your own, or drink water coming directly from a spring.

Maps USGS or BLM map Kayenta(1:100,000), or Chaiyahi Rim SE, Oak Springs, Inscription House Ruin, and Shonto NW(1:24,000).

Main Attractions Two very deep scenic canyons, Anasazi ruins, & old Navajo hogans. Within 200 meters, upper Far End has Douglas fir, aspen, birch, piñon, cedar and lots of other alpine vegetation.

Ideal Time to Hike Spring or fall, but with the higher altitude, summers aren't so hot.

Hiking Boots Any dry weather boots or shoes.

Author's Experience He made a loop-hike between the two trailheads in lower Geshi Canyon in 4 1/2 hours. Later he climbed down into upper Geshi to the ruins, and returned in 1 3/4 hours. Later he made another short trip to the 2 waterfalls in Geshi and returned in a couple of hours. He also hiked down into the upper part of Far End Canyon and back in 2 1/2 hours.

A cable & step trail in the wall of upper Geshi Canyon.

Map 78, Far End & Geshi Canyons, Navajo N., Arizona

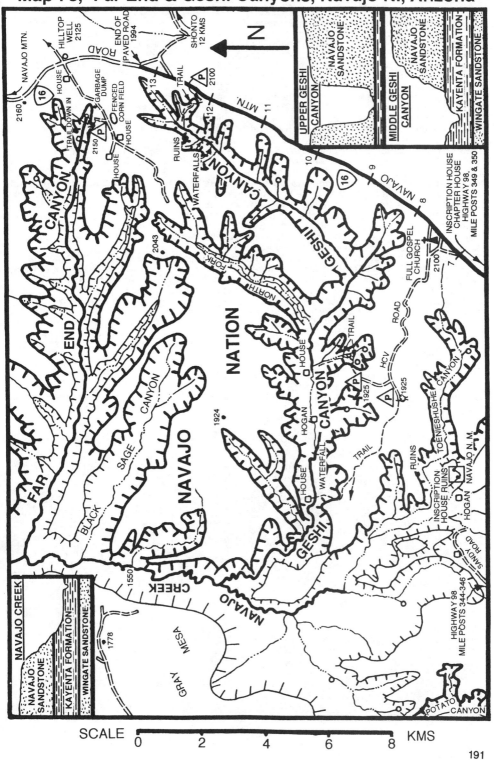

SCALE |0----2----4----6----8| KMS

Potato, Binne Etteni, Toenieshushe & Inscription House Canyons, Navajo Nation, Arizona

Location and Access Shown here are several canyons located about 40 kms south of Navajo Mountain and halfway between Page and Kayenta The map also shows the junction of Highways 98 and 16(Navajo Mtn. Road). To get to Potato Canyon, drive along Highway 98 to a point just east of mile post 344, then turn north onto a moderately good road. Drive about 12 kms to a Navajo home. Park there--or maybe best to park somewhere just south of that house. To get into Binne Etteni and Toenieshushe Canyons, continue north on the same road you use to reach Potato Canyon--but be aware it'll begin to get sandy right away. If you have a 2WD, best to stop at the 1875 meter car-park. A second route into Binne Etteni and Toenieshushe is via the normal route into Inscription House Canyon. To get there, drive north on the Navajo Mtn. Road to the Inscription House Chapter, which is just north of mile post 5. To get from there to the trailhead, proceed north a ways, then turn left on one of several roads running west to the trailhead at 1975 meters. There are lots of roads and homes in that area, so be patient. You might also stop at the chapter house for more information.

Trail or Route Conditions Potato Canyon; from the car-park at 1925 meters, walk west, aim for a rounded hill, then veer north and head down a drainage. Look for a stock trail leading to the bottom. There are a number of ruins as shown on the map. There's a dryfall in the head of the canyon, so you have to return the same way. From the Inscription House Trailhead, walk west on a very good and well-used trail. It zig zags down to the bottom of Inscription House Canyon, then disappears into a maze of livestock trails. From there walk down-canyon to find two big ruins shown in the small square. This is a national monument which is closed to the public. Officially, if you cross the fence at either site, you're illegally trespassing. Binne Etteni Canyon has several old hogans and at least one trail out(and likely more). Toenieshushe Canyon has at least one set of ruins. It could be interesting up-canyon. Better have a Navajo permit to hike here.

Elevations The trailheads, 1925, 1875 & 1975 meters; Inscription House Chapter, 2050 meters.

Hike Length and Time Needed It's only a couple of kms from the trailhead to the rim of Potato Canyon, then perhaps another 8 or 9 kms to the head of the gorge. An all-day hike. From the trailhead at 1875 meters, it's about a dozen or so kms to the head of Binne Etteni or Toenieshushe. Another all-day hike. The distance down to the Inscription House Ruins is about 7 kms, one-way. Another all-day hike, especially if you want to visit other canyons or sites.

Water Little or no water in Potato Canyon, but several seeps near the Inscription House Ruins and elsewhere. Better take all the water you'll need because there's livestock in all these canyons.

Maps USGS or BLM map Kayenta(1:100,000), or Two Red Mesas, Square Butte, Inscription House Ruin & Whirlwind Rock(1:24,000).

Main Attractions Potato Canyon has Anasazi ruins. Inscription House sites are very good.

Ideal Time to Hike Spring or fall, but with the moderate altitude, summers aren't too hot

Hiking Boots Any dry weather boots or shoes.

Author's Experience He spent 5 hours in Potato Canyon; about 6 1/2 hours in Inscription House Canyon and part way up Binne Etteni. Later, he walked from the 1875 meter car-park, into Toenieshushe Canyon and back; 3 1/4 hours round-trip.

The main Anasazi ruins in an alcove in Inscription House Canyon.

Map 79, Potato, Binne Etteni, Toenieshushe, & Inscription House Canyons, Navajo Nation, Arizona

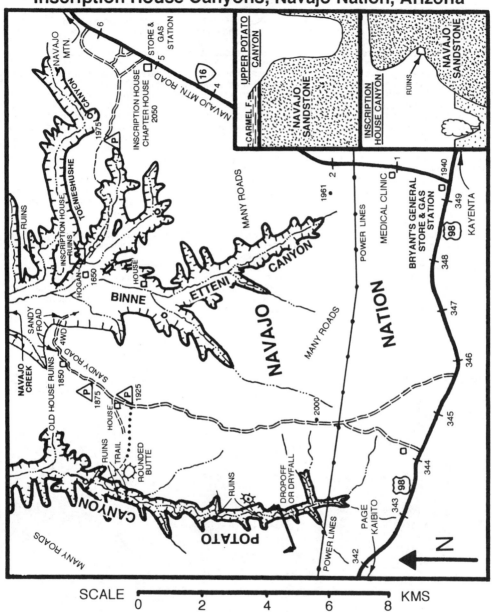

SCALE

0   2   4   6   8   KMS

Navajo National Monument, Navajo Nation, Arizona

Location and Access The two tiny separate parts of the Navajo National Monument featured here are located due west of Kayenta and near the middle of the Navajo Nation. To get there, drive northeast from Tuba City on Highway 160; or southwest from Kayenta on this same road. When you reach the junction called Black Mesa, between mile posts 374 and 375, turn north and drive 14 kms on a paved road to the visitor center. Be sure and stop there for more precise and updated information. In this park you must be escorted by a national park ranger to the Betatakin Ruins, and you must have a permit to visit the Keet Seel Ruins. Both services are free, but you have to make arrangements at the visitor center. You can camp at Keet Seel, but you'll need a free camping permit to do that. From the visitor center you then drive nearly a km to the north and park at the gate next to the campground. It's possible to hike other canyons via the bottom of Tsegi Canyon from mile post 382, but you'll need a Navajo hiking permit to do that.

Trail or Route Conditions The first part of the route to both ruins is on an old road, then at the rim of the canyon, it turns into a heavily used trail. At the bottom of the Navajo Sandstone, the trail separates. One runs west to Betatakin Ruins along the canyon bottom. It's a nice hike in a deep red rock canyon. The other trail follows Keet Seel Canyon bottom all the way to Keet Seel Ruins. Navajo 4WD's make it part way up this canyon. The author counted 32 stream crossings on his way up, but the stream is small and shallow, and you can usually step or hop across. This too is a nice hike between high Navajo Sandstone walls. There are vehicle tracks and livestock trails going up Dowoziebito Canyon, which has a similar appearance to the Keet Seel drainage.

Elevations Visitor center, 2225 meters; Keet Seel Ruins, 2200; low point on trail, 1925 meters.

Hike Length and Time Needed The one-way distance to Betatakin is about 4 kms, but the escort goes slow, making it a 3-4 hour hike, round-trip. From the car-park to Keet Seel is about 12 kms(13 kms from the visitor center). To do this hike round-trip in one day is easy for some; hard for others. Or you can do it in two days, camping one night near Keet Seel.

Water There's year-round water in Long, Keet Seel & Dowoziebito, but there are lots of cattle to muddy it up. So on either hike carry your own water, or take it directly from a spring along the way. Just below Keet Seel Ruins is where the resident ranger gets his water.

Maps USGS or BLM map Kayenta(1:100,000), or Betatakin Ruin and Tall Mtn.(1:24,000).

Main Attractions Two big well-preserved Anasazi ruins. Dowozhiebito has minor sites.

Ideal Time to Hike Keet Seel is open from Memorial Day to Labor Day, and sometimes later into September, while Betatakin is open from about May 1 through September. For the latest information & regulations, call 1-602-672-2366/7. The higher altitudes make summer hiking pleasant.

Hiking Boots Use dry weather boots or shoes to Betatakin, waders if going to Keet Seel, Long or Dowoziebito Canyons.

Author's Experience One afternoon he did the Betatakin hike; next day he did the round-trip hike to Keet Seel in 6 1/2 hours. Years later he hiked up to the ruins in Dowozhiebito in 6 1/2 hours, round-trip.

One of the largest and best preserved of all Anasazi ruins is here at Keet Seel.

Map 80, Navajo National Monument, Navajo N., Arizona

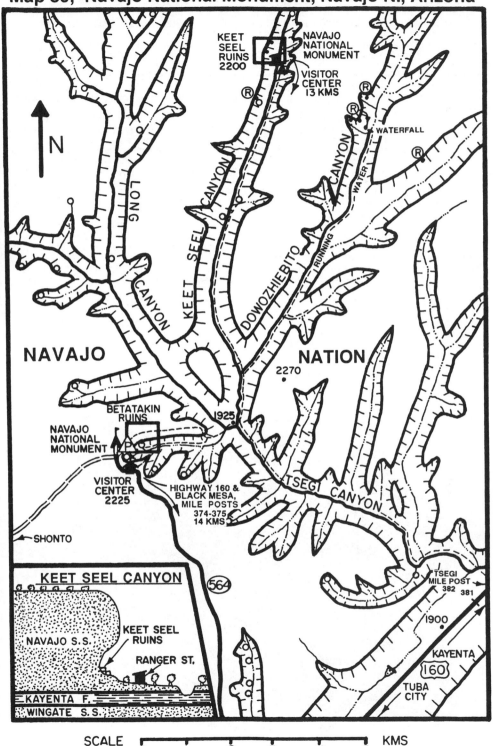

KEET SEEL RUINS 2200
NAVAJO NATIONAL MONUMENT
VISITOR CENTER 13 KMS
N
LONG CANYON
KEET SEEL CANYON
DOWOZHIEBITO CANYON
RUNNING WATER CANYON
WATERFALL
NAVAJO
NATION
2270
BETATAKIN RUINS
1925
NAVAJO NATIONAL MONUMENT
VISITOR CENTER 2225
HIGHWAY 160 & BLACK MESA, MILE POSTS 374-375 14 KMS
SHONTO
TSEGI CANYON
564
TSEGI MILE POST 382 381
1900
KAYENTA
160
TUBA CITY

KEET SEEL CANYON

NAVAJO S.S.
KEET SEEL RUINS
RANGER ST.
KAYENTA F.
WINGATE S.S.

SCALE
0 1 2 3 4 5   KMS

195

Canyon De Chelly Nat. Mon., Navajo Nation, Arizona

Location and Access Canyon De Chelly National Monument is located in the northeastern corner of Arizona and in the heart of the Navajo Nation. The monument consists of several drainages, but the two main canyons are Canyon De Chelly and Canyon Del Muerto. To reach the area from Interstate Highway 40 to the south, exit at Chambers and drive north on Highway 63 past Ganado and to the small community of Chinle. You can also drive in from the north and Mexican Water. As you drive east from Chinle, you'll soon come to the visitor center and campground; then the road splits. One paved highway goes along the rim of Canyon Del Muerto, while the other follows the rim of Canyon De Chelly. There are a number of overlooks along each canyon usually with some ruins in sight. A pair of binoculars would be handy at these viewpoints. Stop at the visitor center before going into the park for further information on hiking possibilities, trails, maps, etc.

Trail or Route Conditions Although all canyons on this map are under control of the National Park Service, the land still belongs to the Navajos, and they continue to farm and graze sheep and goats in the canyon bottoms. There's a sandy 4WD-type track up each of the canyons to summer hogans and fields, but all non-Indians are required to have a Navajo guide when entering the canyon bottoms. There are many sheep trails down into these well-watered canyons from the mesa top, but the author has opted not to show them on this map, because we can't use them anyway. There is one exception however, and that's the trail to White House Ruins. This trail begins at White House Overlook and zig zags down to the bottom lands, then it heads downstream while crossing the creek. You'll have to wade the creek but it's not deep water. Some people walk barefoot in the sandy wash. The White House Overlook is about 10 kms from the visitor center.

Elevations White House Overlook, 1850 meters; canyon bottom, 1725; visitor center, 1700 meters.

Hike Length and Time Needed From White House Overlook to the ruins is about 4 kms, round-trip. This means the total walk-time is only 1 or 1 1/2 hours, but you'll want to spend a little time in the canyon, so plan on maybe 3 hours or half a day for the trip.

Water There are high mountains to the east and an impervious layer of Supai Formation at the bottom of the drainages, so there is flowing water here year-round, or at least most of the time. But it's best take your own drinking water.

Maps USGS or BLM map Canyon de Chelly(1:100,000), or Canyon Del Muerto(1:62,500).

Main Attractions Some of the best preserved Anasazi Indian ruins anywhere.

Ideal Time to Hike Spring or fall are best, but the roads are open year-round. Because of the higher elevations, summers aren't as hot here as in other canyons of the Colorado Plateau.

Hiking Boots Dry weather boots or shoes, which can be removed easily to make one crossing of the shallow stream. Or use wading-type shoes.

Author's Experience The author camped at the campground(it was free in 1986), then made two trips along the South Rim Drive. Once he walked down to the White House Ruins and returned, all in less than 3 hours.

Near the center of the foto are the White House Ruins in Canyon De Chelly.

Map 81, Canyon de Chelly N. M., Navajo Nation, Arizona

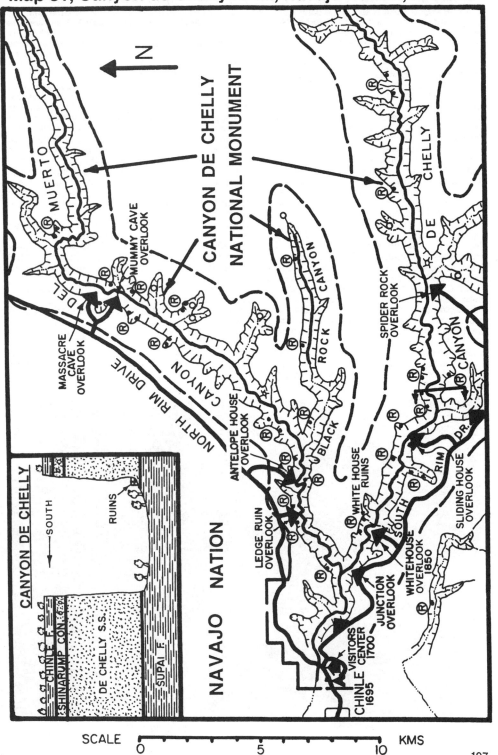

N

CANYON DE CHELLY
NATIONAL MONUMENT

MUERTO

DEL

CANYON

MUMMY CAVE
OVERLOOK

MASSACRE
CAVE
OVERLOOK

NORTH RIM DRIVE

CANYON

ANTELOPE HOUSE
OVERLOOK

LEDGE RUIN
OVERLOOK

ROCK CANYON

BLACK

SPIDER ROCK
OVERLOOK

DE CHELLY

CANYON

WHITE HOUSE
RUINS

WHITEHOUSE
OVERLOOK 1850

JUNCTION
OVERLOOK

SOUTH RIM DR.

SLIDING HOUSE
OVERLOOK

CHINLE VISITORS
CENTER 1700

CHINLE 1695

NAVAJO NATION

CANYON DE CHELLY

← SOUTH

RUINS

CHINLE F.

SHINARUMP CON. Z.

DE CHELLY S.S.

SUPAI F.

SCALE

0   5   10   KMS

Soap and Jackass Creek Canyons(Navajo Nation), Arizona

Location and Access There are 4 canyons on this map, but Jackass and Soap Creeks are the best. All are located on either side of the Colorado River just downstream from where Highway 89A crosses the Navajo Bridge which spans Marble Canyon(in March, 1994, they were building a second new bridge). East-side canyons are on the Navajo Nation; the west-side hikes are on BLM land. To enter Jackass Creek, drive to a point near mile post 532; for Salt Water Wash stop and park near m.p. 529. To get into the south fork of Soap Creek, turn east and drive through a gate at m.p. 548. There's a tap with water year-round, corral & campsites nearby. For Badger Canyon, stop and park between m.p. 542 & 543.

Trail or Route Conditions As you begin walking down Jackass the drainage is shallow, but soon deepens. After 2-3 kms you'll be in some pretty good narrows. In the middle part of these narrows is a steep place and a dryfall. In the past this used to be a problem for hikers, but in recent years someone has chopped steps in a couple of places and has anchored a rope to the wall which can be used to climb up or down. The author guesses some kind of rope will always be there, as Navajos go down to the river to fish. However, most hikers should be able to make it up or down without the rope. To get into Salt Water, walk due west from the car-park, across one wash, then as you meet the second dry creek bed, walk down this to the Colorado. This route is easy & uninteresting. From the car-park near Soap Ck., walk east into a shallow drainage. It soon gets deeper. Walk past an area of rockfalls on a trail to the right or south side. There are narrows in the lower section, but nothing difficult. Badger Canyon is more difficult, because there are 3 dropoffs of 10, 20, and 40 meters. Unless things have changed in recent years, you'll need a rope, or two, to get down this one. You might also bench-walk along the Colorado from Soap Ck. and get into the bottom end of Badger(?).

Elevations Jackass Creek Trailhead, 1200 meters; Salt Water Wash Trailhead, 1350; Soap Creek Trailhead, 1275; Badger Canyon Trailhead, 1150; and the Colorado River, about 925 meters.

Hike Length and Time Needed Down Jackass or Salt Water to the river is about 5 kms each, an easy half-day hike round-trip. It's about 7 kms from the car-park to the Colorado via Soap Ck., and about 5-6 hours, round-trip. The walk/climb down Badger to the river is 3 or 3 1/2 kms, but with all the rope-work and jumaring back up, it will likely be a long all-day adventure.

Water Carry water in your car and in your pack. You may find pothole water in Jackass & Soap Creeks.

Maps USGS or BLM map Glen Canyon Dam(1:100,000), or Lee's Ferry & Tanner Wash(1:62,500).

Main Attractions Four routes into Marble Canyon and the Colorado, and some short interesting narrows in lower Jackass and Soap Creeks. Apparently good narrows in Badger(?).

Ideal Time to Hike Spring or fall, or perhaps in winter warm spells. Hike mornings in summer.

Hiking Boots Dry weather boots or shoes.

Author's Experience The author walked down Jackass to the river and back in 3 hours. Down Salt Water Wash and back, about 4 hours. That included a skinny dip in the ice cold Colorado. He hurried down the south fork of Soap Creek to the river and back in 4 1/2 hours. He has yet to make it all the way down Badger Canyon.

This is the narrows and the one steep place in Jackass Creek.

Map 82, Soap and Jackass Creek Canyons(Navajo Nation), Arizona

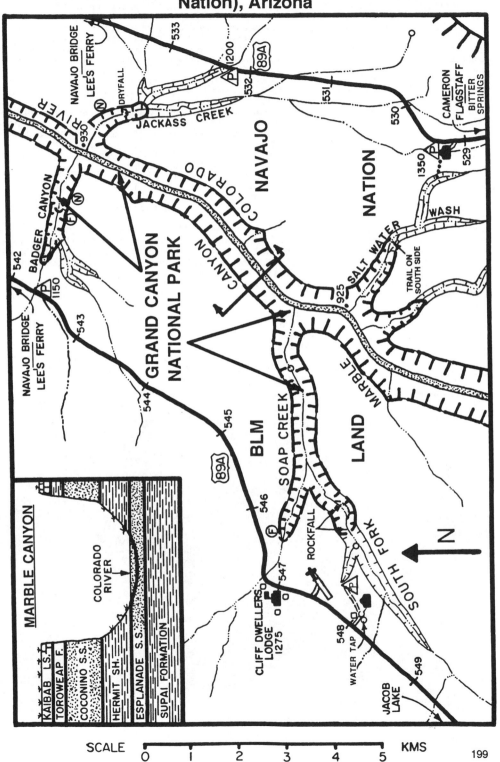

SCALE 0 1 2 3 4 5 KMS

Rider Canyon & Tanner Wash(Navajo Nation), Arizona

Location and Access Tanner Wash flows northwest from the community of Bitter Springs which is at the junction of Highways 89 & 89A. Bitter Springs is east of the Colorado River and is part of the Navajo Nation. Tanner Wash, along with Rider Canyon, end at the Colorado River, which in this region runs through Marble Canyon. Rider Canyon in on BLM land on the west side of Marble Canyon. Below the rim, on both sides of the river, you'll be in Grand Canyon National Park. To reach Tanner, park just off the highway someplace at Bitter Springs. To get to Rider, first get one of the USGS maps below; then drive to a point just west of mile post 557 on Highway 89A, the main road linking Jacob Lake and Navajo Bridge over Marble Canyon. Drive southeast on a dirt road 8 kms to the Cram(Kram) Ranch, always bearing left at junctions. From the ranch head south for a km then the road gradually veers left or east. Always turn left until you're going due east at km 15(mile 9) or so, then turn right. This part is rougher than at the beginning, but with slow going you can get out there with most cars. The last part of the route is shown on this map.

Trail or Route Conditions From Bitter Springs, walk west a short distance and into upper Tanner Wash. Halfway to the river you'll come to a huge dropoff. At that point, retreat 150 meters or so and walk along the west side of the canyon on a terrace just above the Coconino Sandstone. After about a km, you can get down into the canyon again. The route is down a steep rock slide. Most people would consider this slightly challenging. Right under the dryfall is a pool of water called Hermit Spring. Near the river is some running water and a section of narrows. For Rider, walk due north from the car-park about 1 1/2 kms to the edge of the canyon, then look for a break in some NE-to-SW-tending cracks in the Kaibab Limestone. The author went down one route, but came out another, then marked both with a cairn. There are likely other routes too. From the rim to the bottom it's steep, but route-finding is easy. In the bottom, walk the dry creek bed to the Colorado. Near the bottom end you'll walk through some moderately good narrows with several potholes.

Elevations Bitter Springs, 1560 meters; Rider Canyon car-park, 1435; Colorado R., about 920 meters.

Hike Length and Time Needed From Bitter Springs down Tanner to the river and back will take a strong hiker all day. Less energetic hikers can make it to Hermit Spring and back on an easy day-hike. The Rider Canyon hike is about 7 kms from the car-park to the Colorado River. Strong hikers can do the round-trip in a bit more than half a day, but for others the hike will last a full day.

Water Always carry water, but there's a spring below the dryfall in Tanner, and in the lower canyon. Rider Canyon is dry, except for potholes. Lake Powell water in the river is usually drinkable--believe it or not!

Maps USGS or BLM map Glen Canyon Dam(1:100,000), or Tanner Wash & Emmett Wash(1:62,500).

Main Attractions Big potholes, moderate narrows and a grand dropoff in the middle of Tanner Wash.

Ideal Time to Hike Spring or fall, or maybe some winter warm spells. Summers are hot here.

Hiking Boots Any rugged dry weather boots or shoes.

Author's Experience The author went into Tanner Wash on two different days: once in April and again in June. He didn't make it past the dryfall the first time, but did on the second trip. He camped at the Rider Canyon car-park, then walked down to the river and back in 5 1/4 hours.

This is the Coconino Sandstone narrows in the middle part of Tanner Wash.

Map 83, Rider C. & Tanner Wash(Navajo N.), Arizona

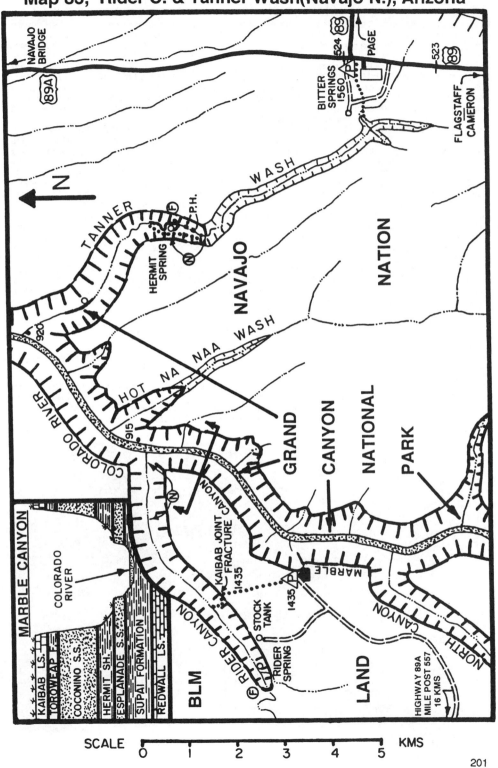

SCALE 0 1 2 3 4 5 KMS

Eminence Break and Tatahatso Point, & Shinumo Wash Hikes, Navajo Nation, Arizona

Location and Access The 3 hikes on this map are located in the southern part of Marble Canyon near Eminence Break. Both are on the east side of the Colorado and on Navajo Nation lands. This is a fault-line escarpment, the result of which are two steep routes down to the Colorado from Tatahatso Point. Also included is a hike into Shinumo Wash and along the Colorado River to the lower terminal of a tramway. This tram was used to transport supplies down to a camp and potential dam site below. To get there, drive along Highway 89 between Cameron and Page. At Cedar Ridge, and just north of mile post 505, turn west onto a good graded road. Drive west, as shown on the *Map 86, Area Access Map--Little Colorado River Gorge.* About halfway to Marble Canyon will be an outcropping called Tooth Rock. Just north of that, the graded road veers northwest and heads straight for a butte called Shinumo Altar. To reach Shinumo Wash, take this road as shown on this map. To reach Tatahatso Point, continue due west on a less-used ungraded road from the stock pond north of Tooth Rock.

Trail or Route Conditions As you near the rim of Shinumo Wash, locate the horse trail which begins right where a dry creek bed from a shallow drainage falls off the canyon rim. It's about 200 meters from the road. Follow this well-used trail down-canyon and along the Redwall bench all the way to the lower end of the tramway. From the car-park on Tatahatso Point, look for an indentation in the canyon rim just downhill to the south. The spot should be marked with a stone cairn. Head straight down this steep gully. When the slope becomes less steep, veer left or east and get out of the ravine. There's a trail of sorts, but watch for stone cairns. Further down it heads southwest on top of the Redwall bench. The last part is used by boaters. A second way to get down to the river from Tatahatso Point, is to head north from the road, as shown on the map. Stay close to the escarpment. After a ways, you'll come to a very steep crack which is on the fault line. Climb down to the bottom(beginners might take a short rope for one steep place) of Tatahatso Wash, then walk down-canyon. To actually reach the river, continue north on the Redwall bench until you find a route down.

Elevations Shinumo W. rim, 1650 meters; Tatahatso Pt. Trailhead, 1725; the Colorado, 970 meters.

Hike Length and Time Needed It's about 12-13 kms down Shinumo Wash to the old tramway campsite and will take all day, round-trip. It's about 3 kms to the river via the trail off the south side of Tatahatso Point, and will take about half a day or a little longer, round-trip. Going down Tatahatso Wash, it's about 11-12 kms to the river and will take all-day, round-trip.

Water Carry plenty in your car and pack. River water is pretty good if you have to drink it.

Maps USGS or BLM maps Glen Canyon Dam & Tuba City(1:100,000), or Emmett Wash, Nankoweap and Shinumo Altar(1:62,500).

Main Attractions Great scenery from the rim & a good look at the spring called Vaseys Paradise.

Ideal Time to Hike Spring or fall. Summers are hot, especially at the bottom.

Hiking Boots Rugged hiking boots would be best.

Author's Experience He walked down to the bench overlooking the river in Tatahatso Wash, then returned, all in 6 1/2 hours(the author didn't actually reached the river, but the route shown is from Butchard's old hiking maps). After lunch he went down the south side *Hiker's Trail* to the river, had a skinny dip, and returned, all in 3 1/2 hours. His trip down Shinumo Wash to the bottom of the tramway took 7 1/2 hours round-trip.

The Redwall Cavern and raft as seen from the trail to the dam builders campsite.

Map 84, Eminence Break and Tatahatso Point, & Shinumo Wash Hikes, Navajo Nation, Arizona

NAVAJO

N

MARBLE CANYON

SHINUMO WASH

TWENTYNINE MILE CANYON

BENCH

REDWALL

HORSE TRAIL

ROUTE DOWN TO RIVER

SOUTH CANYON

RUINS

STANTON CAVE

DAM BUILDERS CAMPSITE
LOWER TRAMWAY SITE

P 1650

VASEY'S PARADISE (SPRING)

REDWALL CAVERN

UPPER TRAMWAY SITE

1674

HOME

HORSESHOE RESERVOIR

BLM LAND

NATION

BREAK

SHINUMO ALTAR 3 KMS

1714

REDWALL BENCH

TATAHATSO WASH

EMINENCE

RIVER

CEDAR RIDGE & HIGHWAY 98--20 KMS
MILE POST 505

STOCK POND

SECOND TRAMWAY SITE

1725

TATAHATSO POINT

1725   1750

CAR-PARK STEEP & ROUGH SPOT

COLORADO

HIKER'S TRAIL

970

REDWALL BENCH

SMALL POINT

STOCK POND

MARBLE CANYON

COLORADO RIVER

KAIBAB LS.
TOROWEAP F.
COCONINO F.
HERMIT SH.
ESPLANADE S.S.
SUPAI F.
REDWALL LIMESTONE
TEMPLE BUTTE LS.
MUAV LS.
BRIGHT ANGEL SHALE

STOCK POND

CEDAR RIDGE & HIGHWAY 98--20 KMS
MILE POST 505

SCALE   KMS
0   5   10

South Canyon and Vaseys Paradise, Arizona

Location and Access South Canyon is located in the northeastern corner of the Grand Canyon and in the southern-most part of the long narrow strip of the national park known as Marble Canyon. To get there, drive along Highway 89A, the main road running between Jacob Lake and the new Navajo Bridge at Marble Canyon. About 32 kms east of Jacob Lake, and between mile posts 559 and 560, turn south on the House Rock Buffalo Ranch Road #445. Drive this well-maintained dirt and gravel road about 32 kms, then turn left or east onto Road #632. After another 2 kms you'll come to the House Rock State Buffalo Ranch. At the ranch, turn right or east, open--then drive through and close--two gates, then drive another half km to the edge of the canyon and park at the trailhead. You can camp at the trailhead which is on BLM land. All the deep canyon bottoms are in the national park, while the rim lands are BLM. Road #445 has some big sharp rocks, so lower the pressure in your tires to avoid a flat.

Trail or Route Conditions From the trailhead, locate the path over the rim, and walk straight downhill to the canyon bottom on a hiker-made trail. Once in the canyon you'll walk in the mostly-dry creek bed. Near the bottom, and in the Redwall Limestone, you'll come to a series of dropoffs or dryfalls, and potholes or pourover pools. At that point you'll have to climb up to the left or north and walk along a bench at or near the top of the Redwall. This trail takes you to an overview of the Colorado. From there, contour along the bench, around the corner to the north and down a gully to the river. Once at river level, walk down-canyon to visit Stanton's Cave and Vaseys Paradise, a spring with a huge volume. This is perhaps the most interesting hike in the Marble Canyon area. For a different view of Vaseys Paradise, use the previous map showing the route down to the river on the opposite side.

Elevations The trailhead is about 1710 meters; the Colorado River, about 875 meters.

Hike Length and Time Needed From the trailhead to the river is about 10 kms. This isn't a long walk, but there's enough to see, that you'll want to make it an all-day hike. Some people do it in two days. Many who visit this area are trout fishing in the frigid Colorado.

Water Plan to carry water to the river, but once there you'll have good water from the gushing spring at Vaseys Paradise. There may be a seep or two and pothole water in the lower end of South Canyon.

Maps USGS or BLM maps Tuba City and Glen Canyon Dam(1:100,000), or Emmett Wash and Nankoweap(1:62,500).

Main Attractions Lush Vaseys Paradise, two Anasazi Indian ruins very near the trail, the almost disintegrated skeleton of a person just south of the best ruins, and Stanton's Cave. In this cave have been found split-willow figurines dating back 3000 to 4000 years, and the remains of giant condors, saber-toothed tigers and other animals that have been extinct for 10,000 years in North America. This cave is now locked tight with a metal gate and bars for protection.

Ideal Time to Hike Spring or fall, or perhaps warm winter dry spells. Morning-hike in summer.

Hiking Boots Any dry weather boots or shoes.

Author's Experience The author camped along Road #445, then with an early morning start hiked down to the sites along the river and back to the car in about 6 1/2 hours.

The giant spring called Vaseys Paradise, as seen from the east side of the Colorado River.

Map 85, South Canyon and Vaseys Paradise, Arizona

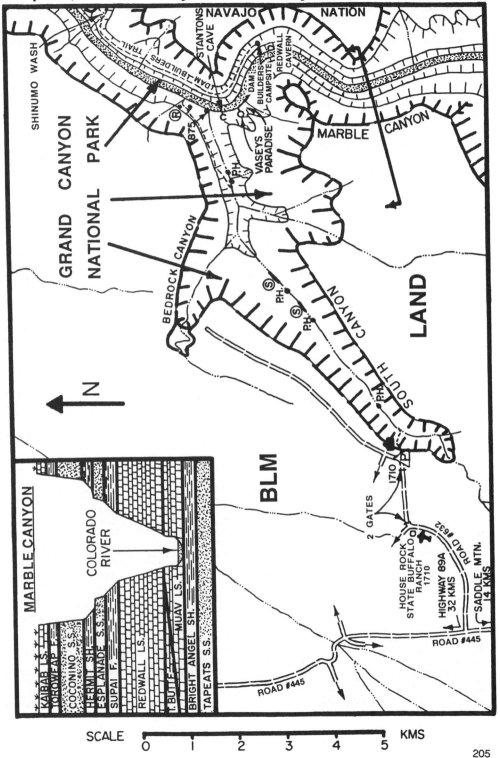

SCALE

KMS

0   1   2   3   4   5

Introduction--Little Colorado River Gorge, Navajo N, Arizona

Introduction and Access The Little Colorado River begins in east central Arizona and flows northwest until it reaches the Colorado River in the eastern part of the Grand Canyon. The lower end of the Little Colorado flows through one of the most spectacular gorges in the world. Up to the present time, this country has been almost unknown to anyone except for a few Navajo herdsmen, but now with some improved roads, the area has moderately easy access.

To get into the upper end of the gorge, you could park and walk from Cameron, but that part of the canyon isn't so interesting, and at times there is quicksand in places not far below Cameron. Perhaps it's best to proceed along State Highway 64, which runs between Cameron and Desert View, just inside the Grand Canyon. Between mile posts 285 and 286, turn off to the north and proceed to the overlook of the upper gorge. There's a steep route down to the bottom just to the east of that viewpoint parking lot. Or to reach a good trail down to a water gauging station in the upper end of the gorge, proceed to a point halfway between mile posts 477 & 478 on Highway 89, not far north of Cameron. From there turn west onto a good dirt road which heads west along the north side of Shadow Mountain, which is an old extinct volcano. With the maps listed below, proceed west to the stock pond shown, then turn south and follow that road all the way to the gorge immediately north of the previously mentioned viewpoint on Highway 64. From there a good trail zig zags down to the river and an old cable car and good campsites.

To reach a couple of routes down to the bottom of the middle gorge from the west, proceed along Highway 64 to a point about halfway between the Grand Canyon entrance gate and the service station(both at Desert View), then turn east onto a good gravel road. After a short distance, the main road turns right into a NPS facility, but you continue east on a rather rough road to the left. Continue following this same road down a steep hill to Cedar Mtn.(around the north side is the better of the two routes), then downhill more until you reach a Navajo home and corral, then turn north and proceed in the direction of Gold Hill. This road is steep in a couple of places with rough spots everywhere, but the author made it OK in his VW Rabbit Diesel with oversized tires. However, that road did test his car pretty good in a couple of places! Park somewhere east of Gold Hill, or try to make it to the point just north of Blue Spring--if you have the right vehicle. From the south side of that point is a trail leading down to Blue Spring, plus another route or two down to the bottom just to the north.

To reach the middle and lower end of the gorge from the east, one way would be to drive to a point along Highway 89 just north of the turnoff to Tuba City, Kayenta and Highway 160. Between mile posts 482 & 483, turn west onto another well-used and maintained dirt road. This road heads west, then the main branch turns northwest staying on the northeast side of Big Canyon. Further along, this main road runs north passing Big Reservoir and ends near Tooth Rock. At that point it intersects another very wide and graded road which begins at Cedar Ridge and very near mile post 505; this one runs west to Tooth Rock before veering northwest and heading for Shinumo Altar and beyond. There are many other good roads in the area, but this one seems to be the best for general access to most of the canyon, and to grazing areas for Navajos. From along this road, there are other short unmaintained roads leading to the rim of Waterhole & Big Canyons, the head of north fork of Big Canyon called Dry Wash, Salt Trail Canyon and to a rugged route down to the lower end of the Little Colorado, called by some, the Walter Powell Route. Be sure you have the two maps listed below in your hands while driving, so you can keep from being lost. The main roads to this area are not highlighted on the USGS maps as they are here. Follow both maps carefully.

Maps USGS or BLM maps Cameron and Tuba City(1:100,000). There are also maps at 1:62,500 and 1:24,000 scale, but for the purpose of access, the two above are the only practical ones to have. For actual hiking, see the next two hiking maps for the larger scale maps available.

Distances The distances in this area are not what they are on the Arizona Strip, but you'll need to pay attention to your fuel tank anyway. Also, for the most part these roads are in pretty good condition, so you'e fuel consumption won't be as great as when you're on really rough, low gear-type tracks. The best places to fuel up nearby are Cameron and Tuba City. There are also pumps at Desert View and The Gap(gasoline only). Page is a larger place with all facilities, but it's quite a ways north. Always start any trip to remote areas with a full tank of fuel.

Road Conditions The main roads mentioned above are graded occasionally and are made from the same material they run over. Close to the rim of the gorge everywhere, and in the northern part of this area, the surface material is what's been eroded from the Kaibab Limestone and Toroweap Formation. This makes for mostly-passable roads in wet weather. In the southern and eastern part, roads are formed from the clay beds of the Chinle Formation. These can be very slick in wet weather. For the most part, if you're out there and a storm comes, all you have to do is wait a while and the desert air will dry the surface enough to drive on. If it's a big storm, you may have to wait overnight, or longer in cooler weather. There are almost no places in this region where deep sand is a problem.

Water Good water is pretty scarce in these parts, so always carry a good supply in your car. The author always has about 8, one gallon jugs full with him at all times! In the gorge, brown flood waters in the Little Colorado comes mostly from lands of the Navajo Nation where grazing is practiced year-round, so you'll always have to purify it. If you arrive at the end of a dry spell, you'll only have puddles of water throughout most of the gorge, then after Blue Springs there will be a clear stream flowing to the Colorado. As for springs, Blue Spring gets the name because of its turquoise blue water. It has a minerally taste, but it's totally different from any other the author has tasted. It's barely drinkable--at least in small amounts! There's also a big spring right at the bottom end of Salt Trail Canyon.

Map 86, Area Access Map--Little Colorado River Gorge, Navajo Nation, Arizona

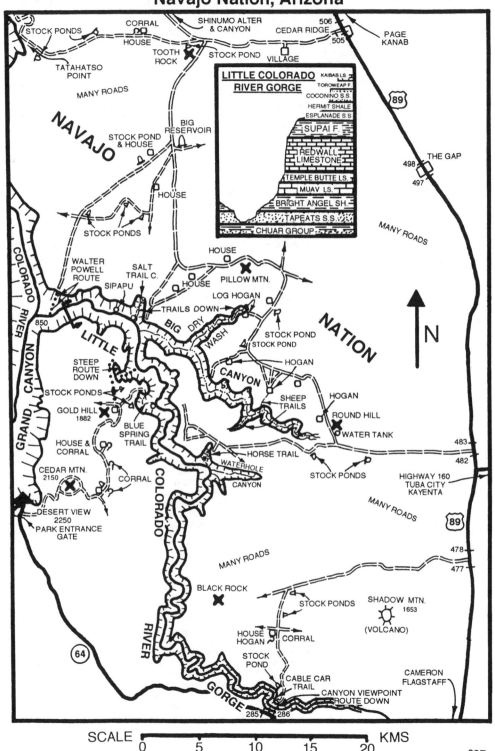

SCALE

0   5   10   15   20   KMS

Upper L. Colorado River Gorge Trails, Navajo Nation, Arizona

Location and Access This map covers the upper 2/3's of the gorge of the Little Colorado River, which starts just below Cameron. You could get down into the canyon right at Cameron, but there are at times some quicksand narrows at about where the gorge is first shown on this map. It you want to get into this upper end, then it's perhaps best to drive west from Cameron on Highway 64 for about 20 kms. Between mile posts 285 & 286 turn north and drive 400 meters to the Canyon Viewpoint, where you can look straight down into the gorge. It's quite a site! Just east of that parking lot is a route down. This make quick access into the gorge. Just across the canyon is a trail leading down to the river and an old cable car. Get there by driving north out of Cameron on Highway 89. About halfway between mile posts 477 & 478, turn west onto a good dirt road. Follow this route on the *Map 85, Area Access Map--Little Colorado River Gorge*. As you near the rim of the gorge, the road gets a little rough, but any higher clearance car can make it to the trailhead.

To get into the middle of the gorge and Blue Spring, drive west from Cameron toward Desert View, the first stop in Grand Canyon N. P. Immediately after you pass through the entrance gate, turn right or east, and drive down a good dirt road for 300 meters or so, then go straight ahead instead of turning south into a NPS complex. You'll be on a rather rough road as you head down the steep grade past Cedar Mtn., then finally north toward Gold Hill. Just after you pass the house east of Gold Hill, turn northeast and drive across a valley to the rim just north and above Blue Spring. There are rough places all along this route, so only higher clearance cars can make it.

To reach the horse trail down into the gorge just northwest of Waterhole Canyon, leave Highway 89 between mile posts 482 & 483, and drive west. Follow the route on the *Area Map*, and the Tuba City USGS map carefully. The last part of this drive is on a little-used road heading south. Look for a little side road running west to the lip of the shallow upper drainage and the trailhead at 1500 meters.

Trail or Route Conditions From the Canyon Viewpoint, walk 100 meters east from the parking lot and look for a steep route down a slit in the first ravine. It's very steep, but easier than it looks. Once at the bottom just walk down-canyon, if the stream is small enough to cross safely. Check out the river first from the rim, or Cameron. From the trailhead on the opposite rim, walk down the partly-washed out trail to the river. It's been washed out in a place or two but it's easy. At the bottom is an old cable car and upstream a ways a gauging station.

From the end of the road near Blue Spring, walk west along the rim. You'll pass a stone fence, then after about 100 meters, you'll climb down a near vertical wall. This too is easier than it looks. Just follow the stone cairns. Only people and maybe mtn. goats can get down this upper part. After a ways, the slope eases up, but you still have to follow the stone cairns and an emerging trail. About 2/3's the way down you'll come to a constructed horse trail. This leads down to Blue Spring. An interesting hike might be to go down to the river, then downstream(north) in the channel or on a bench, then exit the gorge to the west via another minor canyon just to the north of Blue Spring, as shown on the map. At the head of that canyon is an exit on the north.

To reach the bottom of the gorge along the horse trail near Waterhole Canyon, walk into the shallow drainage and head down-canyon. At the big dropoff, head to the right or west side and look for a constructed horse trail. It heads down-canyon, but is visible in only a few places. On the bench not far above the river, the trail is built-up into a mini-dugway as it goes down over a dropoff, but part of it has now collapsed and you'll have to climb down. It's easy, but livestock can't do it anymore. Below this ramp, head east on a bench for half a km, then scramble down a rock slide to the river.

For the long-hike enthusiasts, it's possible to get into the gorge in the upper end, and walk all the way to the Colorado River. Walking is fairly easy most of the way, but only if there isn't too much water in the river. It would be best to do this hike when there's just a trickle of water in the canyon.

Elevations Trailheads, 1500 to 1550 meters; Blue Spring, about 1000 meters.

Hike Length and Time Needed To reach the river via the upper canyon routes will take only about half an hour from either trailhead. For many people, the hike down to Blue Spring and return would be nearly all-day, but strong hikers can do it in half a day, round-trip. Going down to the river via the horse trail near Waterhole Canyon will take most people about half a day or a little more. To hike from the Canyon Viewpoint in the upper gorge, to the Colorado River, then up the Tanner Trail to Lupan Point(just west of Desert View), will take 4 to 6 days. See Map 110, Tanner Trail.

Water Take your own on all day-hikes, and treat river water. The turquoise blue water from Blue Spring has a strange mineral taste, but it seems drinkable(?). A small amount didn't bother the author.

Maps USGS or BLM maps Cameron & Tuba City(1:100,000), or Vishnu Temple, Blue Spring, & Coconino Point(1:62,500).

Main Attractions Super scenery, great gorge, interesting routes, few if any other hikers.

Ideal Time to Hike If you're just hiking the trails down to the river, then spring or fall. If hiking up or down the river, then whenever there is little or no water in the Little Colorado--as seen from the Cameron Bridge. Fall might be good, or late winter or early spring--before spring runoff. Summer is very warm. Butchart did it once in January, but others have hiked it in June and August.

Hiking Boots Rugged hiking boots for hiking the trails down, but waders if walking the gorge.

Author's Experience He hiked down to the river via the horse trail west of Waterhole Canyon and returned in less than 3 hours. He walked from his campsite near the house east of Gold Hill, over to and down the Blue Spring Trail, and returned, in 4 hours. From the same campsite, he also walked halfway down to the river in the canyon north of Blue Spring, then returned, in about 3 hours. He hiked down to the river along the old cable car trail, and returned, in one hour. He has also been down to the river several times from the Canyon Viewpoint, each in about an hour round-trip.

Map 87, Upper Little Colorado River Gorge Trails, Navajo Nation, Arizona

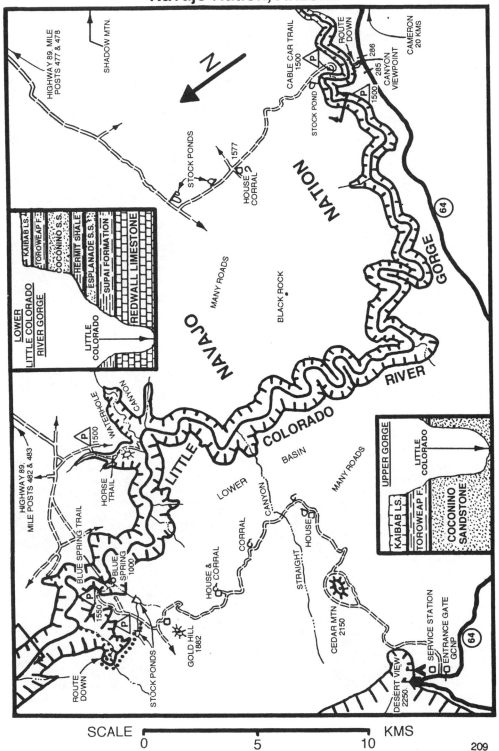

SCALE

0   5   10   KMS

The middle part of the Little Colorado River Gorge as seen from the canyon rim.

The turquoise-blue waters of Blue Spring flows into the flooding and muddy Little Colorado River.

Part of the old horse trail going down to the river near the mouth of Waterhole Canyon.

Typical modern Navajo hogan and sheep & goat corrals. This one is located at the upper end of Big Canyon(seen next map).

Lower L. Colorado River Gorge Trails, Navajo Nation, Arizona

Location and Access This map covers the lower part of the Little Colorado River Gorge and several of its tributaries, including Big and Salt Trail Canyons. Access is via a long drive on an improved graded dirt road, then some driving on secondary dirt roads. To start, drive along Highway 89 between Cameron and Cedar Ridge. Locate the road running west from between mile posts 482 & 483, as shown on *Map 85, Area Access Map--Little Colorado River Gorge*. Drive west until you're near the head of Big Canyon, then at the fork in the road, veer right or northwest. You'll pass a rounded hill, a hogan, then turn westward. At one point will be three secondary roads running south to a hogan and corral at 1540 meters. Park on the hill above the hogan. If you continue on the main road, you'll end up veering north. Soon you'll turn west on a secondary road from on top of a bench heading for the upper end of Dry Wash. Park at an old log hogan at 1550 meters.

From the Big Canyon area, continue on the main graded road to the north, then west. You'll soon pass Pillow Mountain, then on to Salt Trail Canyon, but for now, let's go to the other end of this main road, which will be the easiest way to the lower end of the gorge. Drive to Cedar Ridge, a small Navajo community on Highway 89. Just north of mile post 505, turn west onto a good graded dirt road. Where you leave the highway will be 5 small Chinese elm trees and the foundation of an old building. Drive this improved road due west until you reach Tooth Rock. Just before this landmark, turn left or south onto another graded road. This one runs due south past Big Reservoir. Finally, when the main road turns east, look for another secondary road continuing south. Follow this until you see a large parking place to the right, or west, just east of the upper end of Salt Trail Canyon.

To do a rugged route down to the very end of the Little Colorado, go back to Big Reservoir and look for one of two secondary roads heading to the southwest. Take either one, as shown on this map and the *Area Access Map*. Make your way to the rim of the canyon at about 1805 meters. All these secondary roads get a little rough, but any higher clearance car, driven with care, can make it to the trailheads. Just in case of a wash out, you've got a shovel in the trunk of your car, right?

Trail or Route Conditions From the hogan & corral at 1540 meters, walk south beginning west of the corral. When you reach the canyon rim, walk west 600-700 meters until you come to a sheep & goat trail heading down. This trail gives access to the head of Big Canyon. From the log hogan at 1550 meters, look for a trail going down off the little rim to the north. This trail crosses the upper end of Dry Wash, just before it drops off, then it follows the north rim to the west. After walking on the rim about one km, you'll come to where a wide constructed trail begins to zig zag down into the canyon just below a big dropoff. At the bottom, the trail vanishes. The upper end of this canyon is impressively deep.

From the parking lot at the head of Salt Trail Canyon, locate the trail which zig zags down off a minor rim to the west; then at the next level, and on the very rim of the gorge itself, look for a cairn and trail heading down. Head straight down the very steep gully, watching for stone cairns marking the route. Halfway down, you'll have to stay on the Redwall bench on the right or west side. Finally, as you near the main canyon, the trail drops down off the Redwall and zig zags to the bottom. At the bottom will be a large spring to the right of a leveled place, which at times has been a helicopter landing site for fish researchers. To walk downriver, either walk right along the river bed or bank; or look for a pretty good trail going downstream behind the tamaracks near the canyon wall. Downstream about 4 kms is what is known to the *Hopi Indians* as the *Sipapu*. This is a sacred place to the Hopi. They believe their ancestors came out of that hole when they first came to be. It's actually an old mineral hot spring, or perhaps a geyser(?). It's a large mound of deposited minerals 6 or 8 meters high, with a hole in the middle. Now however, the mineral water comes out on the riverside of the mound. *Please don't desecrate this site in any way, because it could lead to the closing of this canyon to hiking!*

From the car-park at 1805 meters, walk due east to find a steep ravine heading down off the rim. You'll have to route-find this one, but by using it you can reach the lower end of the Little Colorado. According to John W. Powell's book, *The Colorado River and its Canyons*, Walter H. Powell, cousin of John, climbed out of the canyon in this area. This may have been the route he took, thus the name.

Elevations Trailheads, 1540, 1550, 1700 and 1805 meters; mouth of Little Colorado, about 850.

Hike Length and Time Needed In half an hour you can get down into the upper end of Big Canyon from the hogan at 1540 meters; and about the same time to get to the bottom of the trail in upper Dry Wash. With two cars, or one mtn. bike, it's possible to make a loop-hike using these two canyons & trailheads and do it in one full day-hike. It looks like easy walking. Plan on taking all-day to walk down Salt Trail Canyon to the Sipapu and back, which is about 8 to 10 kms, one way. It isn't far, but it'll take most of a day to reach the Little Colorado and perhaps Colorado, and return to the rim via the Walter Powell route.

Water Take plenty in hot weather. Also a spring at the mouth of Salt Trail Canyon. Purify river water.

Maps USGS or BLM map Tuba City(1:100,000), or Vishnu Temple & Blue Spring(1:62,500).

Main Attractions Magnificent gorge, rugged hikes, and solitude.

Hiking Boots Rugged boots for the Salt Trail & Powell routes; any boots for the Big Canyon trails.

Author's Experience The author found the trail and reached the bottom of Big Canyon, then returned; 2 hours. He hunted for the trail, then made it down Dry Wash about 2 kms, then returned, all in less than 2 1/2 hours. The trip down Salt Trail Canyon to the Sipapu and back, took over 7 hours on a hot spring day. He made it down Walter Powell Route to the confluence of the Colorado & Little Colorado in 6 3/4 hours, round-trip.

Map 88, Lower Little Colorado River Gorge Trails, Navajo Nation, Arizona

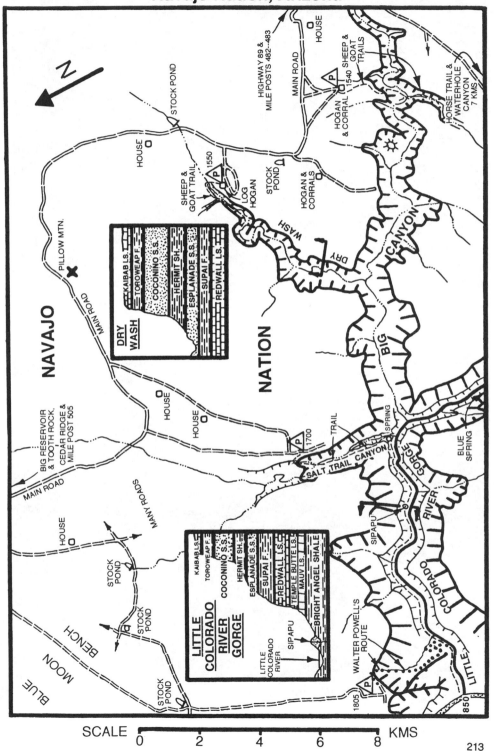

SCALE

0   2   4   6   8   KMS

An Arizona game & fish research helicopter lands at the mouth of Salt Trail Canyon.

Confluence of the Little Colorado(left) and the Colorado Rivers. Notice how muddy the Little Colorado is.

The Sipapu, a sacred Hopi Indian site. It's actually a mineral spring, and perhaps an old geyser.

This foto is from the south side of Tatahatso Point and Eminence Break. The hiker's trail leads down to the left to the sand bar on river in the middle part of the picture.

Nankoweap Trail, Grand Canyon, Arizona

Location and Access The Nankoweap Trail is but one of many ways to reach the bottom of the Grand Canyon and the Colorado River. This one is located on the eastern side of the central portion of the Grand Canyon. Nankoweap Creek flows east and into the Colorado River, which at that point is running south. To reach the trailhead in winter when snow covers the Kaibab Plateau, use the northern or low altitude route. This one begins about 32 kms east of Jacob Lake on Highway 89A and between mile posts 559 and 560. Turn south on the House Rock Buffalo Ranch Road, #445. This road has lots of big sharp rocks, so lower the pressure in your tires to avoid a flat. Follow it south past the Buffalo Ranch and on to the trailhead at the base of Saddle Mtn. Park and camp there, which is at the beginning of the old and now-blocked off Road #445G, and Trails #31 & 57. In summer you can use the road beginning about 1 km south of Kaibab Lodge, which is located about halfway between Jacob Lake and the North Rim Village. On the east side of the highway is Road #424(some maps say it's #442, and some people say #610?). Drive east for one or two kms, then turn south on Road #610. Stay on this route until you reach the Nankoweap Trailhead, a total distance of 20 kms from the highway.

Trail or Route Conditions Other writers have stated it's hard to locate and follow Trail #57 from the upper trailhead, but for the most part, that's not quite true. The only place route-finding might be a problem is on the plateau rim in the first 2 or 3 kms. When you reach the flat area of 2500 meters, veer to the north a bit and drop off the rim on the trail as shown. A little further on, veer to the right and keep just to the right of the ridge. Once you reach the saddle at 2275 meters, it's an easy trail to follow down to the creek. Much of the first part of the trail is on a bench just below the Coconino Sandstone cliff. If you've driven in from the north, walk up the old road to the south until Trail #31 & 57 veer to the left. It runs down into a canyon, then Trail #31 turns east toward Marble Canyon, but you continue south on #57 until it meets the Nankoweap Trail at the 2275 meter-high saddle. If you follow Trail #31 east, you will end up at the rim of Marble Canyon. At that point, according to other information sources, you can route-find straight down to the Colorado River; or bench-walk to the south, then west, and enter Little Nankoweap Canyon. Walk even further to the west and you'll intercept the Nankoweap Trail. The author has yet to check this new route out.

Elevations Rim trailhead, 2680 meters; northern trailhead, 2100; Colorado River, 841 meters.

Hike Length and Time Needed From the rim trailhead to the Colorado is about 23 kms(rim to saddle, 5 kms; saddle to Nankoweap Spring, 13 kms; spring to river, 5 kms; saddle to north trailhead, 3 kms). Most people can make it from the top to bottom in one day, but the return trip will take some hikers two days. A 2-3 day round-trip hike for most people.

Water The author found a good stream of water at 1125 meters; and the Colorado River. Consider caching some water on the way down(near the 1900 meter point) for the return trip, especially if it's warm weather. If it's really hot, getting back up may be a problem.

Maps USGS or BLM maps Grand Canyon and Tuba City(1:100,000) or Grand Canyon N.P.(1:62,500), and for the approach route, the Kaibab National Forest North map(1:125,000).

Main Attractions An interesting route to the Colorado, and one Anasazi ruin near the river.

Ideal Time to Hike Spring or fall, or perhaps in winter dry spells. Carry lots of water in summer!

Hiking Boots Any dry weather boots or shoes; rugged boots if using Trail #31 down to Marble Canyon.

Author's Experience The author camped at the rim trailhead, then took 4 hours to walk to the spring at 1125 meters. He had lunch and a skinny dip, then returned to his car. Total walk-time was just under 9 hours. Years later, he drove to the Saddle Mtn. Trailhead and walked up to the pass and back in just over 2 hours.

This is a view you'll get near the trailhead as you begin the hike to Nankoweap Canyon.

Map 89, Nankoweap Trail, Grand Canyon, Arizona

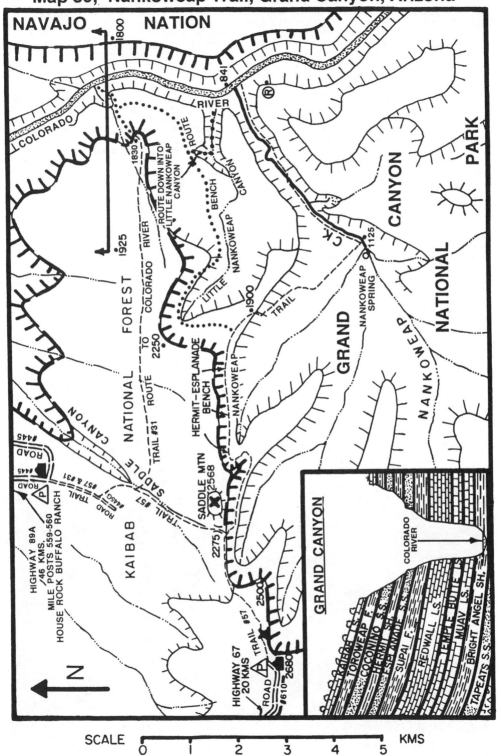

SCALE

0   1   2   3   4   5   KMS

North Kaibab Trail, Grand Canyon, Arizona

Location and Access The North Kaibab Trail runs from the Grand Canyon North Rim, down Roaring Springs Canyon to Bright Angel Creek and Canyon and on to the Colorado River. To get there, drive south from Jacob Lake, which is at the junction of Highways 89A and 67, to the North Rim Village. It's 72 kms from Jacob Lake to the North Rim on a good paved road. The highest point on this highway is 2690 meters, thus it's closed by snow from about November 1 to mid-May. The normal official opening date for this road is May 15 each year. Some roads branching off this paved highway are open for an even shorter period of time, but each year is different.

Trail or Route Conditions Before hiking be sure to stop at the North Rim visitor center to get any last minute information about the trail, camping, water, etc. To reach the trailhead, drive north from the North Rim Village about one km to a big "S" curve. The signs will tell you where it is and where to park. The North Kaibab Trail is one of the most-used trails in the Grand Canyon. It's well-maintained, and in the minds of some, over-used by mule trains. However, the mule caravans go only to Roaring Springs and return. Beyond that it's only hikers. The trail heads down Bright Angel Canyon to the Colorado River. An alternate route from the same trailhead, would be to walk a rim trail to the northeast and down into upper Bright Angel Creek. This is the Old Kaibab Trail. From just north of the Phantom Ranch, another trail runs east to Clear Creek and to a waterfall in its upper canyon. Keep in mind, if you're just out day-hiking, then no permit is needed. If you plan to camp one or more nights, then you must have a free camping permit.

Elevations North Rim Village, 2500 meters; North Rim Trailhead, 2511; Cottonwood Camp, 1000; Colorado River, 731 meters.

Hike Length and Time Needed The distance from the trailhead to the Colorado River is about 22.5 kms one-way. To walk from the rim to Bright Angel Campground next to the Colorado, is a full-day for most anyone. Going down is the easy part. The return-trip is a very long day-hike. Most do it in 2-3 days round-trip. Very strong hikers--or runners, can make it to the river and back in one very long day. An alternate to going all the way to the river would be to use the Old Kaibab Trail down to the mouth of Roaring Springs Canyon, then back up to the North Rim Trailhead. This is about 23 kms round-trip, or an all-day hike for most. Still another trail runs from just north of the Phantom Ranch to Clear Creek. From Bright Angel Creek to Clear Creek and the waterfall is 14 kms one-way, or an all-day hike round-trip from the Bright Angel Campground.

Water Phantom, Bright Angel and Clear Creeks all flow year-round.

Maps USGS or BLM map Grand Canyon(1:100,000), or Grand Canyon N.P.(1:62,500).

Main Attractions A great hike along a well-watered canyon in the heart of the Grand Canyon.

Ideal Time to Hike Because of the high altitude of the region and cold season entry problems, one can hike here only from mid-May through October.

Hiking Boots Any dry weather boots or shoes.

Author's Experience In March, 1972, the author hiked from the South Rim to Roaring Springs and back. That trip lasted 3 days. His last experience was in summer with a round-trip loop-hike down the Old and up the newer Kaibab Trails. This was completed in 5 1/2 hours total walk-time.

At the bottom of the Grand Canyon and on the North Kaibab Trail is the Phantom Ranch.

Map 90, North Kaibab Trail, Grand Canyon, Arizona

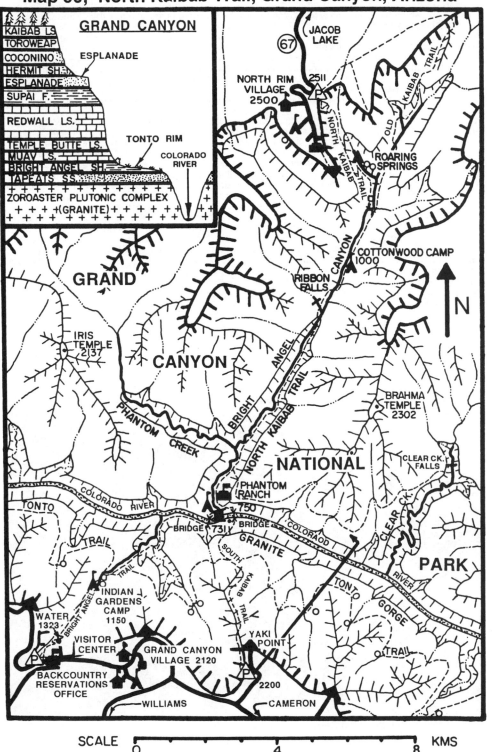

GRAND CANYON

KAIBAB LS.
TOROWEAP
COCONINO
HERMIT SH.
ESPLANADE — ESPLANADE
SUPAI F.
REDWALL LS.
TEMPLE BUTTE LS.
MUAV LS.
BRIGHT ANGEL SH.
TAPEATS SS.
ZOROASTER PLUTONIC COMPLEX
+ + + + (GRANITE) + + + + +

TONTO RIM
COLORADO RIVER

JACOB LAKE
67
NORTH RIM VILLAGE 2500
2511
ROARING SPRINGS
OLD KAIBAB TRAIL
NORTH KAIBAB TRAIL
COTTONWOOD CAMP 1000
RIBBON FALLS
BRIGHT ANGEL CANYON

GRAND
CANYON
IRIS TEMPLE 2137
PHANTOM CREEK
BRAHMA TEMPLE 2302
CLEAR CK. FALLS
NATIONAL

N

COLORADO RIVER
TONTO TRAIL
PHANTOM RANCH
750
BRIDGE 731
BRIDGE
COLORADO
GRANITE
SOUTH KAIBAB TRAIL
CLEAR CK. RIVER
PARK
TONTO GORGE
TRAIL

WATER 1323
BRIGHT ANGEL TRAIL
INDIAN GARDENS CAMP 1150
VISITOR CENTER
GRAND CANYON VILLAGE 2120
YAKI POINT
2200
BACKCOUNTRY RESERVATIONS OFFICE
WILLIAMS
CAMERON

SCALE   0   4   8   KMS

Shinumo Ck.(North Bass Trail), Grand Canyon, Arizona

Location and Access This hike is often called the North Bass Trail, but for much of the way it follows White Creek, which flows through Muav Canyon. As the trail nears the bottom of the Grand Canyon, it follows the lower part of Shinumo Creek. This trail is located 20 to 25 kms west of North Rim Village of the Grand Canyon. The entire mapped area is within the boundaries of Grand Canyon National Park. To get to the trailhead at Swamp Point, drive along Highway 67, the road linking Jacob Lake with the North Rim Village. About one km south of Kaibab Lodge, turn west on Forest Service Road #422. After about 2 kms, turn left or south on Road #270. After 3 more kms turn west on Road #223, and after 10 more kms, turn south or left on Road #268. Finally, after only one km, turn left and onto #268B. This road passes a turnoff to Tipover Spring, then continues on to Swamp Point. This last section of road is about 16 kms of slow travel, but the author got there in his VW Rabbit. You must use the Kaibab National Forest North map when you make this drive.

Trail or Route Conditions The North Bass Trail is one of the least used in the park. So if you'd like to reach the bottom of the Grand Canyon and off the beaten path, this is the one for you. The trail is easy to follow in most places, but in others it requires attention to stay on track. In some places it branches out to several trails, but you can't go too far wrong. As you leave the trailhead, you'll walk west down to a cabin, then the trail heads southeast towards some springs. Between the two springs, it then heads straight down and into the bottom of Muav Canyon. In the upper half of the canyon there is some brush which may scratch bare legs so have a pair of long pants handy. Further down the trail runs across the Supai-Esplanade bench, then drops into lower Shinumo Creek. Further on it rises upon a bench before it reaches the river.

Elevations Swamp Point Trailhead is 2291 meters, the Colorado River 675 meters.

Hike Length and Time Needed From trailhead to river is about 22.5 kms. Because this trail is not as well established as others, it'll take you a full day just to reach the river. Most people should be able to make the return trip easily in two days, or about 3 days round-trip. Some can do it in 2 days round-trip, with a light pack. In warm summer weather get an early start when returning to the trailhead parking place.

Water There are two good springs about 1 km below the trailhead, and running water at various places in Muav Canyon. Shinumo Creek runs year-round. Carry water in your car.

Maps USGS or BLM map Grand Canyon(1:100,000), or Grand Canyon N.P.(1:62,500), and for the approach roads, the Kaibab National Forest North map(1:125,000).

Main Attractions A quiet hike in a remote corner of the Grand Canyon. Somewhere near the bottom of Shinumo Creek are the remains of a camp and garden of an early settler, W. W. Bass.

Ideal Time to Hike Because of high altitude roads, late spring and fall. Avoid mid-summer.

Hiking Boots Any dry weather boots or shoes.

Author's Experience The author camped on the rim, then went quickly down to a viewpoint of the confluence of White and Shinumo Creeks, then hurried back to the rim in a total of 8 hours.

In all parts of the Grand Canyon, you'll come across cactus with fruit (ripe in the fall).

Map 91, Shinumo Creek(North Bass Trail), Grand Canyon, Arizona

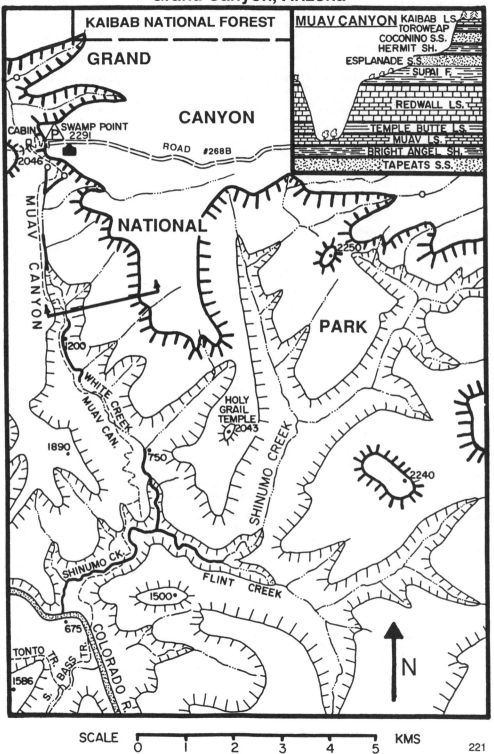

KAIBAB NATIONAL FOREST

GRAND

CANYON

MUAV CANYON

KAIBAB LS.
TOROWEAP
COCONINO S.S.
HERMIT SH.
ESPLANADE S.S.
SUPAI F.
REDWALL LS.
TEMPLE BUTTE LS.
MUAV LS.
BRIGHT ANGEL SH.
TAPEATS S.S.

CABIN
2046

SWAMP POINT
P 2291

ROAD #268B

NATIONAL

2250

PARK

MUAV CANYON

1200

WHITE CREEK

MUAV CAN.

1890

750

HOLY GRAIL TEMPLE
2043

SHINUMO CREEK

2240

SHINUMO CK.

FLINT CREEK

1500

675

COLORADO R.

BASS TR.

TONTO TR.

S. BASS TR.

1586

N

SCALE   0   1   2   3   4   5   KMS

221

Thunder River, and Tapeats & Deer Creeks, Grand Canyon, Arizona

Location and Access These three streams are located in the northwest portion of the central part of the Grand Canyon and on the north side of the Colorado River. The easiest way to get to the Monument Point Trailhead and the Bill Hall Trail, is to first drive to the Kaibab Lodge, located about halfway between Jacob Lake and the North Rim Village. About a km south of the lodge turn west on Forest Service Road #422. Drive about 32 kms and turn left on Road #425. Follow this about 16 kms to Big Saddle Camp very near the edge of the plateau rim. Just past the camp and at the 4-Way, turn west on Road #292A, which is the Monument Point Road. Drive 2 1/2 kms to the end of that road and the beginning of the Bill Hall Trail. The easiest way to get to the Thunder River Trail and the Indian Hollow Campground, is to again drive to Kaibab Lodge, about halfway between Jacob Lake and the North Rim Village. About a km south of the lodge turn west on Forest Service Road #422. Drive about 32 kms and turn left on Road #425. Follow it about 13 kms(or to within 3 kms of Big Saddle Camp), then turn west on Road #232, which is the Indian Hollow Road. From there to the campground and trailhead is about 10 kms. While planning this trip, keep in mind the highway from Jacob Lake to the North Rim is closed from sometime in November until May 15 each year. If you're there earlier than that, you'll have to use Road #422, which begins east of the sawmill in Fredonia. While in Fredonia, stop at the ranger station and pick up the Kaibab National Forest North map in order to follow these roads to either trailhead. Remember, while you're on the plateau or rim, you'll be on Forest Service land and you can camp anywhere you like. Just don't leave behind a trashy campsite. Once you walk over the rim, they you'll be in the national park.

Trail or Route Conditions From Monument Point, the Bill Hall Trail drops down through the Kaibab Limestone, then rounds the corner and finally cuts through the Toroweap, Coconino and Hermit Formations to the Esplanade bench. It then intersects the Thunder River Trail, which begins at the Indian Hollow Campground. For a ways it runs nearly level until it again drops off the Supai and Redwall to a place called Surprise Valley. At the trail junction, turn left and continue to the next rim where you'll first hear, then see, Thunder Spring, the beginning of Thunder River. This river is only about a km long, while dropping 350 meters. It's the biggest spring the author has ever seen. Walking downhill you'll reach Tapeats Creek, which may be difficult to cross in May or June because of high water. There are campsites on both sides of Tapeats. To make it to the Colorado, first travel the east side of Tapeats, then cross over to the west side for the final steep descent to the river. Other writers have mentioned Indian ruins on the east side, and a route down to the river on the west side of Tapeats(?).

Most people now use the Bill Hall to get down to the bottom of the canyon, but the Thunder River Trail offers a less-crowded alternative. From the Indian Hollow Campground, first walk west about one km along the rim, then drop down through the Kaibab, Toroweap and the Coconino Formations and head east along the contact point between the Hermit Shale and the Esplanade Sandstone. After 8 kms, you'll reach the Bill Hall Trail, then it's to the next escarpment and down the Supai and Redwall to Surprise Valley. To reach Deer Creek, turn right at the trail junction and continue on to Deer Creek Valley and the Colorado River. This trail is easy to follow all the way. Some people may want to descend Deer Creek, walk along a bench above the Colorado, and ascend Tapeats Creek, making a loop-hike(or do it in reverse order). This is not an easy hike however, and recommended only for experience and tough hikers only.

Elevations Monument Point, 2150 meters; Indian Hollow Campground, 1900; Thunder Spring, 1150; campsites on either creek, about 750; and the Colorado River, about 600 meters.

Hike Length and Time Needed It's about 19-20 kms from Monument Point to the Colorado via Tapeats Creek, which can be a short day-hike for some people, an all-day hike for others. The return trip up to the rim will nearly double the descent time, so start early and carry water. From the Indian Hollow Campground to the Bill Hall Trail, 8 kms; from the same trailhead to Deer Creek, 20 kms; trailhead to lower Tapeats, 22 1/2 kms. Most people will want a full day to reach the campsite on Deer Creek, but faster hikers can make it in 4 or 5 hours. About 3 days will allow you time to visit both major stream valleys and return to the rim.

Water Carry water in your car, there's none on the rim. Carry water for the descent, and consider caching some halfway down for the return trip up. It's hot as hell in the lower canyon in summer. Thunder River, and Tapeats and Deer Creeks each have good year-round flowing streams. This water is normally good to drink as is(?).

Maps USGS or BLM map Grand Canyon(1:100,000), or Grand Canyon N. P.(1:62,500), and for the driving route to the trailheads the Kaibab National Forest North map(1:125,000).

Main Attractions A less-crowded route to the bottom of the Grand Canyon; Thunder Spring, one of the biggest anywhere; and Deer Creek Falls, best in the Grand Canyon.

Ideal Time to Hike Because of moderately high altitude roads and the possibility of snow, late spring or mid-fall. Avoid mid-summer if possible.

Hiking Boots Any dry weather boots or shoes.

Author's Experience The author made it from Monument Point to Tapeats Creek in 3 hours, but the river was too high to cross in late May, 1985. He ended up camping on Deer Creek at the end of a 7 1/2 hour day. The next day he walked back to Monument Point in just under 4 hours. On another occasion he drove to the Indian Hollow Campground and walked along the rim a ways observing the trail below.

222

Map 92, Thunder River, and Tapeats & Deer Creeks, Grand Canyon, Arizona

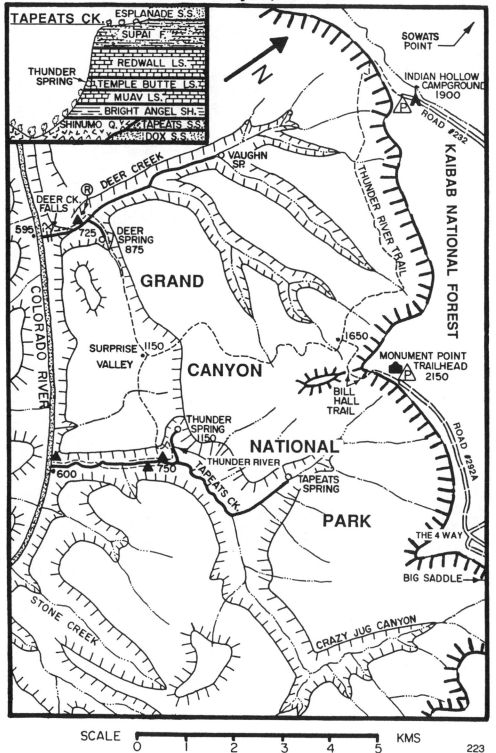

TAPEATS CK.
- ESPLANADE S.S.
- SUPAI F.
- REDWALL LS.
- THUNDER SPRING
- TEMPLE BUTTE LS.
- MUAV LS.
- BRIGHT ANGEL SH.
- SHINUMO Q. TAPEATS S.S.
- DOX S.S.

N

SOWATS POINT

INDIAN HOLLOW CAMPGROUND 1900

ROAD #232

KAIBAB NATIONAL FOREST

DEER CREEK

VAUGHN SP.

THUNDER RIVER TRAIL

DEER CK. FALLS

595

725

DEER SPRING 875

GRAND

SURPRISE VALLEY 1150

CANYON

1650

MONUMENT POINT TRAILHEAD 2150

BILL HALL TRAIL

COLORADO RIVER

THUNDER SPRING 1150

NATIONAL

ROAD #292A

600

750

THUNDER RIVER

TAPEATS CK.

TAPEATS SPRING

PARK

THE 4 WAY

BIG SADDLE

STONE CREEK

CRAZY JUG CANYON

SCALE
0   1   2   3   4   5   KMS

One of the grandest sites in the Grand Canyon; the spring at the head of Thunder River.

Tapeats Creek, just below where Thunder River joins it.

Another site to see is the Deer Creek Falls, as its water tumbles into the Colorado River.

The Colorado River looking east, from the bottom of Deer Creek Falls. In the middle of the foto is the bench just above the river, which you would use if walking from the mouth of Deer Creek to Tapeats Creek.

Snake Gulch, Arizona

Location and Access Snake Gulch is located directly west of Jacob Lake, and south of the small town of Fredonia, both of which are located on what is called the Arizona Strip. This drainage flows west from the Jacob Lake area and eventually empties into Kanab Creek. In summer you can drive west from Jacob Lake, then south on Road #422 for about 3 kms to the Oak Corral, then west for 1 1/2 kms. If you have a high clearance car or pickup, you can then drive north to the trailhead near an old stone cabin. The trailhead marks the beginning of the Snake Gulch Primitive Area. You can also drive south out of Fredonia(which has a Forest Service Ranger Station) on logging Road #422, and reach the same area. For those who want to make a loop-hike, you may want to leave a vehicle or mtn. bike at the Slide Canyon Trailhead. The road to that location is good for all vehicles except for maybe one or two bad spots, which can be fixed with a shovel. One last alternative is to turn west from Road #422 onto the Gunsight Point Road and drive about 14 kms. At that point you'll be at the head of the Swapp Trail #50. All of the main roads on this map are graded perhaps once a year and are in generally good condition for all cars--except during or just after heavy rains.

Trail or Route Conditions There has been an old 4WD track all along the bottom of Snake Gulch, but it's now blocked off and will slowly fade away with time. About 5 kms below Stone Cabin Trailhead you'll begin to see pictographs along the canyon walls. There are also some Anasazi ruins. Halfway down-canyon is the old Swapp Trail #50 coming in from the north. This is an old cattle trail. Lower Snake is an open valley. You can exit up Slide Canyon where there's a very old vehicle track up to Slide Spring. You'll then have to route-find up to the rim and to Road #267C.

Elevations Stone Cabin Trailhead, 1750 meters; mouth of Slide Canyon is 1200; Oak Corral, 1925 meters.

Hike Length and Time Needed From the Stone Cabin Trailhead, one can do a day-hike down to as far as the pictographs and ruins, then return the same way. But if a longer hike is wanted, it's recommended you do as the author did: walk down Snake to Slide Canyon, up a short trail to the car-park at 1695 meters, then walk back along Roads #267C, #267, #235, and #423 to the car-park near Oak Corral. Because it's easy walking it can be done in 2 long days, or 3 easy ones. The distance covered is about 58 kms, round-trip. A mtn. bike or other vehicle left at one end would save a lot of road-walking.

Water All named springs have year-round flows, including two with metal tanks. In wetter years Kanab Creek can flow year-round. Carry water in your car, as all other creeks are dry.

Maps USGS or BLM map Fredonia(1:100,000), Jumpup Canyon and Big Springs(1:62,500), and for the access route, the Kaibab National Forest North map(1:125,000).

Main Attractions Second best place on the Colorado Plateau to see pictographs.

Ideal Time to Hike Spring or fall, or morning-hike in summer.

Hiking Boots Any dry weather boots or shoes.

Author's Experience The author parked just west of Oak Corral and walked down Snake, up Slide, and by road back to his car. He made the 58 km hike in two days, and in about 15 hours total walk-time.

Pictographs in Snake Gulch.

Map 93, Snake Gulch, Arizona

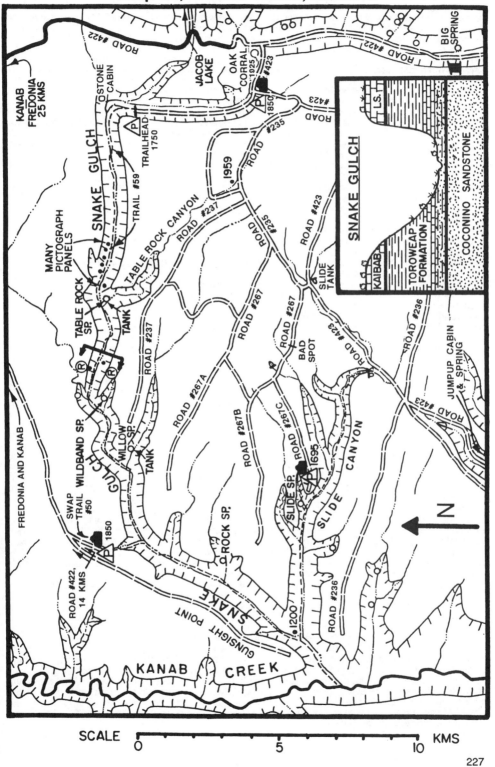

SCALE 0 5 10 KMS

227

Introduction--The Western Grand Canyon-Arizona Strip

Location and Access This map shows the western half of the Arizona Strip, which is the land south of the Utah-Arizona state line and north of the Grand Canyon. It includes the area south of St. George, Kanab and Fredonia, and all of the Bundyville, Mt. Trumbull and Shivwits Plateau country. All access is from St. George and Fredonia. To get there, drive south out of St. George on I-15 to the Bloomington Exit #4. Leave the freeway and drive east through a suburb, then turn south on the paved Mt. Trumbull Road. A few kms away the pavement ends at the state line. From Fredonia, drive west toward Pipe Springs and Colorado City. Between mile posts 24 and 25 turn south onto the Mt. Trumbull Road. The Mt. Trumbull Road makes a loop to the south and passes Mt. Trumbull and the abandoned townsite of Bundyville(sometimes called Mt. Trumbull). This is a good well-maintained road, with a fair amount of traffic, but in places it has large sharp rocks, so you better release a little air from your tires to avoid a flat.

Here is the land ownership situation. As you leave St. George and Fredonia, you'll be driving across BLM land. Once you get east and south of the Mt. Trumbull area, you'll come to the Grand Canyon National Park. To the west of that and in the Parashant Canyon--Shivwits Plateau country, you'll be in the upper part of Lake Mead National Recreation Area. Below the rim will always be the Grand Canyon N. P. again. South of the Colorado River is the Hualapai Nation lands

Distances It's a long way from this area to the nearest gas station so fill up and take extra fuel. Remember, on rough dirt roads you'll consume a lot more fuel than on a paved highway. It's about 125 kms(78 miles) from St. George to the Shivwits Fire Camp. On one trip the author drove from St. George to Twin Point, then the Shivwits F.C., and on to Shanley Tanks; then back to the Mt. Trumbull Road, to the bottom of Whitmore Canyon, to Trail Canyon on the Parashant rim, then back to St. George. Round-trip was 517 kms(322 miles). Excuse the bragging, but his VW Rabbit Diesel used about 23 liters(6.1 gallons) and got 52.92 MPG. You'll need a lot more fuel than that with your gas hog, so go well-prepared with plenty of fuel and anything else you think you might need in a far away and isolated region!

Road Conditions The Mt. Trumbull Road is well signposted by the BLM at every major turnoff. Some parts of this road may be considered all-weather, while other parts can be slick and muddy during and right after heavy rains. It can be traveled by any vehicle, except for the section running up to the east side of Mt. Trumbull which has some steep curves and is not well suited for trailers. You can bypass this steeper section by taking a lower road around the north side of the mountain. To reach Toroweap Point, follow the main road south as shown. This is also a very good road for all vehicles, but it deteriorates a little at the very southern end. To reach the Whitmore Trail, turn south at the Bundyville School and follow this moderately good road down into Whitmore Canyon. Cars can usually make it to the Bar 10 Lodge, but it gets rough near the ranch(take a shovel). Below the lodge, the road deteriorates rapidly, but any HCV should be able to make it to the trailhead overlooking the river. One way to reach Parashant Canyon is to drive west from Bundyville, then southwest and south to the head of Trail Canyon. This road is very good all the way to where it drops down into Trail, but don't go further unless you have a 4WD and like washed-out roads. It's terrible driving in the bottom of Parashant. To reach the Shivwits country, turn off the Mt. Trumbull Road at Diamond Butte. This is a very good road to the Shivwits Fire Camp and can be traveled by any car, but like all area roads, it can be slick in places during or after heavy rains. From near the fire camp the road to Twin Point is good for any vehicle, but much of the road to Kelly Point is for HCV's only. Always carry a shovel in your car, and if caught by heavy rains, just stop and wait an hour or two before driving on.

Water There is no running water anywhere in this area except in the very bottom of several canyons near the Colorado River. Take water for the duration of your trip! Once on the Strip, there are very few reliable places to get good water at or near any of the main roads. One water tap is on the road immediately north of the BLM Ranger Station on the south side of Mt. Trumbull. It's just west of the trailhead for the hike up the mountain. It comes from Nixon Spring higher up-slope, but its storage capacity is only one 55 gallon drum, so please don't waste water. There is also a tap next to the BLM workshop west of Poverty Mountain. This is at a major junction with a side road running southeast to the upper Parashant. The workshop is in the forest behind a green gate. This piped water comes from Poverty Spring. The author has filled up his water bottles there several times and never gotten a bad batch. The Bar 10 Lodge in Whitmore Canyon has a good spring and the owner, Tony Heaton of St. George, is there year-round. In emergencies only, you may be able to get a little water at the Toroweap Ranger Station and the Shivwits Fire Camp. These people have to haul their water in, so try and avoid intruding on them(Shivwits is open only from mid-May to late September). You should also be able to get water in an emergency at Buster Esplin's Wildcat Ranch. He uses the well water only for showering and washing dishes, but it will likely be OK to drink as is? His metal stock tank has gold fish in it, so you can get wash water there anytime. All around this area you will observe earthen dams called stock ponds or tanks. They will have water for perhaps washing after good rains. These tanks have muddy water much of the year, but are not reliable sources for water. Two stock tanks which seem to have water all the time will be the ponds at Lake Flat(north of Shivwits Fire Camp) and the Shanley Tanks on the Kelly Point Road. You'll have to treat all water from any of these tanks. In addition, there is a good year-round flow at Green Spring west of Mt. Dellenbaugh, but the road getting there is rough as hell. Just south of Green Spring are the Ambush Tanks. These are lava potholes, but likely have water year-round. Also the Parashant Ranch south of Oak Grove has the Mociac Well, which will normally have water. The BLM people pump it to a tank, then it flows down to a second tank just above the brick house. You'll have to turn the facet on at the second tank to get water to the outside tap at the house. The BLM says it may not be drinkable, but the well is capped, so it couldn't be all that bad(?). The VT and Waring Ranches are now abandoned.

Map 94, Introduction to the Western Grand Canyon and the Arizona Strip, Arizona

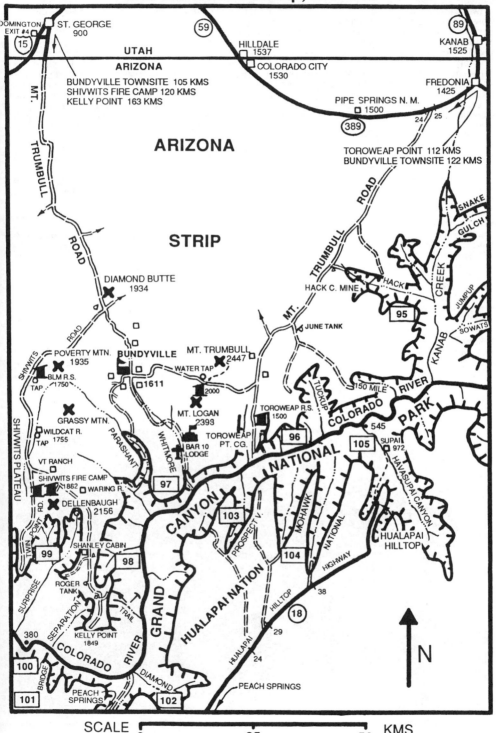

BLOOMINGTON EXIT #4

15

ST. GEORGE 900

UTAH

59

ARIZONA

BUNDYVILLE TOWNSITE 105 KMS
SHIVWITS FIRE CAMP 120 KMS
KELLY POINT 163 KMS

HILLDALE 1537

COLORADO CITY 1530

89

KANAB 1525

FREDONIA 1425

PIPE SPRINGS N. M. 1500

389

24 25

ARIZONA

MT. TRUMBULL ROAD

TRUMBULL ROAD

STRIP

TOROWEAP POINT 112 KMS
BUNDYVILLE TOWNSITE 122 KMS

SNAKE GULCH

DIAMOND BUTTE 1934

HACK C. MINE

HACK

CREEK

JUMPUP

95

SOWATS

MT.

ROAD

JUNE TANK

POVERTY MTN. 1935

BUNDYVILLE

MT. TRUMBULL 2447

WATER TAP

SHIVWITS TAP

BLM R.S. 1750

1611

2000

150 MILE

COLORADO

RIVER

PARK

GRASSY MTN.

WILDCAT R. TAP 1755

MT. LOGAN 2399

TUCKUP

TOROWEAP R.S. 1500

545

SUPAI 972

SHIVWITS PLATEAU

VT RANCH

PARASHANT

WHITMORE

BAR 10 LODGE

TOROWEAP PT. CG.

96

105

HAVASUPAI CANYON

SHIVWITS FIRE CAMP 1862

WARING R.

97

NATIONAL

HUALAPAI HILLTOP

DELLENBAUGH 2156

TWIN POINT RD.

99

SHANLEY CABIN

98

103

PROSPECT V.

MOHAWK

NATIONAL

104

HIGHWAY

CANYON

SURPRISE

ROGER TANK

TRAIL

GRAND

HUALAPAI NATION

HILLTOP

38

18

SEPARATION

KELLY POINT 1849

29

380

COLORADO

RIVER

HUALAPAI

24

100

BRIDGE

DIAMOND

PEACH SPRINGS

N

101

PEACH SPRINGS

102

SCALE

0   25   50   KMS

229

Lower Kanab Creek, & Tributary Canyons, Grand C., Arizona

Location and Access This map features the lower part of Kanab Creek where it enters the Grand Canyon. This area is due south of Kanab and Fredonia. One normal route to lower Kanab Creek is via one of two roads in the upper end of Hack Canyon. To get there, drive west out of Fredonia on Highway 389. Between mile posts 24 & 25, turn south onto the Mt. Trumbull Road. Drive south about 36 kms and turn left or southeast at the sign stating Hack Reservoir. Drive a sometimes-maintained road down Hack Canyon to an old uranium mine and beyond to near Willow Spring. Another way to Willow Spring is to regress about 12 kms to the northeast from the beginning of Hack Canyon to where some power lines cross the Mt. Trumbull Road. At that point, a good road heads south past a corral & windmill, a fence & cattle guard, the Sunshine stock tank(pond), 4 metal tanks, and finally to the rim of Hack. From where a side road turns left and heads north, go about 400 more meters and look for a minor track on the right. Walk or drive this about 200 meters to a cedar post & wire fence corral. Just below this is the beginning of a good trail heading down to Willow Spring. This road is good for cars. Another route in is from Jumpup Cabin & Spring and Jumpup Canyon. A fourth way is to drive to Sowats Point. To do these routes get the *Kaibab National Forest North map* at the Fredonia Ranger Station and have someone there point out the better roads. These road are occasionally graded and in good condition for all cars--in dry weather conditions.

Trail or Route Conditions From Willow Spring, down to about Jumpup Creek, there are old cattle trails, then you walk in the dry creek bed or the small stream. From Sowats Point, there's a good well-used trail down into Sowats, Kwagunt, lower Indian Hollow and Jumpup Canyons. You can use either of these canyons to reach lower Kanab Creek. If roads to Sowats Point are dry, this is the recommended access route. As you get further into Kanab Canyon, the walls get higher and water begins to flow. Below Shower Bath Spring there's an area of boulders which makes for slow walking, but it's not so difficult.

Elevations Willow Spring Trailhead, 1625 meters; Willow Spring, 1175; Sowats Point, 1800; the Colorado River, 529 meters.

Hike Length and Time Needed From the Hack Canyon trailheads to the Colorado, is about 48 kms, or 4-5 days round-trip. From Sowats Point to the Colorado, 35 kms one-way, or about 3-4 days round-trip.

Water Kanab Creek has a year-round flow in its lower end. Jumpup Spring is troughed, and Lower Jumpup Spring has a flowing stream for half a km. Sowats and Kwagunt have flowing water in places, as shown. Kwagunt has 3 nice *tricklefalls* lower down. Willow Spring has water year-round.

Maps USGS or BLM maps Fredonia & Grand Canyon(1:100,000), and Kaibab National Forest North(1:125,000).

Main Attractions Very deep moderately narrow canyon, solitude, and Shower Bath Spring.

Ideal Time to Hike Spring or fall, or winter dry spells(Hack Canyon route only). From Sowats Point, late spring or early fall, because of higher altitude roads. Avoid hiking in summer if you can!

Hiking Boots Dry weather boots or shoes, but you'll want waders for the bottom end of Kanab Creek.

Author's Experience In 1986, the author walked from the Hack Canyon Mine to the Colorado River and back in a period of 4 days, or about 26 hours total walk-time. He later drove out to the Willow Spring Trailhead and walked down the trail to the bottom of Hack Canyon. Still later, he drove to Sowats Point and made a one day loop-hike down to Lower Jumpup Spring, then down Sowats and up Kwagunt. Total walk-time was 9 1/2 hours. He has also driven to the Jumpup Cabin and walked down below Upper Jumpup Spring a ways.

John W. Powell was the man who named this feature called Shower Bath Spring in the lower end of
Kanab Creek.

Map 95, Lower Kanab Creek, and Tributary Canyons, Grand Canyon, Arizona

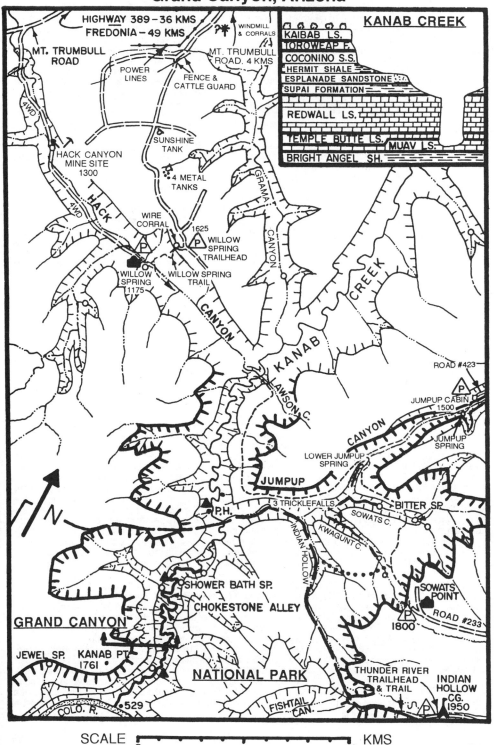

HIGHWAY 389 – 36 KMS
FREDONIA – 49 KMS

WINDMILL & CORRALS

MT. TRUMBULL ROAD

MT. TRUMBULL ROAD. 4 KMS

POWER LINES

FENCE & CATTLE GUARD

4WD

HACK CANYON MINE SITE 1300

SUNSHINE TANK

4 METAL TANKS

HACK

4WD

WIRE CORRAL

1625

P

WILLOW SPRING TRAILHEAD

P

WILLOW SPRING TRAIL

WILLOW SPRING 1175

CANYON

GRAMA CANYON

KANAB CREEK

KAIBAB LS.
TOROWEAP F.
COCONINO S.S.
HERMIT SHALE
ESPLANADE SANDSTONE
SUPAI FORMATION
REDWALL LS.
TEMPLE BUTTE LS.
MUAV LS.
BRIGHT ANGEL SH.

KANAB CREEK

ROAD #423

P

JUMPUP CABIN 1500

JUMPUP SPRING

LAWSON C.

CANYON

LOWER JUMPUP SPRING

JUMPUP

P.H.

3 TRICKLEFALLS

SOWATS C.

BITTER SP.

N

INDIAN HOLLOW

KWAGUNT C.

SOWATS POINT

SHOWER BATH SP.

CHOKESTONE ALLEY

P
1800

ROAD #233

GRAND CANYON

JEWEL SP.   KANAB PT. 1761

NATIONAL PARK

529

COLO. R.

FISHTAIL CAN.

THUNDER RIVER TRAILHEAD & TRAIL

INDIAN HOLLOW CG. 1950

P

SCALE   0   5   10   KMS

231

The Tuckup, Cove Canyon and Lava Falls Trails, Grand Canyon, Arizona

Location and Access These three trails are located in the western part of the Grand Canyon and on the north side of the Colorado River. To get there, first stock up on fuel, food and water in Fredonia, then drive west on Highway 389 to between mile posts 24 & 25, and turn south onto the Mt. Trumbull Road. Drive this very good gravel & improved road for about 89 kms to the Toroweap Ranger Station. Stop there to get updated information about the area, then continue south another 10 kms to the canyon rim overlook at Toroweap Point. This is one of the best places in the park to view the Colorado River.

To hike the Tuckup Trail, drive back up the Toroweap Point Road about 1 km, and turn east. From the main road, a 4WD track runs east for about 5 kms, then the actual trail begins. Cars can go only a km or two on this last section. To reach the trailheads at the heads of Tuckup and 150 Mile Canyons, turn south off from the Mt. Trumbull Road just south of Findlay & Heaton Knolls and drive 200 meters to June Tank. From there the road branches into two. By following a good topo map carefully and always staying on one of these two main roads, you can make it to either trailhead easily, even in a car--if road conditions are dry! The author's VW Rabbit made it to both trailheads.

If you're going to do the hike down Lava Falls, drive south from the Toroweap Ranger Station, and instead of turning left to reach the Tuckup, turn right not far north of the cinder cone known as Vulcans Throne, and drive about 4 kms to the end of a very rough road.

Trail or Route Conditions The Tuckup is an old livestock trail that begins near Toroweap Point and ends at the mesa top above Buckhorn Spring at the head of 150 Mile Canyon. At the halfway point there's an entry/exit beginning above Schmutz Spring. The Tuckup is used very little today, but even though it may fade in places, it's easy to follow, because it contours along the Esplanade Sandstone which forms a natural bench. Few people do the entire trail, but some are going to the Tuckup & 150 Mile Canyon trailheads and taking day-hikes down to the next level, or attempting to reach the Colorado. To do that requires about 4 rope pitches in either canyon. There is a very good pictograph panel and old copper mine(just below Cottonwood Sp.) in Tuckup; and an old stockman's or hermit's camp just above Hotel Spring in 150 Mile.

There's an interesting hike down to the river from very near the end of the road leading to the beginning of the Tuckup Trail. At the sharp corner where the road comes close to the rim of Cove Canyon, walk east instead of following the road north. There's a faint vehicle track there, which eventually turns to a trail that drops off the rim of Cove, then contours around to the south above the river. The trail soon disappears, then you simply head down a steep ravine to the west, while slowly veering south. Follow this gully to the river. The Cove Canyon Trail itself covers only the first third of this route down.

From or near the Lava Falls Trailhead, which is just southwest of the Vulcans Throne cinder cone, a hiker-made trail drops off the edge and zig zags down the steep lava-covered face of the canyon wall. This trail is marked with stone cairns and is easy to follow, but it sometimes branches into two or three routes. At a point in time after the Grand Canyon was formed, a volcanic eruption of Vulcans Throne and other volcanos occurred, and the result was a lava flow pouring over the Esplanade and into the river. There have been several temporary lava dams at this site in the past million years or so.

Elevations The Tuckup Trail, from 1250 to 1350 meters; except the plateau trailheads which are at 1750 & 1700; Cove Canyon & Lava Falls Trailheads, 1280 & 1300; Colorado River, about 500 meters.

Hike Length and Time Needed From Toroweap Point to Buckhorn Spring is about 100 kms, but you'll have to have two vehicles(or one mtn. bike) to make this hike practical. Most people would want about 4-5 days for the hike, but that would depend on the time of year. In cooler weather, there's less water to carry. Getting out at the head of Tuckup Canyon would cut the time in half. To hike the Cove Canyon Trail(and route) down to the river and back, will take no less than half a day; but for some it'll take all-day. The time it takes will depend on where you park, or if you have a mtn. bike. From the Lava Falls Trailhead to the river is only about 2 1/2 kms, but it's not an easy hike. There are lots of loose rocks and scree. For most people it's a half-day hike round-trip. Another hike you might try is to walk the Tuckup Trail to Tuckup Canyon, with ropes drop down into lower Tuckup by way of Cottonwood Spring, then walk along the Colorado to the Lava Falls Trail, and back up from there. About 100 kms, and maybe a 5 day trip, or longer.

Water Water is your biggest problem in the Toroweap region, so carry plenty in your car. If you're there after recent rains, you'll be in luck, because you'll find seeps and pothole water all along the rim of the Esplanade. After long dry spells, water can be a real problem. The author never did find Schmutz Spring, but Cottonwood, Buckhorn and Hotel Springs each have good water, along with a nice little stream in the middle part of 150 Mile Canyon as shown. Colorado River water should be treated before drinking, but if it's clear, you could probably drink it as is, and survive.

Maps USGS or BLM maps Grand Canyon, Mount Trumbull, Fredonia(access route), or the BLM's Visitor Map--Arizona Strip District(1:100,000), or Tuckup Canyon & Kanab Point(1:62,500).

Main Attractions Wild solitude, and great views of the inner canyon gorge. Also, Vulcans Throne and lots of Recent or Pleistocene volcanic cinder cones to the west, and south across the river.

Ideal Time to Hike Spring or fall, or maybe some winter warm spells. Summers are extra hot.

Hiking Boots Rugged boots or shoes, especially on the Lava Falls & Cove Canyon Trails.

Author's Experience The author hiked several kms along the Tuckup Trail for fotos of the inner gorge, then returned the same way. Later that same day he walked the last km of road to the actual beginning of the Lava Falls Trail, then worked his way down to the Colorado and had a skinny dip before climbing back up. The total round-trip time was 2 1/2 hours. On another trip, he mtn. biked to the end of the road leading to the Tuckup Trail; then found and hiked the Cove Canyon Trail down to the river, had a skinny dip, then returned, all in 4 1/2 hours. Later, he hiked down into Tuckup and 150 Mile Canyons from each trailhead in 5 1/2 and 7 hours each, round-trip.

Map 96, The Tuckup, Cove Canyon and Lava Falls Trails, Grand Canyon, Arizona

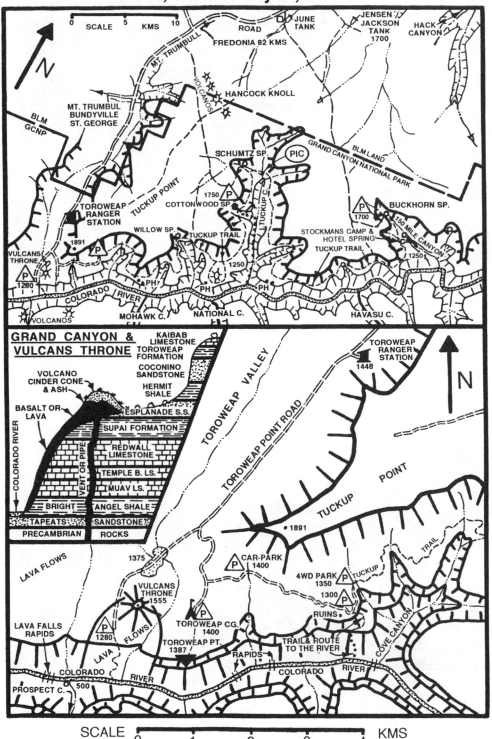

An old cattleman's camp down in 150 Mile Canyon. Please leave everything in place.

From the south side of Toroweap Point, looking down at the Colorado. In the distance is the rockslide near the mouth of Cove Canyon, where a route leads down to the river.

From the south side of the river looking west, one can see Vulcan's Throne to the upper right, and the lava flows and trail leading down to the river.

Pictographs in upper Tuckup Canyon.

Parashant & Whitmore Canyons, Grand Canyon, Arizona

Location and Access Parashant & Whitmore Canyons are located in the western Grand Canyon on the north side of the Colorado. Both are due south of St. George. The easiest way to get there is to locate the Mt. Trumbull Road at the southern end of St. George and drive 95 kms south to the old townsite of Bundyville. From the Bundyville School, get to Parashant by turning west, then south, then west. At the first major junction after a gate, stay left; a km after that turn right. From there stay on the main road heading south, then southeast for a total of 16 kms to a parking area at the top of Trail Canyon. For this drive have the Mount Trumbull map listed below. The Trail Canyon Road down to the Copper Mtn. Mine is first steep, then rough where it crosses the dry creek bed. This road is for 4WD's or mtn. bikes only.

To reach the Whitmore Trail, drive south from Bundyville School staying on the main road. After 23 kms you'll come to the Bar 10 Lodge. Most cars can make it there. From the lodge it's 16 more kms to the Whitmore Trail on a rough-in-places HCV road. Those with higher clearance cars might make it, but with the help of a shovel.

To reach the trailhead above Frog Spring in the lower end of Parashant, turn west off the road in lower Whitmore Canyon just north of the cinder cone. This 4WD road runs southwest, then north, for about 20 kms, before ending in two places on the rim of lower Parashant Canyon above Frog Spring.

Trail or Route Conditions One can reach the bottom of Parashant and the Colorado by walking straight down the dry creek bed; or into the canyon east of the Copper Mtn. Mine. From the mining cabin on top, walk east down the mine road a ways and to the bottom of the tributary, then through 2 sets of good limestone narrows(bypass the first narrows on the west) and into the main drainage. From there, it's an easy/fast walk to the river in the dry creek bed. Along the way are good Redwall Limestone narrows.

An alternate way into lower Parashant is via the Whitmore Point Road to the area above Frog Spring. Near the end of the road in Parashant, take the track veering right or northeast, and go to its end to a bulldozer-cleared site. From there back up 200 meters and walk due north in the drainage just west of the end of the road. After a ways you should start to see a trail. It runs north to the rim, then zig zags down to Frog Spring on the next level below, which is the Redwall bench. From there you'll have to walk down-canyon to a point 300 meters above, or below, the mouth of the Copper Mtn. Mine tributary canyon to find a route down to the bottom. You can also climb down from the Whitmore Point Road to Frogy Fault Canyon, walk northwest to near Cedar Spring, then use one of the two routes mentioned above to get into Parashant's narrows. There is still another way to the Colorado in this area. Observe on the map the route leaving the Whitmore Point Road and running out onto Lone Mountain. Follow this route to the rim, then route-find down a steep ravine to the river.

From about halfway along the Whitmore Point Road, is an old cattle trail running all the way to the Colorado. The Cane Spring Trail starts about two kms southwest of the cinder cone, and runs down toward Cane Spring. Instead of going up-canyon toward this spring, head east down the dry creek bed. After a ways, get upon the bench to the right or south, and follow the route shown on the map. You'll lose the trail in the flats, but when it starts over the last rim, it's easy to find & follow. You can also get to the lower end of the Cane Spring Trail from the cabin near the beginning of the Whitmore Trail.

The Whitmore Trail is well-constructed. It's easy to follow past columnar jointing created by thick lava flows. In his book, *Grand Canyon Treks II*, Butchart says you can walk from the mouth of Parashant along the Colorado to either the Cane Spring or Whitmore Trail. *This is for tough hikers only.*

Elevations Trail Canyon Trailhead, 1645 meters; Whitmore Trailhead, 781; Frog Spring Trailhead, 1150; Cane Spring Trailhead, 1075; Colorado River, between 475 & 495 meters.

Hike Length and Time Needed From the Copper Mtn. Mine to the Colorado is about 16 kms. It's about the same distance from the Frog Spring Trailhead. From either trailhead, it's a one-day hike, round-trip. The route to the river via the Whitmore Point Road and Lone Mtn. is 8 or 9 kms, and an all-day hike & climb. The Cane Spring Trail is roughly 10 kms long, and is also an all-day trip. The Whitmore Trail is 1.5 kms long, or an easy half-day hike round-trip from the trailhead at 781 meters altitude.

Water Carry plenty in your car! Cedar Spring is the only reliable water in the Parashant. At one time the copper miners ran a hose from there across the narrows to the north side. You'll walk under this hose going down-canyon. Normally there's water at the stock pond, tank & trough in Parashant, and sometimes at Cow Tanks(?). Also at Cane Spring, which is about half a km from the trail. Look for two large, pink plastic tanks. Also at the Bar 10 Lodge, next to the airstrip below the lodge, and the Colorado River, which has better water than most people realize.

Maps USGS or BLM maps Mount Trumbull, or Visitor Map--Arizona Strip District(both 1:100,000), or Whitmore Rapids, Whitmore Point, and Whitmore Point SE(1:24,000).

Main Attractions Wild isolation, deep narrows in lower Parashant Canyon, several old cattle trails and routes down to the Colorado River, and volcanic columnar jointing along the Whitmore Trail.

Ideal Time to Hike Spring or fall, or perhaps some warm & dry spells in winter. Summers are hot!

Hiking Boots Any dry weather boots or shoes.

Author's Experience On his third try, the author rode a mtn. bike down to the Copper Mtn. Mine in about 4 hours, then set up camp in the canyon east of the mine. Next day he made it to the river & back in 7 1/2 hours. He returned to his car in 4 hours on Day 3. He made it to the cinder cone in Whitmore in his VW Rabbit, then walked to the river(Whitmore Trail), had a skinny dip and walked back to his car in a total time of 2 3/4 hours. Another trip, he camped just north of the cinder cone and rode a mtn. bike to the Cane Spring Trail. He hiked, and hunted for Cane Spring, then made it to the river for a skinny dip, then returned to his car in a total time of 8 1/2 hours. He once left his bike on the Whitmore Point Road, went over Lone Mtn. to the river, skinny dipped, then returned to his bike in 8 1/4 hours. He has also mtn. biked to Frog Spring and explored on foot.

Map 97, Parashant and Whitmore Canyon Trails and Routes, Grand Canyon, Arizona

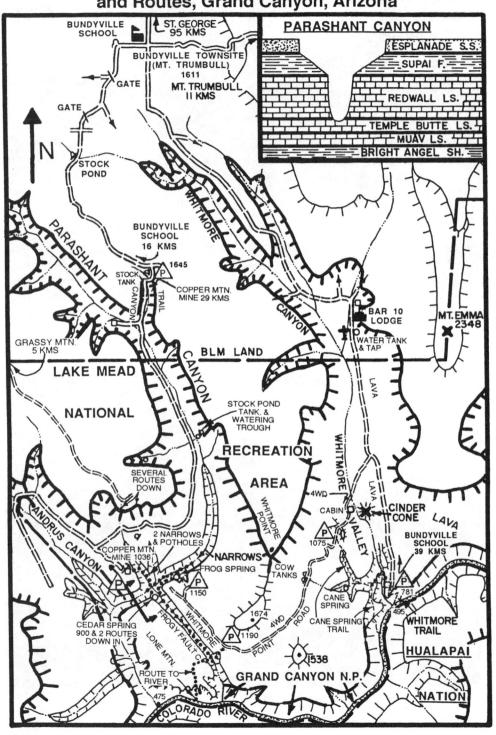

BUNDYVILLE SCHOOL

ST. GEORGE 95 KMS

BUNDYVILLE TOWNSITE (MT. TRUMBULL) 1611

MT. TRUMBULL 11 KMS

GATE

GATE

N

STOCK POND

PARASHANT

PARASHANT CANYON

ESPLANADE S.S.

SUPAI F.

REDWALL LS.

TEMPLE BUTTE LS.

MUAV LS.

BRIGHT ANGEL SH.

WHITMORE

BUNDYVILLE SCHOOL 16 KMS

P 1645

STOCK TANK

TRAIL

COPPER MTN. MINE 29 KMS

CANYON

CANYON

BAR 10 LODGE

WATER TANK & TAP

MT. EMMA 2348

GRASSY MTN. 5 KMS

BLM LAND

LAKE MEAD

NATIONAL

CANYON

LAVA

STOCK POND TANK, & WATERING TROUGH

RECREATION

AREA

SEVERAL ROUTES DOWN

WHITMORE POINT

WHITMORE

4WD

CABIN

LAVA

CINDER CONE

LAVA

ANDRUS CANYON

2 NARROWS & POTHOLES

COPPER MTN. MINE 1036

P

P 1075

VALLEY

BUNDYVILLE SCHOOL 39 KMS

NARROWS

FROG SPRING

P 1150

COW TANKS

P 781

495

CEDAR SPRING 900 & 2 ROUTES DOWN IN

FROGY FAULT C.

WHITMORE

LONE MTN.

1674

P 1190

4WD ROAD

POINT

CANE SPRING

CANE SPRING TRAIL

WHITMORE TRAIL

ROUTE TO RIVER

475

1538

GRAND CANYON N.P.

HUALAPAI

NATION

COLORADO RIVER

SCALE   KMS

0   5   10

The cabins at the Copper Mtn. Mine. Probably the best way into this rugged region is with a mtn. bike.

The limestone narrows in Parashant Canyon.

Looking east up the Colorado River from the top of the Whitmore Trail. Notice the black lava coming down from the left which once flowed into the river.

This picture shows the plastic tanks and watering trough below Cane Spring.

Spring, Indian, 209 and 214 Mile, Trail and Separation Canyons, Grand Canyon, Arizona

Location and Access This area is in the western Grand Canyon on the north side of the Colorado due south of St. George. It's also just north of where Peach Spring Canyon enters the Colorado. Getting to this isolated corner of the Grand Canyon is easy for most of the way, then it gets difficult as you go beyond the Shivwits Fire Camp. You can get there from Fredonia, but by far the best way is to drive south out of St. George on the Mt. Trumbull Road. Just south of Diamond Butte, turn southwest onto the Shivwits(or Parashant) Road #103, and follow the signs to Mt. Dellenbaugh(see Map 99). Near the Shivwits Fire Camp, continue southeast toward the old Waring Ranch, then on to the old sawmill at Green Spring. You'll need a HCV to go beyond the Waring Ranch. The road is over basalt boulders from the Dellenbaugh volcano. From Green Spring, continue south to Ambush Canyon, then the road improves as it continues south to Shanley Tanks & Cabin. From Shanley Tanks(ponds) continue south. Southeast of Blue Mtn., the road improves and is pretty good all the way to Kelly Point. If going down 209 Mile Canyon, park at Roger Tank(all the stock tanks on this road are signposted). If you're going into Separation, park at Roger Tank, or drive 2 kms beyond and park where the road turns east. If you're heading for the Snyder Mine & Trail or 214 Mile Canyons, continue east to Kelly Tanks, and about 200 meters beyond, turn left or southeast and continue to the east rim. You can also drive or mtn. bike from Shanley Tanks toward Amos Spring, for a hike down to that spring and points beyond. For people with cars, it's best to park near the Waring Ranch and mtn. bike to the Shanley Tanks, perhaps using the cabin there as a base*(please leave it clean)*.

Trail or Route Conditions To reach the Colorado via either Spring or Indian Canyons, walk east from Shanley Tanks to a point about 200 meters south of Price Point, then head straight down a big rock slide. From there head east again as shown on the map. Both canyons have dryfalls, so you have to walk along their south rims to just southwest of their mouths, then route-find down to the river.

From the end of the road at 1825 meters, walk 400-500 meters north along the rim, then head east down a gully and into the head of Trail Canyon. Walk down the dry creek bed to the first big dryfall. Immediately below you at that point will be Shanley Spring. Head right 75 meters and walk down a trail to get to the spring itself, which is in two different little caves. *This is the only life-saving water around!* Continue straight down Trail Canyon, skirting to the right to get around one dropoff, until you reach the river. There's no trail in Trail Canyon, but there is in 214 Mile. Head north from Shanley Spring to find the tailings from the Snyder Copper Mine on a southeast facing slope. From there, head northeast to the canyon rim, and to a point just north of a side drainage dryfall, to find the beginning of the Snyder Mine Trail. It zig zags straight down, then veers right or east to almost the bottom, then bypasses a dryfall, and finally ends up in the bottom less than 2 kms up from the river. From near the end of Kelly Point, you can head southeast to find a route down to the river as shown.

The route down into the east fork of Separation is probably the same route taken by three of J. W. Powell's men who abandoned his expedition in August, 1869. They were later killed by Shivwits Indians just south of the Wildcat Ranch(see Arizona Strip map). From the road & trailhead shown, walk due south about 500 meters to the rim and route-find down past Kelly Spring into the upper east fork(from Roger Tank as well). Once down in, it's an easy and fast walk to the river.

From the end of Amos Spring Road, rim-walk west to find many easy routes down to the bottom, then walk the dry creek bed. Amos Spring has good water. The trail below it is marked with stone cairns.

Elevations Shanley Tanks & Cabin, 1842 meters; Roger Tank, 1810; Snyder Mine, 1125; Shanley Spring, 950; the river(lake) at Separation, 373; the Colorado River at Fall Canyon, 435 meters.

Hike Length and Time Needed From Shanley Cabin to the Colorado via Spring or Indian Canyons will take most people one day to reach the river, then a second day coming out. Strong hikers who know either route can do it round-trip in one very long day. It'll be about the same amount of time to go from Roger Tank down 209 Mile Canyon and back. From a camp on the rim at 1825 meters, strong hikers could reach the river via 214 Mile or Trail Canyons and return in the same day, but that's too much for most. Best to set up camp near Shanley Spring, then go to the river via either canyon and return, taking several days. Down Separation from either car-park to the river is about 19-20 kms, and will take one day down, then another very long day back up.

Water Shanley Spring, Amos Spring, Kelly Spring(maybe at a seasonal spring on top of the Esplanade rim below), most of the time at Shanley & Ambush Tanks(natural potholes), and the Colorado.

Maps USGS or BLM maps Peach Springs & Mount Trumbull, and Visitor Map--Arizona Strip District(1:100,000), or Price Point, Granite Park, Amos Point, Separation Canyon & Travertine Rapids(1:24,000).

Main Attractions A true wilderness experience and historic copper mine(before 1910?).

Ideal Time to Hike March & April, and October & early November. The Shivwits Plateau can have deep snow in winter, and the lower canyons are unbearably hot in summer.

Hiking Boots Rugged dry weather hiking boots.

Author's Experience On a hot August 1, the author made it to Ambush Tanks with his VW Rabbit with oversized tires, and camped. Next day he rode a mtn. bike to the rim of Trail Canyon, hiked down and found Shanley Spring and the mine, and made it back to Ambush in 11 1/2 hours, round-trip. On the next trip, he drove to a site 2 kms south of Shanley Tanks and camped. From there he mtn. biked to Trail Canyon, then hiked to Shanley Spring, and to within 2 kms of the river in 214 Mile Canyon before returning in 11 1/2 hours, round-trip. Once he hiked from Shanley Cabin to halfway down the last slope to the river in Indian Canyon and returned in 11 hours round-trip. He also mtn. biked the Amos Spring Road and got to the area near the stone cabin and returned in 8 1/2 hours. On a boating trip up from Lake Mead, he walked up Separation to where the route exits east fork. That hike took 9 1/2 hours round-trip.

Map 98, Spring, Indian, 209 and 214 Mile, Trail and Separation Canyons, Grand Canyon, Arizona

SHIVWITS PLATEAU & FALL CANYON

KAIBAB L.S.
TOROWEAP F.
COCONINO S.S.
HERMIT SH.
ESPLANADE S.S.
SUPAI SHALE
REDWALL LIMESTONE
TEMPLE BUTTE L.S.
MUAV LIMESTONE
BRIGHT ANGEL SHALE
TAPEATS SANDSTONE
PRECAMBRIAN GRANITES & GNISSES

SANUP PLATEAU

TONTO RIM
COLORADO RIVER

CABIN

ST. GEORGE 130 KMS
SHIVWITS FIRE CAMP & RANGER STATION 15 KMS

SHIVWITS PLATEAU

KELLY POINT ROAD

1851

SPRING

CANYON

COLORADO

RIVER

DOWN

INDIAN CANYON

DOWN

SHANLEY CABIN
SHANLEY TANKS
1842

PRICE POINT 1975

ROUTE

DROPOFF

PRICE CANYON

DOWN

209 MILE CANYON

GREEN
SPRING CANYON

SPRING

ROAD

AMOS SPRING
AMOS SPRING
STOCK TANK
AMOS POINT

BLUE MTN. 2006

KELLY

ROUTE

FALL CANYON

435

GRANITE

STONE CABIN

1404

SANUP

FORK

WEST FORK

CANYON

RUNNING WATER

ROGER TANK 1810

KELLY SP.

SEASONAL SPRING?

POINT

KELLY TANKS

1825

214 M. CANYON TRAIL

TRAIL

SNYDER MINE 1125

DROPOFF

SHANLEY SPRING 950

CANYON

PARK

GORGE

SEPARATION

DRYFALL

EAST FORK

PH.

ROAD

1849

KELLY POINT

RUNNING WATER

ROUTE

1406

DOWN

220 MILE CANYON

NATIONAL

HUALAPAI NATION

COLORADO RIVER

PLATEAU

GRAND

CANYON

GRANITE GORGE

PEACH SPRINGS CANYON
PEACH SPRINGS & HIGHWAY 66

N

SCALE

| 0 | 3 | 6 | 9 | 12 | KMS |

241

Shovel and wheel barrow in one shaft of the Synder Mine.

Shanley Spring is hidden under the dryfall at the center of the foto.

Looking north across the Colorado at the mouth of Separation Canyon. Notice the sand or silt bars created by high waters in upper Lake Mead.

The Shanley Tank or stock pond, and the Shanley Cabin in the background.

Burnt and Surprise Canyons, Grand Canyon, Arizona

Location and Access These two canyons are located on the southern fringe of the Shivwits Plateau which is due south of St. George, Utah, on the Arizona Strip. Both drain south into the Colorado River, which in these parts is at the upper limits of Lake Mead. Much of the land on the plateau top is part of the Lake Mead National Recreation Area; land below the rim is Grand Canyon N.P. In this same general area is the Shivwits Fire Camp and Mt. Dellenbaugh, an old volcano. To get there, one could jump in a boat and come up from Lake Mead and Pearce Ferry launching site, but this book is aimed more for land lovers. Therefore the normal route is via St. George. Drive south out of town on the Mt. Trumbull Road. See the Introduction to the Arizona Strip for more details. Drive about 74 kms until you're just south of Diamond Butte, then turn southwest onto the Shivwits(or Parashant) Road. Stay on this main road signposted for Mt. Dellenbaugh and the Shivwits Fire Camp. At a place called Oak Grove on the BLM maps, turn to the right and head for Twin Point. After passing through two gates you'll be at the Parashant Ranch, which is now a BLM workshop; then on past another stock tank(pond), a chained area, and a gate at the LMNRA boundary. Open, then close, all gates as you drive on. About half a km past this last gate, turn left and drive another 1 1/2 kms to reach the head of Twin Creek Canyon. This is where you park if you're heading down Surprise Canyon. If you're going into Burnt Canyon, continue on the Twin Point Road for 4 or 5 kms past the Twin Creek Canyon turnoff, then immediately after passing through a shallow drainage, turn right or west, and drive past a stock tank to the rim of Burnt Canyon just east of Burnt Spring.

Trail or Route Conditions There is some kind of old trail going down into the very upper or northern end of Burnt Canyon, but the author didn't find it and Buster Esplin wasn't aware of its location. So the best route down into Burnt is at the car-park shown. You can camp there under some large cedar trees. From the car-park, walk south along the rim for about 500 meters, then turn right or west and follow a game trail over the cliff and down a steep slope. There are about 3 routes through the cliffs in this area. Once over the cliffs, head straight down and into the obvious drainage running west into the bottom of Burnt Canyon. At the bottom, you'll have to climb down over several ledges into the main drainage. Then turn south--down-canyon, and walk about 300-350 meters and on the right will be Burnt Spring in a small cement box surrounded by cattails. About 200 meters below the spring and on the west wall is a 15-20 meter-long petroglyph panel, the style of which the author has not seen before. About 50 and 200 meters below the petroglyph panel are two minor cow trails on the east side skirting some minor ledges. From there on down to the Colorado River the walking is easy and fast. You can also walk down an old cow trail at the end of Twin Point which runs out to Neilson Spring.

For Surprise Canyon, you can enter via Green Spring, Ambush Canyon, and other points south, but the easiest and fastest way to the river is via Twin Creek and Twin Spring Canyons. From the car-park, just walk to the canyon rim and make your way down in. After 100 meters or so, you'll be at the bottom, then it's clear sailing all the way to Surprise Canyon. About 1 km up from The Confluence, water begins to flow in lower Twin Spring. There are some good limestone narrows there too, including a chokestone lodged about 5 meters above the creek. The surprise in Surprise is that there's a nice stream of water running all the way down to the Colorado River. In 1991 or 1992 there was a big flood which took out all the willows, tamaracks and trees which used to make walking very slow. But in the spring of 1993, it was a very fast and easy walk. It should be an easy hike for many years to come.

Elevations Both car-parks, 1825 meters; Sanup Plateau, 1400; High water--Lake Mead, 373 meters.

Hike Length and Time Needed Burnt Canyon to the river(lake) is about 19 or 20 kms one-way. Best to arrive at the car-park in the afternoon, take empty jugs down to the spring, fill-up, then walk a ways and camp. Second day, walk to the Colorado and back. Third morning; walk back up to your car in the cool morning shade. To the river and back should take about 16 to 20 hours total walk-time(?). It's about 33-35 kms to the river through Twin Spring and Surprise Canyons. In the past it would have taken 4-5 days round-trip because of all the bushwhacking, but since the last big flood reamed out the canyon, it's now possible for a fast hiker with light pack to make it to the Colorado in one very long day. Then maybe two easy days for the return trip to the top of Twin Creek Canyon. Should be an easy 3-4 day hike, round-trip.

Water Burnt Spring, lower part of Burnt Canyon, lower Twin Spring Canyon, all or most of Surprise Canyon, Green Spring, Ambush Tanks(?), and the Colorado and/or upper Lake Mead.

Maps USGS or BLM maps Mount Trumbull, Peach Springs, and the Visitor Map--Arizona Strip(1:100,000), or Tincanabitts Point, Mount Dellenbaugh, Amos Point & Devils Slide Rapids(1:24,000).

Main Attractions A wilderness experience. Wild and woolly canyons seldom visited.

Ideal Time to Hike From about early or mid-April through May, and October to early November. Snows can be deep on top of the Shivwits in winter. Summers are depressingly hot at the bottom! If you can manage it, come up from Lake Mead in a boat. This could be down throughout most of each winter.

Hiking Boots Dry weather boots in Burnt & Twin Spring Canyons, waders in Surprise Canyon.

Author's Experience In the summer of 1991, the author took a big pack down Twin Creek, Twin Spring and into Surprise Canyon to about what he calls Halfway Canyon, then realized he couldn't make it to the river and back with 2 1/2 days of food, so he returned the same day. Total walk-time, 12 hours! Next day, July 31, he day-hiked down Burnt Canyon to Gate Rock near the bottom end of east fork and back in 8 hours. He drank 7 3/4 liters of liquids that day, perhaps more the day before! In late March of 1993, he boated up from Pearce Ferry on Lake Mead. He hiked up to Gate Rock in Burnt Canyon in 2 1/2 hours, then returned. Round-trip was about 5 hours. He also walked up Surprise to Halfway Canyon in 4 1/2 hours. That round-trip hike was under 9 hours.

Map 99, Burnt and Surprise Canyons, Grand Canyon, Arizona

VT RANCH
WARING RANCH
SHIVWITS ROAD
WELL
ST. GEORGE 120 KMS
1882
SHIVWITS FIRE CAMP RANGER STATION
OLD SAWMILL
GREEN SPRING
OAK GROVE 1800
STOCK TANKS
AMBUSH C.
POT HOLES
SHIVWITS
PLATEAU
MT. DELLENBAUGH 2156
TRAIL
PARASHANT RANCH
BLM WORKSHOP
GREEN
SPRING
CANYON
SHIVWITS
HOME RANCH TANK
KELLY
SPENCER TANK
SPRING
LMNRA BOUNDARY
TWIN SPRING
TWIN
SHANLEY TANKS
TWIN CREEK CANYON
SPRING
SUICIDE
POINT
SHANLEY CABIN 1842
CABIN
P 1825
MATHIS SP.
CANYON
POINT
KELLY POINT
STOCK POND OR TANK
COTTONWOOD SPRING
SPRING
PLATEAU
BURNT CANYON SP. 1250
P 1825
TWIN
POINT
ROAD
POINT
THE CONFLUENCE
AMOS SP.
ROAD
PET
RED POINT 1573
1825
RUNNING WATER
AMOS SPRING
GATE ROCK
CANYON
CANYON
HALFWAY CANYON
AMOS POINT
BURNT
EAST FORK
1381
P
SURPRISE
SANUP
PLATEAU
STONE CABIN
SANUP
COW TRAIL
RUNNING WATER
CORRAL
RUNNING WATER
NEILSON SPRING
SALT
CREEK
PLATEAU
TWIN POINT SALT CREEK
KAIBAB LS.
TOROWEAP F.
COCONINO S.S.
SANUP PLATEAU
HERMIT SH.
ESPLANADE S.S.
SUPAI SHALE
1473
REDWALL LIMESTONE
HWM 373
LAKE MEAD
SURPRISE
COLORADO RIVER
TEMPLE BUTTE LIMESTONE
HUALAPAI
LAKE MEAD HWM 373
N
GRANITE
GORGE
MUAV LIMESTONE
LAND
BRIGHT ANGEL SHALE
COLORADO RIVER
TAPEATS SANDSTONE
PRECAMBRIAN GRANITES & GNISSES

SCALE
0   3   6   9   12   KMS

Typical scene in the middle part of Surprise Canyon. A giant flood in the early 1990's reamed out this canyon making walking very easy and fast.

The narrows of lower Twin Spring Canyon.

Burnt Spring in the upper end of Burnt Canyon.

These petroglyphs are just below Burnt Spring in Burnt Canyon.

Quartermaster, Horse Flat and Clay Tank Canyons, Hualapai Nation, Grand Canyon, Arizona

Location and Access The three canyons featured here are at the extreme western end of the Grand Canyon and at the upper limits of Lake Mead. All are located south of the Colorado River-Lake Mead and on lands of the Hualapai Nation. This means you'll have to have a permit to hike and camp in this area. In 1994, the Hualapais were charging $7 a day for camping and hiking. To get this permit, drive to Peach Springs located on old Highway 66 between Seligman and Kingman. Stop at the river runners office, which is also their wildlife office, located in the center of town on the main highway just west of a small supermarket and across the street from the only gas station around. On weekends you can get a permit at that gas station. They can give you the latest information on road conditions, but usually not too much concerning hiking. Throughout the Hualapai Nation, you'll run into people looking after cattle or gathering wood, so it's wise to have a permit handy and on your dashboard, because they all seem to know you're supposed to have a one--unlike the Navajos! For more information call 1-602-769-2210 or 2227, which are the fone numbers for the river runners and wildlife offices. You can also get a permit at French's service station in Meadview.

There are two ways to get to this area. First, from Peach Springs, drive west on Highway 66 for 5 kms. Between mile posts 100 and 101, turn north and west on Buck & Doe Road. Drive west on this good graded road until you're in this area. Be sure to have the Peach Springs map handy for this drive. A second way in is to head for Kingman, then turn north on Highway 93. After 43 kms, turn north onto the Pearce Ferry Road running past Dolan Springs toward Meadview. After 47 kms, and just past mile post 29, turn right onto the Diamond Bar Road and head in the direction of Grand Canyon West, 21 miles(34 kms) ahead. When you reach the Buck & Doe Road, turn south toward the canyons as shown.

To get to Quartermaster Canyon, turn off Buck & Doe Road as shown, and proceed to Jeff Tank. From there, a rougher road heads over a low divide and down a drainage, ending near the rim and the beginning of a trail into the canyon. Cars can make it to the area around Jeff Tank OK, but maybe not much beyond that(?). A mtn. bike is handy here. To get into Horse Flat or Clay Tank Canyons, turn off Buck & Doe Road and head for Clay Tank. Not far beyond that are the two parking places and the beginning of these two hikes. Cars can make it to the first carpark, but may not get over the divide and to the second one at the stock pond & metal tank. Car drivers may want a mtn. bike.

Trail or Route Conditions From the end of the road, which is 100 meters from the lip of Quartermaster Canyon, walk west up the slope, then look down for the trail below. It should be visible. Head down to the trail and follow it around the corner to the main canyon. This is an old homesteader's trail, but it's visible only higher up as it zig zags down the steep canyon wall. Further down it disappears, so just walk along the dry creek bed all the way to the Colorado. Just before the river(lake) is a very large spring and signs of an old homestead. Just around the corner from that is another large spring which cascades into the river(lake). To get into Horse Flat Canyon, stop and park at the first car-park beyond Clay Tank, then walk down-canyon in the dry creek bed to the fork as shown. From there route-find up the slope to the north and to the rim of the canyon, then again route-find down a steep route to the bottom. Now for Clay Tank Canyon. From the car-park at the stock pond and metal tank, walk down-canyon, first on a vehicle track, then a cow trail. The drainage slowly deepens, then be looking for a low divide to the left or north, which is where you get down into Clay Tank. This divide is only 20 meters above the creek bed, and the divide itself may be only 50 meters away. Clay Tank Canyon will surely capture this first drainage in just a few thousand years. From the divide, just head down the steep slope to the bottom. Once there you can walk further down-canyon on wild donkey trails. About 2 kms from the Colorado is running water, lots of riparian vegetation, and calcite deposits(like what you find in limestone caves).

Elevations Lake Mead high water mark, 373 meters; Buck & Doe Road, 1500 to 1600; car-parks, 1450 to 1500 meters.

Hike Length and Time Needed It's only about a km to the bottom of Quartermaster, but about 12-13 kms to the river(lake). Strong hikers who know where the trail is, and with an early morning start, should be able to reach the river(lake) and return in one long day. Getting down into Horse Flat Canyon to the Colorado and back may well be too much for most hikers in one day. It's about 13 or 14 kms to the river, but the route off the rim is not always easy to locate and follow. At the lower end of Horse Flat Canyon is running water and lots of bushwhacking. The route down into Clay Tank and to the river(lake) is easier and less complicated than the other two, so it can be done by most hikers in one long day.

Water Good running water is found in the lower ends of all three canyons as shown on the map.

Maps USGS or BLM map Peach Springs(1:100,000), or several other maps at 1:24,000 scale.

Main Attractions Wild solitude and challenging route-finding along trails and obscure routes.

Ideal Time to Hike Spring or fall, or some winter warm and dry spells. Summers are too hot.

Hiking Boots Rugged boots, except waders in the lower end of Clay Tank and Horse Flat Canyons.

Author's Experience He wandered around the upper end of Quartermaster all day. In the afternoon he found the right road and trail, but just walked down to near the canyon bottom and back in a couple of hours. Once from his boat, he explored the very bottom end of the canyon with the large spring and waterfall. He also walked to within half a km of the river(lake) in Clay Tank Canyon, then returned, all in less than 8 hours. After spending all day looking for the right route into Horse Flat, he finally found the spot, but didn't go down. That was at the end of a 10 1/2 hour day. He bushwhacked up from the river(lake) once too, but for less than 2 kms.

Map 100, Quartermaster, Horse Flat and Clay Tank Canyons, Hualapai Nation, Grand Canyon, Arizona

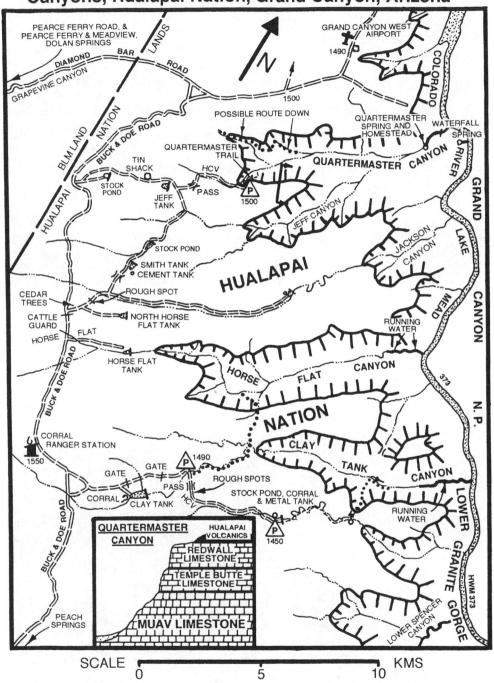

PEARCE FERRY ROAD, & PEARCE FERRY & MEADVIEW, DOLAN SPRINGS

GRAND CANYON WEST AIRPORT

1490

DIAMOND BAR LANDS ROAD

GRAPEVINE CANYON

1500

BLM LAND

NATION

HUALAPAI

BUCK & DOE ROAD

POSSIBLE ROUTE DOWN

QUARTERMASTER SPRING AND HOMESTEAD

WATERFALL SPRING

QUARTERMASTER TRAIL

QUARTERMASTER CANYON

COLORADO RIVER

TIN SHACK

HCV

STOCK POND

JEFF TANK

PASS

P 1500

JEFF CANYON

JACKSON CANYON

STOCK POND

SMITH TANK

CEMENT TANK

HUALAPAI

LAKE MEAD

GRAND CANYON N.P.

CEDAR TREES

ROUGH SPOT

NORTH HORSE FLAT TANK

RUNNING WATER

CATTLE GUARD

HORSE FLAT

BUCK & DOE ROAD

HORSE FLAT TANK

HORSE FLAT CANYON

NATION

373

CORRAL RANGER STATION

1550

GATE

GATE

P 1490

CLAY TANK CANYON

ROUGH SPOTS

LOWER GRANITE GORGE

GATE

CORRAL

PASS

HCV

CLAY TANK

STOCK POND, CORRAL & METAL TANK

RUNNING WATER

HWM 373

QUARTERMASTER CANYON

HUALAPAI VOLCANICS

P 1450

REDWALL LIMESTONE

BUCK & DOE ROAD

TEMPLE BUTTE LIMESTONE

PEACH SPRINGS

MUAV LIMESTONE

LOWER SPENCER CANYON

SCALE

0   5   10   KMS

249

A waterfall in the lower end of Quartermaster Canyon.

This is one of two stone buildings at Indian Gardens, or Meriwhitica Spring. This was once a homestead of some kind.

These are the ruins of an old camp between Bridge and Separation Canyons. It was probably built and used by the surveyors who were looking for potential dam sites.

Looking east from the viewpoint west of Bridge Canyon and at the western end of the Grand Canyon.

Meriwhitica, Spencer, Milkweed and Bridge Canyons, Hualapai Nation, Grand Canyon, Arizona

Location and Access All canyons on this map are located in the western Grand Canyon south of the Colorado River. Land north of the river is part of Grand Canyon N. P., while all areas you'll be driving and walking on are part of the Hualapai Nation lands. Because of this, first drive to Peach Springs located between Kingman and Seligman, and pay $7-a-day for hiking and camping. Get this permit at the river runners and wildlife office, on the main highway in town just west of the little supermarket. On weekends you can get a permit at the gas station across the street. Call 1-602-769-2210 or 2227 for more information. Then drive west on Highway 66 for 5 kms. Between mile posts 100 and 101, turn north and west onto Buck and Doe Road. If you're going to Bridge Canyon, drive about 10 kms on this road, and about a km after you pass under the large power lines, turn right or north and drive past an airstrip tower, a stock tank(pond), then to the rim of Hindu Canyon(marked 1426 meters), which is a good place to park your car and camp. This car-park is about 28 kms from Peach Springs. People with a mtn. bike can continue another 23 kms to an overlook of the Colorado River. A 4WD may or may not make it past the washed out section(?). If you're heading for Milkweed and Spencer Canyons, drive to Milkweed Spring, the best entry point to that canyon. To reach the Meriwhitica Canyon Trailhead, drive 11 kms along Buck and Doe Road from the turnoff to Milkweed Spring. At that point turn right onto Meriwhitica Road at the sign pointing to North Tank(just before the turnoff is a metal water tank). Drive another 13 kms to the trailhead.

Trail or Route Conditions For Bridge Canyon, walk or ride a mtn. bike along an old 4WD track running into Hindu Canyon, then north to a low pass between Hindu and Bridge. From the pass at 1274 meters, walk north down the dry creek bed and soon you'll see a developing trail veering to the east side. The trail contours along the east wall of Bridge, then zig zags down past a minor spring and disappears into the creek bed. At that point you can continue down the creek to the Colorado. The last km is along a stream with some wading and a little bushwhacking. If you stay on the Tonto Bench and follow a rim trail, you'll end up in the southern branch of Separation Canyon and at the river. Along this trail you can get down to the river in a number of places, including one good trail leading down to what appears to be a camp built and used by people who were testing bedrock to determine the feasibility of making a dam in these parts. An alternate to walking down Bridge, would be to continue riding your mtn. bike to the end of the road at 1405 meters, for an excellent view of the western Grand Canyon.

In Milkweed Canyon, simply walk downs along the small stream, or get up on the bench on the west side. If you stay along the creek, you'll have to climb down one old tree trunk leaning up against a cliff, in order to get over a dropoff 3 kms below the spring. If you stay on the canyon rim, walk along one of many wild donkey trails which bypass the dropoff. Further down Milkweed, you'll be in a granite gorge. In it is one narrow place with a chokestone. Two people may be able to get over this and the deep pool below with an air mattress, but you can also retreat 100 meters and climb up and around it on the west side(the author went around on the east side). A little below that, it's a fast jwalk down to Spencer Spring, then the stream starts flowing again. From there you'll have willows and tamaracks all the way to the Colorado, unless a flash flood has reamed it out, as it periodically does. From the Meriwhitica Trailhead, walk down a very good, wide trail to the bottom of the canyon, then look for wild donkey trails which take you to Indian Gardens and Meriwhitica Spring. There you'll find two old stone houses. From the spring, stay on the east side to find a trail going down to the lower end of Spencer Canyon and the Colorado.

Elevations Each trailhead, about 1450 meters; Lake Mead high water mark, 373 meters.

Hike Length and Time Needed Down Milkweed and Spencer Canyons to the Colorado is about 30 kms, one-way. This is a long day-hike for some, a day and a half for others, and about 3 days round-trip. From Meriwhitica Trailhead to the spring is about 8 kms, then another 8 or so to the Colorado. To the spring and back is an easy day-hike, and strong hikers might be able to reach the river and return in one very long day(walking is slow in lower Spencer). A loop-hike, down either canyon to the river and up the other, is about 48 kms, or about a 3 day hike. Use a mtn. bike as a shuttle back to your car. Going down Bridge Canyon to the river and back is a one-day hike; two long days or more, if you go to Separation and back.

Water Always have water in your car. At Milkweed Spring, and in Milkweeds' pink granite gorge; from Spencer Spring to the Colorado; and Meriwhitica Spring. Bridge Canyon has one minor spring, and running water near the river, plus a small stream at the Dam Builder's Camp.

Maps USGS or BLM map Peach Springs(1:100,000), and several maps at 1:24,000 scale.

Main Attractions A true wilderness experience, and no other hikers.

Ideal Time to Hike Spring or fall, or some winter warm spells. Summers are too hot.

Hiking Boots Dry weather boots or shoes, but waders down near the Colorado in either canyon.

Author's Experience The author left a mtn. bike at Milkweed Spring, then drove to Meriwhitica Trailhead. He hiked down to the river that day and camped near Spencer Spring. Next day he hiked up Spencer and through Milkweed to the road and spring, then biked back to his car. Total walk-time for the two days was 14 1/2 hours, plus 1 2/3 hours of biking. He also walked from the campsite at 1426 meters to the bottom of Bridge, up along the Tonto a ways, then back to the car-park in just under 9 hours(no bike). Another time he biked from the car-park to the viewpoint at 1405 meters, plus took a short hike in lower Hindu to the big dropoff. Round-trip was about 5 1/2 hours. On still another trip he came up from Lake Mead in a small boat with 10 hp motor, and hiked halfway to Bridge Canyon from Separation Canyon's southern branch. That took 7 hours round-trip. When water in Lake Mead reaches its highest level at 373 meters, lake water backs up to about Bridge Canyon. Coming into this region by boat is the best way if doing lots of hiking. And nobody is there to check for a hiking permit!

Map 101, Meriwhitica, Spencer, Milkweed and Bridge Canyons, Hualapai Nation, Grand Canyon, Arizona

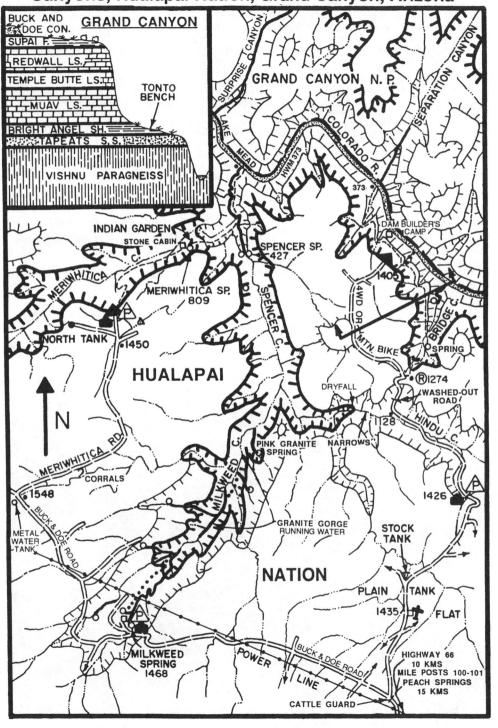

GRAND CANYON

BUCK AND DOE CON.
SUPAI F.
REDWALL LS.
TEMPLE BUTTE LS.
MUAV LS.
BRIGHT ANGEL SH.
TAPEATS S.S.
VISHNU PARAGNEISS

TONTO BENCH

GRAND CANYON N. P.

SURPRISE CANYON

SEPARATION CANYON

LAKE MEAD

HWY 373

COLORADO R.

373

DAM BUILDER'S CAMP

INDIAN GARDEN
STONE CABIN

SPENCER SP.
427

1405

MERIWHITICA

MERIWHITICA SP.
809

SPENCER C.

BRIDGE C.

SPRING

NORTH TANK

1450

HUALAPAI

4WD OR MTN. BIKE

DRYFALL

R 1274
WASHED-OUT ROAD

N

MERIWHITICA RD.

128

HINDU C.

CORRALS

MILKWEED C.

PINK GRANITE SPRING

NARROWS

1548

BUCK & DOE ROAD

1426

METAL WATER TANK

GRANITE GORGE RUNNING WATER

STOCK TANK

NATION

PLAIN TANK

1435

FLAT

MILKWEED SPRING
1468

POWER

LINE

BUCK & DOE ROAD

HIGHWAY 66
10 KMS
MILE POSTS 100-101
PEACH SPRINGS
15 KMS

CATTLE GUARD

SCALE
0   5   10   KMS

253

Diamond Creek Canyon, Hualapai Nation, Grand Canyon, Arizona

Location and Access Diamond Creek Canyon is located in the western Grand Canyon, on the south side of the Colorado River, and in the central part of the Hualapai Nation. The bottom end of Diamond Creek is located about 32 kms due north of Peach Springs, which is on Highway 66 in western Arizona. To get there, exit Interstate Highway 40 at Seligman or Kingman, then drive Highway 66 to Peach Springs. This is where you'll have to pay the $7-a-day fee for camping and hiking on Hualapai lands. Having to pay a fee to hike and camp is a real drag for most of us, but if it's wild, isolated wilderness canyons with no other hikers you're looking for, then the Hualapai Nation is a pretty good place to go. Get this permit at the river runners and wildlife office in the center of town, or on weekends at the gas station across the street. For more information, call 1-602-769-2210 or 2227. With the permit in hand, drive north out of town and into Peach Springs Canyon. This gravel road is well-maintained and frequently traveled, and any vehicle can make it to the Colorado River, a distance of 34 kms.

Trail or Route Conditions There is no trail into Diamond Creek Canyon, but for the most part walking is easy right in the creek itself or along side the stream. There's running water in the lower end of the canyon, but there are no swimming holes in the narrows to hinder progress. In the area of the junction of Blue Mountain Canyon and Diamond Canyon, there are some willows and other brush, but they are very minor obstacles. There are ways of getting into this canyon system from the upper end and from the Hualapai Hilltop Highway, but going up from the bottom seems more practical. The lower end of the canyon has some pretty good narrows in the Vishnu Paragneiss.

Elevations Colorado River, 407 meters; halfway into the drainage it's 792; canyon rim altitude is near 2000 meters.

Hike Length and Time Needed From the car-park at the mouth of Diamond Creek(which can also be used as a campsite) to Diamond Spring, is about 18 kms. To get that far into the canyon would take most people two days for the round-trip. From the car-park to the 792 meter point is about 11 kms. Most people can hike to that point and back in a day. From the car-park to the upper end of the narrows is about 6 or 7 kms. Anyone can get there and back in about half a day.

Water There is year-round water up to the 792 meter mark, and likely further up above that(?). There are several apparent year-round springs in both of the major canyons. The author found Diamond Creek water very good for drinking and he saw no sign of cattle or beaver in the canyon.

Maps USGS or BLM map Peach Springs(1:100,000), or Diamond Peak and Peach Springs NE(1:24,000).

Main Attractions There's a section of moderately good narrows 6 or 7 kms up from the road in the Vishnu Paragneiss. This is unusual because most narrow or slot canyons on the Colorado Plateau are in sandstone.

Ideal Time to Hike Spring or fall are best, but some winter warm spells can be pleasant. Summers are extreme hot in the canyon bottom.

Hiking Boots Wading boots or shoes.

Author's Experience The author camped very near the river, then the next morning hiked up to the 792 meter point and back in 5 hours.

The unusual paragneiss or granite-type narrows in the lower end of Diamond Creek.

Map 102, Diamond Creek Canyon, Hualapai Nation, Grand Canyon, Arizona

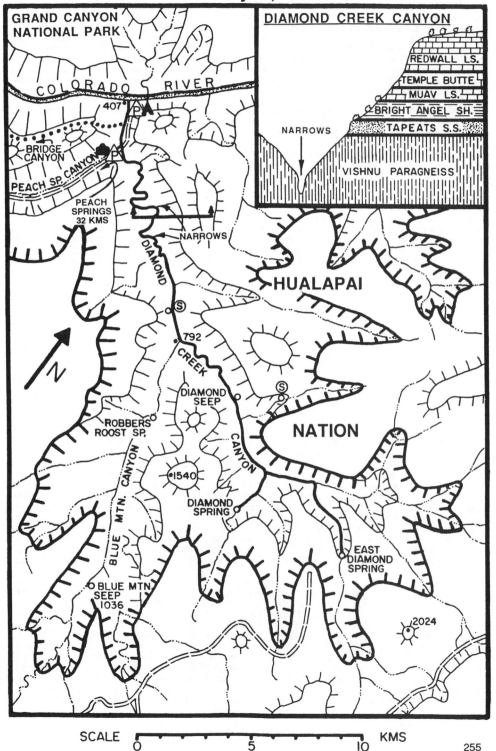

GRAND CANYON NATIONAL PARK

COLORADO RIVER

407

BRIDGE CANYON

PEACH SP. CANYON

PEACH SPRINGS 32 KMS

NARROWS

DIAMOND

CREEK

N

DIAMOND SEEP

ROBBERS ROOST SP.

792

1540

DIAMOND SPRING

BLUE MTN. CANYON

BLUE MTN SEEP 1036

HUALAPAI

NATION

EAST DIAMOND SPRING

2024

DIAMOND CREEK CANYON

REDWALL LS.
TEMPLE BUTTE
MUAV LS.
BRIGHT ANGEL SH.
TAPEATS S.S.

NARROWS

VISHNU PARAGNEISS

SCALE

0   5   10   KMS

Ridenour Mine, and Prospect Valley and Canyon Hikes, Hualapai Nation, Grand Canyon, Arizona

Location and Access Featured on this map is an interesting hike, mtn. bike or 4WD vehicle ride down to an old mine; and a hike down to the Colorado River via a canyon cut out of old volcanic lava. The most prominent feature here is Prospect Valley, which lies just south of the Colorado River, Vulcans Throne and Toroweap Point. It's also on lands of the Hualapai Nation, northeast of Peach Springs and just west of Mohawk Canyon. Being on Hualapai lands means you'll have to get a permit costing $7-a-day for hiking and camping. Get it at the river runners and wildlife office in the center of Peach Springs, just west of a small supermarket; or on weekends at the gas station across the street. Call 1-602-769-2210, or 2227, for more information. The people in that office can give you up-to-date information on road conditions, etc.

To get to the Ridenour Mine and Prospect Valley, drive east out of Peach Springs on old Highway 66 in the direction of Seligman, but turn north onto the Hualapai Highway, between mile posts 110 & 111. This is the road running northeast toward the Hualapai Hilltop and Havasupai Canyon. Proceed to Frazier Wells and mile post 24, then turn north onto a very good and well-graded road. At various junctions, continue north on the road signposted for Mexican Tank, which is situated at the head(south end) of Prospect Valley. You must have the two metric maps(1:100,000 scale) listed below for this drive. From Mexican Tank, continue north and either toward the hilltop above the Ridenour Mine, or down Prospect Valley toward the Colorado. The author made it to the hilltop and car-park above the mine in his VW Rabbit with oversized tires, but walked from there to save pushing a bike back up. However, a low-geared 2WD HCV might be able to make it back up that steep hill(?).

If you continue north down Prospect Valley, you'll need a 4WD for sure. And it better be a good running one too, because it's a long way to the nearest garage! The author was told there are a couple of washed-out places where the road crosses the dry creek bed, so instead of going down the canyon, he drove north along the Prospect Point Road and managed to reach the car-park at 1950 meters, as shown. Here's the simplest way to get to Prospect Point. Drive along the Hualapai Highway to a point just before mile post 29, then turn north on another good graded road. This is the same road you'd take if going to Mohawk Canyon, but continue going straight ahead instead of turning off the right toward Mohawk. Continue straight north on a pretty good road, passing the landmarks shown. You'll eventually land at a place on the rim at a cool 1950 meters altitude.

Trail or Route Conditions From the car-park on top of the hill above the Ridenour Mine, walk, or ride a mtn. bike or 4WD down the moderately good but steep road to where there's an old camp with several old stone buildings. From there it's just a short walk down to the old copper mine and several tunnels. This mine appears to be in the Esplanade Sandstone. Interesting place. While in the area, you could look for a route down to the river, or head south to get a drink from the Ridenour Spring, which was once troughed-up. With a cup, you can still get a good safe drink out of it(cattle and/or wild burros are mucking it up pretty bad though!).

If you have a 4WD or mtn. bike to get down Prospect Valley, stop and park just north of the first cinder cone as shown, then walk north and climb down into Prospect Canyon via the right-hand or eastern drainage. It's steep with lots of loose material, so take rugged hiking boots, at least for the upper part. As you'll be able to tell immediately, this canyon cuts deeply into mostly black volcanic rocks. This makes a contrasting scene in red rock country. You'll end this hike at the river and apparently warm spring across from the trail coming down from Vulcans Throne. If you're the adventurous kind, you could drive to Vulcans Throne, walk down the trail to the river, then float across to the bottom of Prospect Canyon on an air mattress or inner tube. Don't swim, the water is too cold!

If you have only a car and can manage to get out to the end of Prospect Point, park in the shade of piñon or cedar trees, then simply head off the rim to the northwest, route-finding down over the ledges. At the road you can then go down Prospect Canyon, or in the direction of Hongo Spring and some good views into the main canyon opposite Toroweap Point. Being on the south side of the canyon, and looking north at the sunny side across the way, you'll have better opportunities for good fotos than from Toroweap Point.

Elevations Ridenour Mine Trailhead, 1925 meters; Prospect Point Trailhead, 1950; Prospect Canyon Trailhead, 1275; Colorado River, about 500 meters.

Hike Length and Time Needed You can spend a half, or a full day down around the Ridenour Mine. From Prospect Point down through Prospect Canyon to the river and back is a day-hike; perhaps a very long day-hike for some. Or backpack in and stay a night or two.

Water Always carry lots of water in your car. Also, at Ridenour Spring, but take a cup to dip water out with; maybe at the Hongo Spring, if you can find it; and there's supposed to be water somewhere around Frazier Wells or the youth camp nearby(?). And the Colorado River.

Maps USGS or BLM maps Mount Trumbull and Peach Springs(1:100,000), or Vulcans Throne, Vulcans Throne SE, Vulcans Throne SW, and Whitmore Rapids(1:24,000).

Main Attractions Wild solitude, interesting old mine buildings, cinder cones & a route to the river.

Ideal Time to Hike Spring or fall, but beware of the possibility of snow in early spring or late fall!

Hiking Boots Rugged hiking boots for the route down off Prospect Point and on to the river.

Author's Experience The author walked down to the Ridenour Mine and spring, then returned, all in just under 4 hours. He also walked down from Prospect Point to the middle part of Prospect Canyon, and then out to the overlook as shown, then had to return to his car, all in 6 hours. In warm weather, that was a tiresome 6 hours--so be sure you don't over-extend yourself! Getting back out is the hard part.

Map 103, Ridenour Mine, and Prospect Valley and Canyon Hikes, Hualapai Nation, Grand Canyon, Arizona

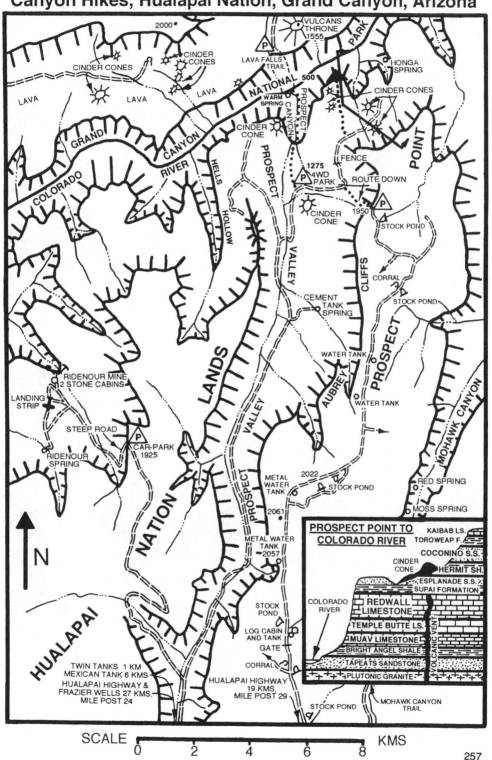

2000

CINDER CONES

CINDER CONES

VULCANS THRONE 1555

LAVA FALLS TRAIL

HONGA SPRING

P

CINDER CONES

LAVA

LAVA

LAVA

LAVA

WARM SPRING

CINDER CONE

GRAND

CANYON

RIVER

COLORADO

HELLS HOLLOW

PROSPECT

CANYON

NATIONAL

PROSPECT

500

1275 4WD PARK

P

FENCE

ROUTE DOWN

P

1950

POINT

PARK

CINDER CONE

VALLEY

STOCK POND

CLIFFS

CEMENT TANK SPRING

CORRAL

STOCK POND

WATER TANK

PROSPECT

LANDS

RIDENOUR MINE 2 STONE CABINS

LANDING STRIP

STEEP ROAD

P

CAR-PARK 1925

RIDENOUR SPRING

VALLEY

AUBREY

WATER TANK

MOHAWK CANYON

NATION

PROSPECT

METAL WATER TANK

2022

STOCK POND

RED SPRING

2061

MOSS SPRING

METAL WATER TANK 2057

N

HUALAPAI

STOCK POND

LOG CABIN AND TANK

GATE

CORRAL

TWIN TANKS 1 KM
MEXICAN TANK 6 KMS
HUALAPAI HIGHWAY &
FRAZIER WELLS 27 KMS,
MILE POST 24

HUALAPAI HIGHWAY
19 KMS,
MILE POST 29

STOCK POND

MOHAWK CANYON TRAIL

PROSPECT POINT TO COLORADO RIVER

KAIBAB LS.
TOROWEAP F.
COCONINO S.S.
CINDER CONE
HERMIT SH.
ESPLANADE S.S.
SUPAI FORMATION
COLORADO RIVER
REDWALL LIMESTONE
TEMPLE BUTTE LS.
MUAV LIMESTONE
BRIGHT ANGEL SHALE
TAPEATS SANDSTONE
PLUTONIC GRANITE
VOLCANIC VENT

SCALE

0   2   4   6   8   KMS

Looking north and down into the upper end of Prospect Canyon at Vulcans Throne across the Colorado River. Notice the old volcano on the left, which has now been eroded away by the slow enlarging of Prospect Canyon.

Lower Mohawk Canyon immediately below the 10 meter-high chokestone and dryfall.

Some of the ruins of cabins and buildings at the old Ridenour Copper Mine.

Just one of the many chokestones and waterfalls in lower National Canyon.

Mohawk and National Canyons, Hualapai Nation, Grand Canyon, Arizona

Location and Access Mohawk and National Canyons are located in the western Grand Canyon south of the Colorado River. Both drain to the north on lands of the Hualapai Nation. Before going hiking, you'll need to pay the $7-a-day fee for camping and hiking on Hualapai lands. Get this permit at the river runners and wildlife office on the main highway in Peach Springs. It's just west of the little supermarket in the middle of town. On weekends get the permit at the gas station across the street. Call 1-602-769-2210 or 2227 for the latest information.

To get to the canyons, drive to a point about 10 kms east of Peach Springs on Highway 66. This is the old road linking Peach Springs with Kingman on the west, Seligman on the east. Between mile posts 110 and 111, turn north on the paved highway running northeast to the Hualapai Hilltop and the entrance to Havasupai and Havasu Canyons.. Take this road and proceed to mile post 29, then turn north. Drive 14 kms on a well-used graded dirt road, then veer northeast into the shallow upper Mohawk drainage. The last 9 kms of road is less-used, but any car can make it. Near the end is a shack on the left, then after another km is the trailhead. For National Canyon, drive along the Hualapai Highway to just past mile post 38 and turn north onto a good dirt road. Drive past several stock tanks(ponds) and an old house, then continue north on a less-used road. Finally the road ends abruptly about a km before the canyon actually begins. Any car driven with care can make it to the trailhead in dry conditions.

Trail or Route Conditions In Mohawk, there's a trail heading down-canyon, mostly on the west side. It passes a cave, then bypasses the Coconino dryfall. This trail leads past Moss and Red Springs, then you'll walk in the dry creek bed all the way down to a 5 meter-high dryfall, then a large chokestone and 10 meter-high dropoff about 3/4 the way through the canyon. At that point, you'll either have to use a long rope and rappel, or bench-walk on the left or west side for about 500 meters. There you'll find a 3 meter-high ledge which may or may not be climbable, so have at least one short rope to get down and back up. If you can make it down this obstacle, you should be able to make it to the river OK.

In National, use a pair of long pants for wading through brush along an overgrown trail for the first two kms to the cave on the right or east. From there, and staying on the east side, walk along a better trail for another km or two until you're well into the upper part of the drainage, then walk the dry creek bed. About 2/3 the way through National, water begins to flow. About 6 kms below that will be some big boulders and waterfalls. Camp 100 meters above this on a sandy bench. You can work your way down through the boulders, only to find about a dozen chokestones and waterfalls in the last 3 or 4 kms to the river. At first there will be several minor falls you can skirt easily, then a short section with 4 chokestones. Skirt these on the east. Then another easy section with deep pools and falls, and finally the last big waterfall. From there you'll have to back-track and look for a way up on the west side and for a trail or route coming up from the river. The author failed to make the final km because of the short mid-November days, but there were many tracks in these narrows, left by boaters coming up from the river. So there is a route through. Better take a short rope to make sure you get to the river. It's important to remember, these canyon bottoms change with every flood, so expect the unexpected.

Elevations Both trailheads about 1700 meters; the Colorado River about 520 meters.

Hike Length and Time Needed From the trailhead to the Colorado is about 25 kms in either canyon. In Mohawk, strong hikers with a rope, can make it to the river in one day, then back out the next. Some may want 3 days round-trip. National Canyon seems a bit longer, and it's slow going in the bottom end, so most will want 3 days round-trip. Perhaps make it to the boulders in one day and camp, then down to the river the next, and all the way back out the 3rd day.

Water Always carry water in your car. In upper Mohawk, you'll find several seeps along the bottom of the Coconino which is Moss Spring. There's a cement trough at one point. Red Spring is found facing northeast in the red Hermit Shale. In 1990, a metal trough was placed there and fenced off with poles. Good water, plus campsite below. Just above the 5 and 10 meter-high dropoffs, you'll likely find potholes with water, and 3 minor seeps in between these two dryfalls. In the lower end of Mohawk is a nice year-round stream, but you have to get down off that one little ledge to reach it. In upper National, there's supposed to be a spring with cement trough a km or two below the cave on the west side, but the author saw no indication of it. However, he never really looked. About 2/3's the way through the canyon water begins to flow. It's intermittent at first, then below the boulders it's a steady stream. Good water.

Maps USGS or BLM maps Grand Canyon, Mount Trumbull and Peach Springs(1:100,000), or Vulcans Throne and Vulcans Throne SE(1 24,000), and National Canyon(1:62,500).

Main Attractions Two challenging routes to the Colorado and interesting narrows, chokestones and waterfalls in lower National Canyon.

Ideal Time to Hike Spring or fall, or some warm dry spells in winter. Summers are hot.

Hiking Boots Any dry weather boots or shoes, but better take waders near the bottom end of both.

Author's Experience The author got to within one km of the river in Mohawk, but couldn't get down the dropoff. He camped in the dry creek bed, then came out the second day. Total two day walk-time was 16 hours. In National, he left his pack at the boulders and got to within one km of the river. Camped at the boulders, then came out the second day. Total two day walk-time was less than 17 hours, which included an hour for lunch and a splash-bath at Red Spring on the way out.

Map 104, Mohawk and National Canyons, Hualapai Nation, Grand Canyon, Arizona

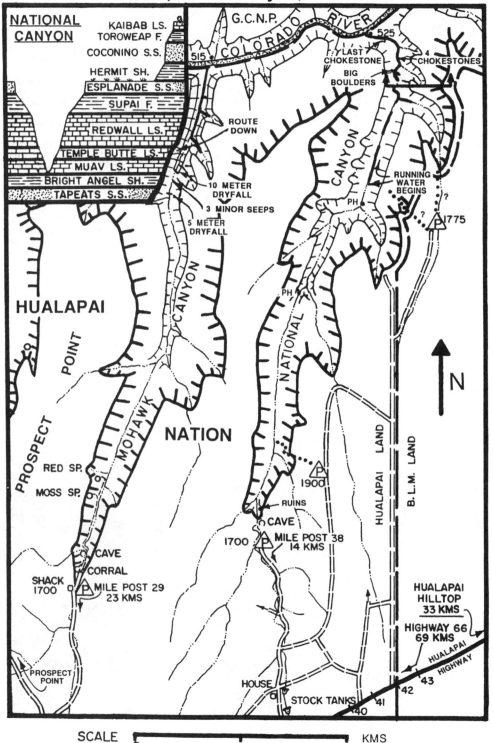

NATIONAL CANYON

KAIBAB LS.
TOROWEAP F.
COCONINO S.S.
HERMIT SH.
ESPLANADE S.S.
SUPAI F.
REDWALL LS.
TEMPLE BUTTE LS.
MUAV LS.
BRIGHT ANGEL SH.
TAPEATS S.S.

G.C.N.P.

COLORADO RIVER

515

525

LAST CHOKESTONE
BIG BOULDERS
4 CHOKESTONES

ROUTE DOWN

10 METER DRYFALL

3 MINOR SEEPS

5 METER DRYFALL

RUNNING WATER BEGINS
?
?
1775

PH

PH

HUALAPAI

PROSPECT POINT

MOHAWK CANYON

NATIONAL CANYON

N

RED SP.
MOSS SP.

NATION

1900
P
RUINS
CAVE
1700
P
MILE POST 38
14 KMS

HUALAPAI LAND

B.L.M. LAND

CAVE
CORRAL
SHACK
1700
P
MILE POST 29
23 KMS

HUALAPAI HILLTOP
33 KMS

HIGHWAY 66
69 KMS

PROSPECT POINT

HUALAPAI HIGHWAY

HOUSE
STOCK TANKS
40
41
42
43

SCALE 0 5 10 KMS

Havasu Canyon, Havasupai Nation, Grand Canyon, Arizona

Location and Access Havasu Canyon is a very long drainage which begins near Williams and flows northwest to the Grand Canyon and Colorado River. Most of the canyon is part of the Havasupai Nation. To reach the trailhead, you must first drive along Highway 66(north of Interstate 40) between Seligman and Peach Springs. At a point 10 kms east of Peach Springs and between mile posts 110 and 111, turn north on the highway running to Hualapai Hilltop. The distance is about 102 kms and is paved all the way. At the Hilltop Trailhead are two parking lots and a tanker truck with water. There are no services whatsoever along the Hualapai Hilltop Highway.

Trail or Route Conditions From the Hualapai Hilltop there's a very good and well-used pack trail leading down Hualapai Canyon to Havasu Canyon and to the village of Supai. The trail is used by mules every day and is extremely dusty. When you arrive at Supai, you'll have to pay an entrance fee. In 1994 it was $9 in the warmest half of the year; $7 during the winter months. In the village of Supai is a lodge, store & cafe where you can sleep and buy basic foods. From Supai, there's a road which accommodates tractors down to as far as Havasu Falls and the campground just below. Camping in the area of Supai is at their campground only. Camping cost $12 per person, per day, in 1994, but was only $8 a night during the winter months. From the campground, there's another trail running downstream along the top of the Redwall and into Beaver Canyon. You can walk this same rim or bench(top of Redwall) on down to an overlook of the Colorado River; or locate a route down to Havasu Creek below Mooney Falls and follow the creek down to Beaver Falls. Pass Beaver Falls on the east side until you find a way down through the Temple Butte Limestone to the creek. You can then walk along the stream down to the Colorado. There's a trail of some kind all the way. Expect lots of riparian vegetation along this route. The lower end of the canyon is part of Grand Canyon National Park and for day-hiking only. Camping is not allowed in this part of the park or canyon. In the fall of 1990, there was a flood in the canyon and things may have changed(?). Ask the person at the tourist office in Supai about some of the other hikes in the area, and buy the book, *Havasu Canyon* by J. Wampler in the same office. For updated information, call their tourist office on their radio-telefone at 1-602-448-2121.

Elevations Trailhead, 1585 meters; Supai, 972; campground, 890; Colorado River, 545 meters.

Hike Length and Time Needed From the trailhead to Supai village is about 13 kms, a very easy half-day hike. The campground is 3 kms below the village, a 45 minute walk. From the campground to the Colorado is another 13 kms. Most people can walk from the campground to the Colorado and back in one day. A suggested hike: Day 1, down to the campground; Day 2, to the river and back; Day 3, back to the trailhead parking.

Water Water flows from Havasu Springs to the Colorado, but treat it below Supai. There is a tanker truck parked at the trailhead with water for the mule traffic, but have your own supply in your car.

Maps USGS or BLM map Grand Canyon(1:100,000), or Grand Canyon N.P.(1:62,500).

Main Attractions Several fotogenic waterfalls and the Havasupai Indians and village.

Ideal Time to Hike Spring or fall, or some winter warm spells. Summers are extra hot in Supai.

Hiking Boots Any dry weather boots or shoes, but waders if you're going to the Colorado.

Author's Experience The author walked down to Havasu Falls and back in 7 1/2 hours.

One of the best attractions in Havasu Canyon is Havasu Falls below Supai Village.

Map 105, Havasu Canyon, Havasupai Nation, Grand Canyon, Arizona

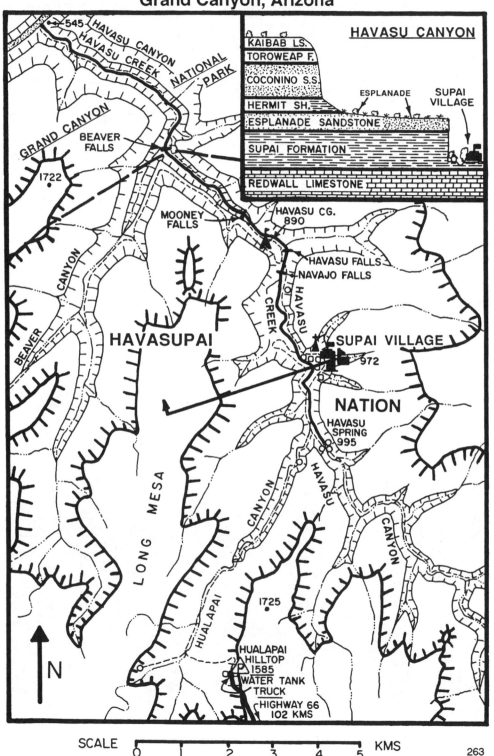

HAVASU CANYON

KAIBAB LS.
TOROWEAP F.
COCONINO S.S.
HERMIT SH.
ESPLANADE
SUPAI VILLAGE
ESPLANADE SANDSTONE
SUPAI FORMATION
REDWALL LIMESTONE

545
HAVASU CANYON
HAVASU CREEK
NATIONAL PARK
GRAND CANYON
BEAVER FALLS
1722
MOONEY FALLS
HAVASU CG. 890
HAVASU FALLS
NAVAJO FALLS
BEAVER CANYON
CREEK
HAVASU
HAVASUPAI
SUPAI VILLAGE
972
NATION
HAVASU SPRING 995
LONG MESA
HUALAPAI CANYON
HAVASU CANYON
1725
HUALAPAI HILLTOP 1585
WATER TANK TRUCK
HIGHWAY 66 102 KMS

N

SCALE 0 1 2 3 4 5 KMS

South Bass Trail, Grand Canyon, Arizona

Location and Access The South Bass Trail which follows Bass Canyon into the Grand Canyon is located in the western part of the national park and on the south side of the Colorado River. The bottom of this trail is just across the river from the end of the North Bass Trail and Shinumo Creek. To reach the trailhead, drive west out of Grand Canyon Village on West Rim Drive(see the Tonto Trail map for a better look at the approach road). About a km west of the Bright Angel Lodge, turn left or south onto Rowe Well Road. After about 3 kms, you'll cross the old railroad tracks, and another 1 1/2 kms more you'll cross back to the west side of the tracks. The first section of road is #328A. When you cross the tracks for the second time you'll be on Road #328. Follow this road west for about 32 kms, until you come to a major junction. At that point you'll be on Havasupai Indian land. Turn right in the direction of the seasonal Pasture Wash Ranger Station, which is 6 or 7 kms from the junction. After the ranger station, it's another 6 or 7 kms to the trailhead on Forest Road #2515. The author got to within 1 km of the trailhead with his VW Rabbit and had to park because of a rough spot at 2019 meters. While at the South Rim's Grand Canyon Village, discuss trail and road conditions with people at the Backcountry Reservations office. This is the place where you'll pick up a permit if you plan to backpack and camp down in the canyon.

Trail or Route Conditions The trail is very good as it first heads down and to the east. After one km you'll see some Anasazi ruins. At the bottom of the Coconino and across the Esplanade, the trail fades in places, so pay attention to the stone cairns. As it goes down Bass Canyon, it's easy to follow. Finally, there are several places in which to get off the Tonto Rim and down to the Colorado. The Tonto Trail to Garnet Canyon is little-used and maybe hard to follow, as are the Seep Spring and the Apache Point Trails. If you're doing the entire Tonto Trail, this is where you'll begin or end that hike.

Elevations The trailhead is 2026 meters, while the Colorado River is 680 meters.

Hike Length and Time Needed From the rim to the river is about 11 kms, and any fit person can do the round-trip in one day. For some it's a very long day. In cool weather it's a lot easier and faster than in summer heat.

Water This country is dry, so carry plenty of water in your car. The author found pothole water in the area where the Tonto Trail crosses Bass Canyon. At the very bottom of Bass Canyon, you may find a small stream. Other than these places the only water is in the Colorado. Consider caching a bottle halfway down for the return trip if it's hot weather.

Maps USGS or BLM map Grand Canyon(1:100,000), or Grand Canyon N.P.(1:62,500).

Main Attractions A quiet and remote hike to the Colorado River, with Anasazi Indian ruins in the lower Coconino Sandstone less than a km from the trailhead.

Ideal Time to Hike Dry weather in spring or fall. Muddy roads may be your biggest concern.

Hiking Boots Any dry weather boots or shoes.

Author's Experience The author stopped, parked and camped at the 2019 meter mark, then walked to the Colorado and back in one day, or in about 7 hours walk-time.

Half way down the South Bass Trail you'll see this gorge with the Redwall Ls. at the bottom.

Map 106, South Bass Trail, Grand Canyon, Arizona

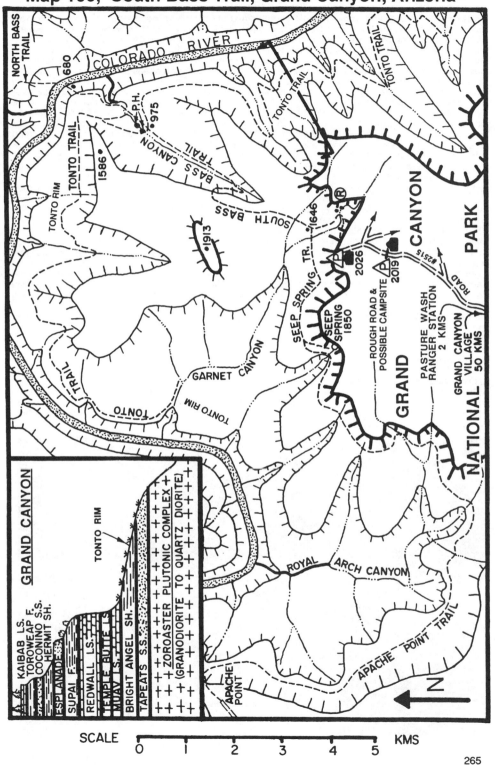

NORTH BASS TRAIL

COLORADO RIVER

680

P.H.

975

TONTO TRAIL

1586

TONTO RIM

TONTO TRAIL

BASS CANYON TRAIL

SOUTH BASS

1913

TONTO TRAIL

TONTO RIM

GARNET CANYON

TONTO RIM

TONTO TRAIL

SEEP SPRING TR.

1646

®

2026

2019

SEEP SPRING 1850

ROUGH ROAD & POSSIBLE CAMPSITE

PASTURE WASH RANGER STATION 2 KMS

#215

ROAD

GRAND CANYON VILLAGE 50 KMS

GRAND CANYON

NATIONAL PARK

GRAND CANYON

TONTO RIM

KAIBAB LS.
TOROWEAP F.
COCONINO S.S.
HERMIT SH.
ESPLANADE
SUPAI F.
REDWALL LS.
TEMPLE BUTTE LS.
MUAV LS.
BRIGHT ANGEL SH.
TAPEATS S.S.
ZOROASTER PLUTONIC COMPLEX
(GRANODIORITE TO QUARTZ DIORITE)

ROYAL ARCH CANYON

APACHE POINT TRAIL

APACHE POINT

N

SCALE   KMS
0   1   2   3   4   5

265

Boucher and Hermit Trails, Grand Canyon, Arizona

Location and Access The Boucher and Hermit Trails are two of the more popular and easily accessible routes down into the Grand Canyon from the south rim. To reach the trailhead, drive west along the West Rim Drive about 13 kms from Grand Canyon Village to a place called Hermits Rest. Located there are toilets, drinking water and a viewpoint. Once at Hermits Rest, locate the actual trailhead parking place.

Trail or Route Conditions The trail leaving Hermits Rest is the beginning of both the Boucher and Hermit Trails. From the trailhead, you first zig zag down through the Coconino Sandstone heading in a southerly direction. Then the trail separates. At the first junction, one path heads southeast and is called the Waldron Trail. A short distance beyond that is the junction of the Hermit and Boucher Trails. If you turn right and follow the Hermit, you'll soon come to the Santa Maria Spring, with toilets and a small cabin rest-stop. The NPS has allowed the original piping and improvements on the spring to deteriorate, so it'll take a little time to fill a water bottle there. Just before the trail begins to drop off the Esplanade bench is Fourmile Spring, which may or may not have water(?). After that it's a steep walk down to the Tonto Rim and Trail. Near the Hermit Creek Campsite are the faint remains of an old tourist camp set up long ago by the Santa Fe Railroad. You can then follow a trail down Hermit Creek to the Colorado River. Back to the trail junction; if you had taken the Boucher Trail to the west, you would have been able to take a short side-trip to Dripping Spring. Just before Dripping Spring, the Boucher turns right, then heads north along the Esplanade bench. Further on it drops down into Travertine Canyon, and finally along Boucher Creek to connect with the Tonto Trail. From the Boucher Creek Campsite, you can walk downstream to the Colorado, or walk in either direction along the Tonto Trail. If you walk east, you can connect with the Hermit and return to the trailhead without back-tracking.

Elevations Hermits Rest, 2069 meters; Dripping Sp., 1700; Santa Maria Sp., 1600; Hermit Ck. Campsite, 900; Boucher Ck. Campsite, 850; Colorado R., between 706 and 747 meters.

Hike Length and Time Needed From Hermits Rest to Boucher Creek Campsite is about 13 kms, then another 3 kms to the Colorado River. That's 32 km round-trip. This hike can be done in one long day, but most people end up making it in two. From Hermits Rest to the Hermit Creek Campsite is about 12 kms. From there to the Colorado is another 3 kms. That makes a total of 30 kms round-trip to the river via the Hermit. Any strong hiker can do this rim-to-river round-trip hike in one day. Or you could go down either the Hermit or Boucher, then walk along the Tonto and ascend the other. This could be done is one very long day, but most would want to do it in two.

Water Santa Maria Spring, maybe at Fourmile Spring(?), Hermit Creek, Boucher Creek, Colorado River, and Dripping Spring(not much water there and you have to catch it in a cup).

Maps USGS or BLM map Grand Canyon(1:100,000), or Grand Canyon N.P.(1:62,500).

Main Attractions Loop-hike into and out of the Grand Canyon using the same trailhead.

Ideal Time to Hike Spring or fall, or possibly during some warm and dry spells in winter. The road to Hermits Rest is open year-round. Summers are very hot at the canyon bottom.

Hiking Boots Any dry weather boots or shoes.

Author's Experience The author hiked from Hermits Rest down the Boucher to the Colorado River, then back the same way in a total time of 8 1/2 hours. The next day, he walked down the Hermit Trail to the river and back in 6 hours round-trip. Years later he took another short hike down to Dripping and Santa Maria Springs and returned in about 3 hours.

This is Hermit Creek below the Tonto Trail.

Map 107, Boucher and Hermit Trails, Grand Canyon, Arizona

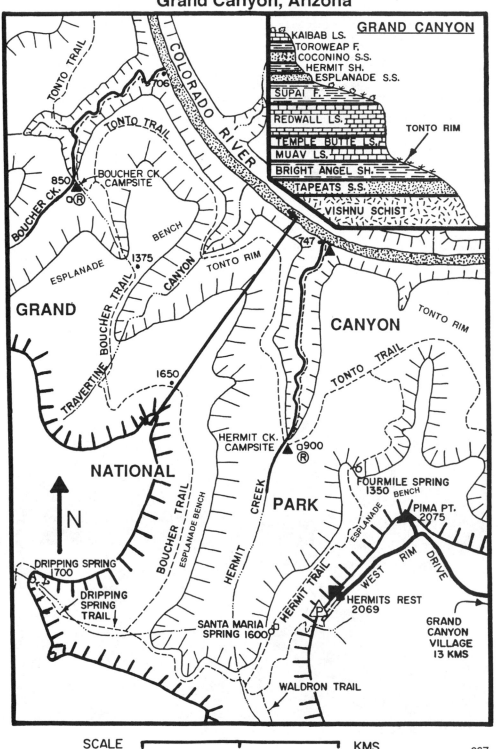

TONTO TRAIL

COLORADO RIVER

706

TONTO TRAIL

GRAND CANYON

KAIBAB LS.
TOROWEAP F.
COCONINO S.S.
HERMIT SH.
ESPLANADE S.S.
SUPAI F.
REDWALL LS.
TEMPLE BUTTE LS.
MUAV LS.
BRIGHT ANGEL SH.
TAPEATS S.S.
VISHNU SCHIST

TONTO RIM

850

BOUCHER CK. CAMPSITE

BOUCHER CK.

Ⓡ

BENCH

1375

TONTO RIM

747

ESPLANADE

GRAND

CANYON

TRAVERTINE

BOUCHER TRAIL

CANYON

TONTO RIM

1650

HERMIT CK. CAMPSITE

900

Ⓡ

TONTO TRAIL

NATIONAL

BOUCHER TRAIL

ESPLANADE BENCH

HERMIT CREEK

PARK

FOURMILE SPRING 1350 BENCH

N

PIMA PT. 2075

DRIPPING SPRING 1700

DRIPPING SPRING TRAIL

SANTA MARIA SPRING 1600

ESPLANADE

HERMIT TRAIL

WEST RIM DRIVE

HERMITS REST 2069

P

GRAND CANYON VILLAGE 13 KMS

WALDRON TRAIL

SCALE

0   1   2   KMS

267

Bright Angel and South Kaibab Trails, Grand Canyon, Arizona

Location and Access The hike featured here is along the two most popular trails into the Grand Canyon and to the Colorado River. Both are located in the middle part of the south rim right where most of the tourist facilities are located. To get to the Bright Angel and South Kaibab Trails, drive north from Williams located on Interstate 40; or drive west from Cameron, located on Highway 89, about halfway between Flagstaff and Page. Your destination is South Rim of the Grand Canyon. To get to the South Kaibab Trailhead, drive west from the Grand Canyon Village in the direction of Cameron, but turn north at the sign to Yaki Point. After a km the trailhead will be on the left. The Bright Angel Trailhead is just west of the historic Bright Angel Lodge right at Grand Canyon Village. Park where ever you can find a place.

Trail or Route Conditions Both these trails are over-used and very dusty. Horses and mules use them both, and the stench of mule urine is overpowering in places. Because the South Kaibab Trail is totally dry except for the Colorado River, it's recommended you descend this route, and return to the rim via the Bright Angel Trail, which has water in several places. In cooler weather you can hike in either direction, but in summer you'll need more water than you can carry. From the South Kaibab Trailhead simply walk down to the river along this zig zag trail. You can cross the river on the first of two bridges, then go downstream a ways, re-cross the Colorado on another foot bridge, and ascend the Bright Angel Trail to the Grand Canyon Village and lodge. You can also complete a loop by walking along the Tonto Trail near the bottom, without crossing the river. Upon returning to the canyon rim, you'll have to use a second car, a bike, or walk 8 kms back to your car.

Elevations Yaki Point, 2200 meters; Grand Canyon Village, 2120; Colorado River, 731.

Hike Length and Time Needed The South Kaibab Trail from Yaki Point to Bright Angel Campground is 11 kms. The Bright Angel Trail from the rim to the Bright Angel Campground is about 15 kms. Added together this makes a loop of about 26 kms. The distance between the Bright Angel Trailhead and Yaki Point is about 8 kms. If you have to walk this too its 34 kms, round-trip. This is a rather long hike, but with two cars, or a car and a mtn. bike, it's not difficult for the fit hiker to do in one day. Many camp in the canyon one night, and return the second day. If you do this, you'll have to get a camping permit well in advance. If you want to camp overnight anywhere in the park, write to the Backcountry Reservations office, Grand Canyon Village, South Rim, Arizona. Write after October 1, for reservations during the next 12 months. That reservations office is shown on this map just west of the campground.

Water Colorado River, the Bright Angel Campground, Garden Creek(below Indian Gardens Camp), and a water tank rest-stop at 1323 meters on the Bright Angel Trail.

Maps USGS or BLM map Grand Canyon(1:100,000), or Grand Canyon N.P.(1:62,500).

Main Attractions Grand Canyon views, the Phantom Ranch, and a great rim-to-river hike.

Ideal Time to Hike Spring or fall. The highway to the South Rim is open year-round, so you can do it anytime. Summers are very hot at the bottom and winters are often cold and snowy on the rim.

Hiking Boots Any dry weather boots or shoes.

Author's Experience He has done it twice, once in March and again at Easter time. The second time he walked the suggested loop-hike in less than 8 hours, which included the tiresome road-walk back to his car.

One of two suspension bridges over the Colorado River between the North and South Rims.

Map 108, Bright Angel and South Kaibab Trails, Grand Canyon, Arizona

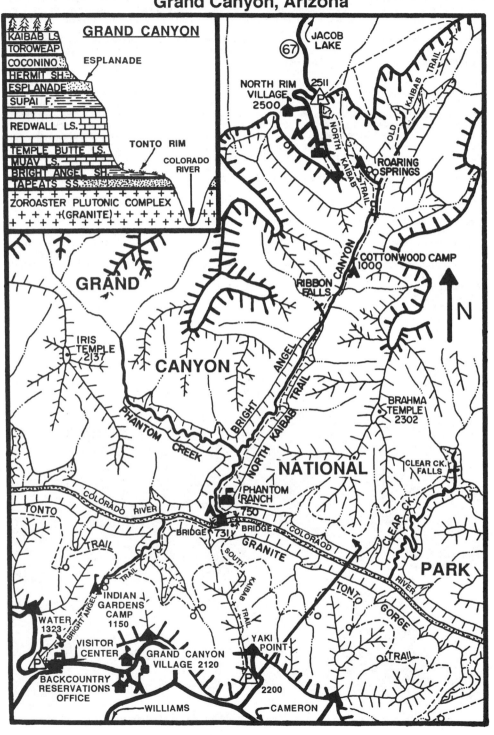

GRAND CANYON

KAIBAB LS.
TOROWEAP
COCONINO
HERMIT SH.
ESPLANADE
SUPAI F.
REDWALL LS.
TEMPLE BUTTE LS.
MUAV LS.
BRIGHT ANGEL SH.
TAPEATS SS.
ZOROASTER PLUTONIC COMPLEX
(GRANITE)

ESPLANADE

TONTO RIM

COLORADO RIVER

JACOB LAKE

(67)

NORTH RIM VILLAGE 2500

2511

OLD KAIBAB TRAIL

NORTH KAIBAB TRAIL

ROARING SPRINGS

COTTONWOOD CAMP 1000

RIBBON FALLS

N

GRAND

IRIS TEMPLE 2137

CANYON

PHANTOM CREEK

BRIGHT ANGEL TRAIL

NORTH KAIBAB TRAIL

BRAHMA TEMPLE 2302

CLEAR CK. FALLS

NATIONAL

PHANTOM RANCH

COLORADO RIVER

TONTO

BRIDGE

BRIDGE 731

750

COLORAOD

GRANITE

CLEAR CK.

PARK

TRAIL

SOUTH KAIBAB TRAIL

TONTO

GORGE

RIVER

BRIGHT ANGEL TRAIL

INDIAN GARDENS CAMP 1150

WATER 1323

VISITOR CENTER

GRAND CANYON VILLAGE 2120

YAKI POINT

TRAIL

BACKCOUNTRY RESERVATIONS OFFICE

WILLIAMS

2200

CAMERON

SCALE 0 4 KMS

Grandview and Hance Trails, Grand Canyon, Arizona

Location and Access The Grandview and Hance Trails are located on the south rim of the Grand Canyon about halfway between Desert View to the east, and the Grand Canyon Village to the west. To reach the trailheads, drive east from Grand Canyon Village about 20 and/or 30 kms, then turn north and drive a short distance to the rim. The Grandview Trail begins at Grandview Point; while the Hance Trail begins near Moran Point. If doing the Hance, you must park at Moran Point, then walk or use a mountain bike to reach the beginning of the trail, which is 1 1/2 kms to the southwest right on the main highway. If you park where the trail actually begins, the NPS will tow your car away.

Trail or Route Conditions The Grandview is a good and well-used trail that was originally built by early-day miners. Now many people use it to get down to the old copper mines in the vicinity of Horseshoe Mesa. When you reach the bottleneck at the south side of the mesa, take the trail which zig zags down to the east into a canyon to one mine with many relics and to a very good, shady, cool spring. From there, the trail winds its way around Horseshoe Mesa connecting with the Tonto Trail part way. On the west side of the mesa, the trail enters Cottonwood Canyon, then ascends the slope back up to the bottleneck. On top of the mesa is an old stone mine building and several prospects or adits where miners removed copper from the Redwall Limestone. The Hance Trail is one of the least-used trails beginning on the south rim, but it's still very good. From the actual trailhead along the highway, the route begins by heading in a northwesterly direction for 200 meters, then enters a shallow drainage near the rim. The trail then zig zags down Red Canyon to the Colorado River. The trail running west from the junction of Red Canyon and the Colorado is the eastern end of the Tonto.

Elevations Trailheads, 2256 & 2176 meters, Colorado River, 795, copper mines, 1500 meters.

Hike Length and Time Needed If you were to go down the Grandview Trail and make a loop-hike around the mesa, then walk back up to Grandview Point, it would be about 21 kms round-trip. The vertical distance from bottom to top is about 1100 meters. Strong hikers can do this round-trip in one day, but many can't. However, it's no problem for anyone to walk down to the mines and return in one day. Overnight trips are popular, with campsites on Hance and Cottonwood Creeks. From the Hance Trailhead parking lot at Moran Point, down to the Colorado River and back is 26 kms. Most can do this in one day with little problem, but for many it's a two-day hike.

Water There's water in Hance and Cottonwood Creeks, and at the spring on the east side of Horseshoe Mesa near the mining area. There maybe some water in Red Canyon, but don't bet on it. Also in the Colorado River.

Maps USGS or BLM maps Cameron and Tuba City(1:100,000), or Grand Canyon N.P.(1:62,500).

Main Attractions Old copper mines and two easy ways to the bottom of the Grand Canyon. The beginning or end of the Tonto Trail which runs all the way through the canyon from east to west.

Ideal Time to Hike Spring or fall, or some warm spells in winter. Summers are hot at the bottom.

Hiking Boots Any dry weather boots or shoes.

Author's Experience The author walked down the Grandview to the mines and had lunch at the spring just below on the east side, then made the loop around the mesa and back to the top in 5 1/2 hours round-trip. The next day he walked down the Hance to the river and back in 6 1/2 hours.

Relics of a bygone era are found along the lower part of Grandview Trail.

Map 109, Grandview and Hance Trails, Grand Canyon, Arizona

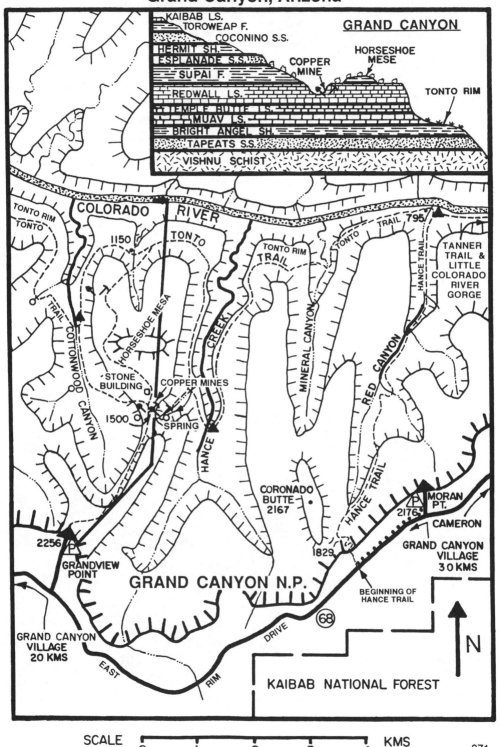

GRAND CANYON

KAIBAB LS.
TOROWEAP F.
COCONINO S.S.
HERMIT SH.
ESPLANADE S.S.
SUPAI F.
REDWALL LS.
TEMPLE BUTTE LS.
MUAV LS.
BRIGHT ANGEL SH.
TAPEATS S.S.
VISHNU SCHIST

COPPER MINE
HORSESHOE MESE
TONTO RIM

COLORADO RIVER

TONTO RIM
TONTO
1150
TONTO
TONTO TRAIL 795
TONTO RIM TRAIL
TONTO TRAIL
HANCE TRAIL
TANNER TRAIL & LITTLE COLORADO RIVER GORGE

TRAIL
COTTONWOOD CANYON
HORSESHOE MESA
CREEK
MINERAL CANYON
RED CANYON

STONE BUILDING
COPPER MINES
1500
SPRING
HANCE

CORONADO BUTTE 2167

HANCE TRAIL
2176
MORAN PT.
CAMERON
GRAND CANYON VILLAGE 30 KMS

1829

2256
GRANDVIEW POINT

GRAND CANYON N.P.

BEGINNING OF HANCE TRAIL

GRAND CANYON VILLAGE 20 KMS

EAST RIM DRIVE
68

N

KAIBAB NATIONAL FOREST

SCALE 0 1 2 3 4 KMS

271

Tanner Trail, Grand Canyon, Arizona

Location and Access The Tanner Trail is located in the southeast corner of the Grand Canyon, about halfway between the Cameron on Highway 89, and Grand Canyon Village. It's the most easterly trail on the south rim of the canyon which descends to the Colorado River. To get to the trailhead located at Lipan Point, drive west from Cameron about 54 kms to a point 3 kms west of Desert View; or drive 37 kms east from Grand Canyon Village on the East Rim Drive. This corner of the canyon is where the Colorado River makes its big sweeping turn to the west after having run in a southerly direction for so long.

Trail or Route Conditions Park your car at Lipan Point parking lot and begin walking east. The trail then heads north over the rim and zig zags down into Tanner Canyon. This route has not seen much in the way of maintenance, but it's used enough so it has become a well-beaten path. There's no way anyone can get lost or miss the trail at any point. Further on down, the trail runs along a kind of ridge to the left, then it finally curves back into the Tanner Canyon drainage as you near the Colorado River. From the river you can walk west towards Red Canyon and the eastern end of the Tonto Trail; or walk north along the Beamer Trail to the mouth of the Little Colorado River. At times, and in some places along the Beamer Trail, you may lose the path, but not for long. There are always stone cairns in critical places. Just north of Palisades Creek, the trail rises to the top of the Tapeats Sandstone bench, until it reaches the Little Colorado. In this 8 km section of trail it's very difficult, if not impossible, to get down to the Colorado River for water. So if you're heading to or from the mouth of the Little Colorado River, carry plenty of water along this section of trail.

Elevations Lipan Point, 2221 meters; the Colorado River, 810 meters.

Hike Length and Time Needed From Lipan Point to the Colorado River is about 13 kms. For a strong hiker this is an easy day-hike, round-trip. Many others however must do it in two days. The trip down is very easy; the return trip up is much more tiring as usual. From the mouth of Tanner Canyon to Palisades Creek is about 7 kms, and from Palisades Creek to the mouth of the Little Colorado about 8 more kms. From the mouth of Tanner Canyon down-river to the actual beginning of the Tonto Trail and Red Canyon is about 18 kms.

Water Always carry water in your car and in your pack because the only water around will be in the Colorado and Little Colorado Rivers. If it's summertime, consider caching a bottle of water on your way down for the return trip.

Maps USGS or BLM map Tuba City(1:100,000), or Grand Canyon N.P.(1:62,500).

Main Attractions A well-used but not crowded route to the Colorado River in the biggest canyon in the world; the ruins of the Tanner Cabin and his old mine; and the route into or out of the Little Colorado River Gorge.

Ideal Time to Hike Spring or fall, or some winter warm dry spells. Summers are very hot at the bottom.

Hiking Boots Any dry weather boots or shoes.

Author's Experience The author made a quick trip to as far as Palisades Creek, then returned the same day. It took about 9 1/2 hours round-trip.

At the bottom of the Tanner Trail are the crumbling remains of the Seth Tanner Cabin.

Map 110, Tanner Trail, Grand Canyon, Arizona

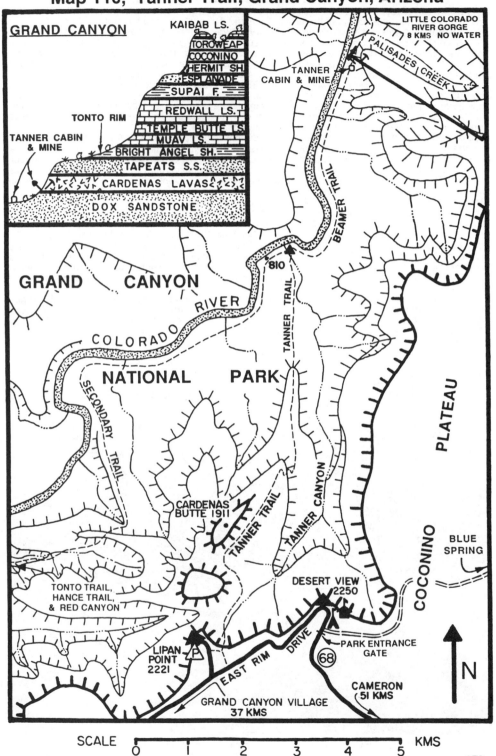

GRAND CANYON

KAIBAB LS.
TOROWEAP
COCONINO
HERMIT SH.
ESPLANADE
SUPAI F.
REDWALL LS.
TEMPLE BUTTE LS.
MUAV LS.
BRIGHT ANGEL SH.
TAPEATS S.S.
CARDENAS LAVAS
DOX SANDSTONE

TONTO RIM

TANNER CABIN & MINE

LITTLE COLORADO RIVER GORGE 8 KMS NO WATER

PALISADES CREEK

TANNER CABIN & MINE

BEAMER TRAIL

GRAND CANYON

COLORADO RIVER

810

TANNER TRAIL

NATIONAL PARK

SECONDARY TRAIL

CARDENAS BUTTE 1911

TANNER TRAIL

TANNER CANYON

PLATEAU

COCONINO

BLUE SPRING

DESERT VIEW 2250

TONTO TRAIL, HANCE TRAIL, & RED CANYON

PARK ENTRANCE GATE

68

LIPAN POINT 2221

EAST RIM DRIVE

N

CAMERON 51 KMS

GRAND CANYON VILLAGE 37 KMS

SCALE
0 1 2 3 4 5 KMS

273

Tonto Trail, Grand Canyon, Arizona

Location and Access The Tonto Trail is an east-to-west track running along the south side of the Colorado River near the bottom of the Grand Canyon. To reach the south rim of the Grand Canyon, drive west from Cameron, which is on the main highway connecting Flagstaff and Page. Or drive north from Williams, which is on Interstate 40 west of Flagstaff. The length of the Tonto Trail is about 145 kms. To shorten this long trek there are a number of access trails from the south rim. They are, from east to west: the Hance, Grandview, South Kaibab, Bright Angel, Hermit, Boucher, and the South Bass Trails. All these are covered elsewhere in this book. All these south-side trails can be approached from the paved highway called the East or West Rim Drive except for the South Bass, and that involves a drive of 50 kms on dirt roads. See the section on the South Bass Trail to get a description of that road. The Tonto officially begins at the bottom end of the Hance Trail in the east, but it's possible to begin at the bottom of the Tanner Trail as well. In the west it dead-ends in Garnet Canyon, but most begin or end the hike at the South Bass Trail.

Trail or Route Conditions From the Hance to the Boucher Trail, the Tonto is well-used, but to the west of Boucher Creek, it isn't used as much. None-the-less, it's not difficult to find and follow. Remember, the name Tonto comes from three geologic formations called the Tonto Group; they are the Muav Limestone, Bright Angel Shale and the Tapeats Sandstone. These three formations form a bench or platform known as the Tonto Rim. This rim or bench runs throughout the canyon and on both sides of the river. If you stay on this bench, you'll always be on or near the trail.

Elevations The South Rim, 2000 to 2200 meters; the Tonto Platform or Rim, roughly 1000 meters.

Hike Length and Time Needed From the bottom of Red Canyon(at the Colorado River) to Cottonwood Ck., 16 kms; Cottonwood Ck. to South Kaibab Trail, 32 kms; South Kaibab to Bright Angel Trail, 8 kms; Bright Angel to Hermit Trail, 20 kms; Hermit to Boucher Trail, 13 kms; and Boucher to South Bass Trail, 35 kms. Better add on 13 kms from Moran Point to the Tonto, and another 8 kms from the Tonto to South Bass Trailhead. This adds up to 145 kms, trailhead to trailhead(this does not include the 24 kms from South Bass to Garnet Canyon). At 20-25 kms a day, that's a 5 to 7 day hike for the average person. Since the hike is mostly level, it's not a strenuous walk. You'll also need a camp stove--no fires-- and some kind of car shuttle.

Water Always in the Colorado River, and in Cottonwood, Hance, Grapevine, Pipe, Bright Angel, Monument, Hermit, Boucher, and Ruby Creeks, and seasonally at various other creeks and springs. Don't do this hike without talking about the water situation with the rangers at the Backcountry Reservations Office in Grand Canyon Village. This is where you'll have to pick up your camping permit. Use the Grand Canyon N.P. map which shows the trails and springs best.

Maps USGS or BLM maps Grand Canyon and Tuba City(1:100,000), or Grand Canyon N.P.(1:62,500).

Main Attractions A long and interesting hike through the heart of the Grand Canyon.

Ideal Time to Hike Spring or fall, or some warm dry spells in winter. Summers are too hot.

Hiking Boots Any dry weather boots or shoes.

Author's Experience The author has not hiked this entire trail, but has been down all the access trails leading to the Tonto and the Colorado River.

From the south side of the Colorado River looking at the mouth of Bright Angel Creek and one of two foot bridges crossing the river.

Map 111, Tonto Trail, Grand Canyon, Arizona

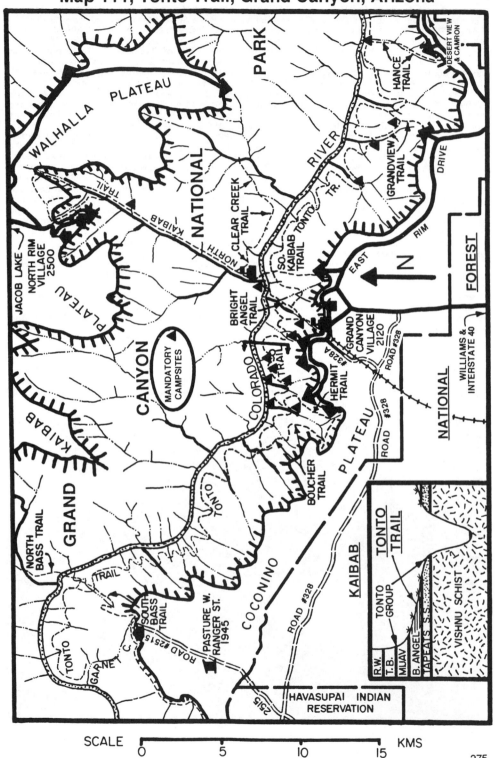

WALHALLA PLATEAU

PARK

NATIONAL

RIVER

DESERT VIEW & CAMRON

HANCE TRAIL

GRANDVIEW TRAIL

NORTH KAIBAB TRAIL

TRAIL

CLEAR CREEK TRAIL

SO. KAIBAB TRAIL

TONTO

TR.

DRIVE

RIM

EAST

JACOB LAKE

NORTH RIM VILLAGE 2500

PLATEAU

CANYON

MANDATORY CAMPSITES

BRIGHT ANGEL TRAIL

N

FOREST

GRAND CANYON VILLAGE 2120

ROAD #328

WILLIAMS & INTERSTATE 40

NATIONAL

COLORADO

TRAIL

HERMIT TRAIL

#328A

ROAD #328

KAIBAB

GRAND

NORTH BASS TRAIL

TRAIL

TONTO

BOUCHER TRAIL

PLATEAU

TONTO TRAIL

TONTO GROUP

VISHNU SCHIST

COCONINO

KAIBAB

R.W.
T.B.
MUAV
B. ANGEL
TAPEATS S.S.

ROAD #328

SOUTH BASS TRAIL

PASTURE W. RANGER ST. 1945

ROAD #2515

TONTO

GARNE C

2515

HAVASUPAI INDIAN RESERVATION

SCALE   0   5   10   15   KMS

275

West Fork of Oak Creek, Arizona

Location and Access The West Fork of Oak Creek is located about halfway between Sedona and Flagstaff and just west of Highway 89A. Normal access is via this old main highway which runs from Flagstaff to Sedona, much of which is in the bottom of Oak Creek Canyon. Highway 89A is now a secondary route, with Interstate 17 running parallel to it just to the east. The area around Sedona is famous for its red rocks and scenery, as well as the West Fork of Oak Creek. The bottom of this canyon has been set aside as a protected natural area with no motorized vehicles allowed. The place is very popular, especially on weekends. To get to the bottom end of the canyon, drive Highway 89A to between mile posts 384 and 385 and park in the narrow space beside the road(it's possible they will have a special parking area built when you arrive). This is about one km north of Don Hoel's Cabins and about 2 kms south of the entrance to Cave Spring Campground. To reach the upper end of the canyon, drive west from the south end of Flagstaff on old Highway 66 about 4 kms, then turn south on Forest Road #231. Drive this very good gravel road 35 kms to where you cross the bridge over upper West Fork at 1980 meters. Be sure and have the Coconino National Forest map in hand before attempting this drive to the upper end of the canyon.

Trail or Route Conditions From the bottom trailhead, a path winds its way up the West Fork for 4 or 5 kms, then the canyon narrows and the foot traffic decreases. From there you must walk in or beside the small stream. For the most part this is easy, but there are a number of deep holes you must wade or swim through. Be sure to have an inner tube or some other float device if you have cameras, maps or wallet you must protect. Perhaps the best way to keep things dry is to line your day-pack with several large plastic bags, then with everything inside, you simply swim through deep pools with the pack on. Add an empty plastic bottle or two for added flotation. The amount of water in the stream is very small, but some pools are rather large and deep and hold sizable trout.

Elevations Upper car-park, 1980 meters; the lower trailhead, about 1625 meters.

Hike Length and Time Needed From the upper car-park to the lower trailhead is about 23 kms. That distance can be walked in one day(if going downstream), but you'd have to have two cars and a shuttle to do it. Some people hike it in two days. Most however just walk up from the bottom trailhead and return the same way.

Water The water in this stream is crystal clear, but there's heavy human traffic in the lower end, so it may be polluted. The higher you go in the canyon the better the water quality will be.

Maps USGS or BLM maps Flagstaff and Sedona(1:100,000), or Dutton Hill, Wilson Mtn., Munds Park(1:24,000), and the Coconino National Forest map if you're driving to the upper end.

Main Attractions Deep, narrow and scenic canyon; and cool clear swimming holes.

Ideal Time to Hike Summer, with temperatures 30° C. or higher(if going beyond the trail's end).

Hiking Boots Wading boots or shoes.

Author's Experience The author came down from the top, but wasn't very well prepared and couldn't keep cameras dry beyond the first big swimming hole. Later he walked up from the bottom to the first group of deep pools and returned. Round-trip was 4 hours. He has not seen the middle third of the canyon but it must be a good one and without a lot of foot traffic.

To hike the entire length of West Fork of Oak Creek, you'll have to pass deep holes like this.

Map 112, West Fork of Oak Creek, Arizona

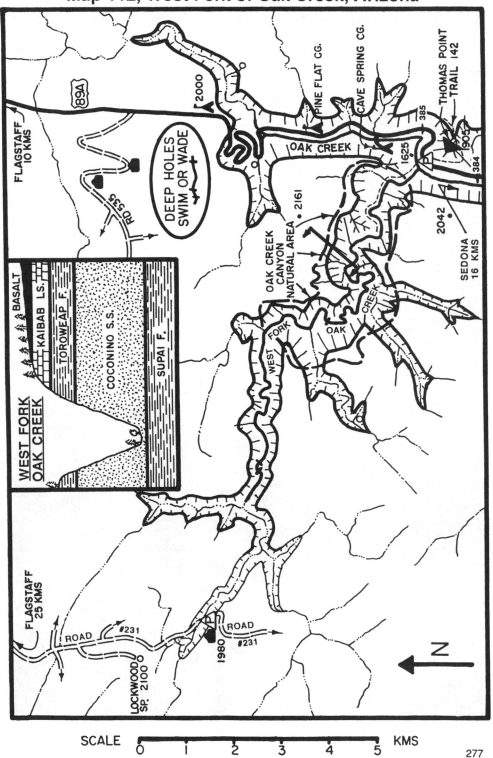

FLAGSTAFF 10 KMS

US 89A

RD 535

2000

DEEP HOLES SWIM OR WADE

PINE FLAT CG.

CAVE SPRING CG.

THOMAS POINT TRAIL 142

OAK CREEK

385

1625

905

384

2161

2042

SEDONA 16 KMS

OAK CREEK CANYON NATURAL AREA

WEST FORK

OAK CREEK

BASALT
KAIBAB LS.
TOROWEAP F.
COCONINO S.S.
SUPAI F.

WEST FORK OAK CREEK

FLAGSTAFF 25 KMS

ROAD #231

ROAD #231

1980

LOCKWOOD SP. 2100

N

SCALE   0   1   2   3   4   5   KMS

277

Wet Beaver Creek Canyon, Arizona

Location and Access Wet Beaver Creek is located south of Flagstaff and Sedona, and just east of the Montezuma Castle National Monument. To get to the Beaver Creek Ranger Station and the beginning of the trail, drive south from Flagstaff or north from Camp Verde on Interstate 17. Exit the freeway near mile post 299 at the Highway 179 exit. This highway runs north to Sedona, but turn south on to the newly paved Forest Service Road #618 instead. Drive 4 kms to the ranger station for last minute hiking information and perhaps maps, then head east about another km to the trailhead.

Trail or Route Conditions From the trailhead, you'll first be walking on an old road which is marked Bell Trail #13. After 3 1/2 kms is a trail junction; straight ahead 1 1/2 kms is Bell Crossing; to the left is the Apache Maid Trail #15. Take Trail #15 to the top of the mesa where you'll find a sign post marking the beginning of where the trail begins to fade. With a compass in hand, head northeast keeping the canyon rim not far to your right. Further along and near the upper end of a shallow canyon, you'll pick up old tire tracks leading to Apache Maid Tanks. Follow this road eastward to where it turns abruptly south. You can then make a shortcut over the north side of Hog Hill, or go on south to the chained area, then route-find east just south of this low hill. All old maps show a road running over the pass north of Hog Hill, but it doesn't seem to exist. The best route is the one north of Hog Hill. In the valley called Waldroup Place, you'll see more roads and a meadow with big ponderosa pines. This is where you'll head south into Waldroup Canyon and over 7 dryfalls or dropoffs(no problems) to the head of Wet Beaver. In upper Wet Beaver, just after you make the first right turn, is where you'll first see water and some big swimming holes. *You must have an air mattress to ferry your pack across a total of 32 such holes, 23 of which you must swim.* You'll emerge from this watery hike at Bell Crossing. Go well-prepared and this will be one of your best hikes ever.

Elevations Ranger station, 1158 meters; Waldroup Place, 1853; the first springs, 1524.

Hike Length and Time Needed The round-trip is only about 35 kms, but the hike is hot and dry on the mesa top, then there's lots of swimming back down through the canyon. Strong and hurried hikers can do it in 2 days, but 3 days is more realistic for the loop-hike.

Water There's no reliable water from the beginning of the Apache Maid Trail #15 to Waldroup Place. There's possibly pothole water in Waldroup Canyon, but the first reliable water is at the top of the Coconino at 1524 meters. Carry lots of water on the Apache Maid Trail #15!

Maps USGS or BLM map Sedona(1:100,000), or Casner Butte, Apache Maid Mtn.(1:24,000), and the Coconino National Forest map.

Main Attractions Clear, pure water and many deep, cold swimming holes.

Ideal Time to Hike In the summer with temperatures of 30° C. or higher.

Hiking Boots Wading boots or shoes in good condition.

Author's Experience With an early morning start the author went up the Apache Maid Trail #15, but had to explore in the heat of the day to reach Waldroup Place. Then into the canyon and out the bottom in 2 full days, or a total walk-time of 16 hours.

The author counted 23 swimming holes like this one on his hike of Wet Beaver.

Map 113, Wet Beaver Creek Canyon, Arizona

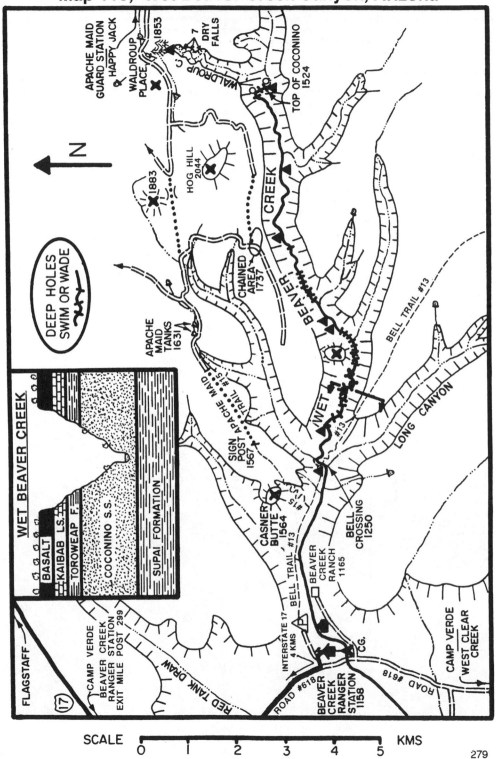

APACHE MAID GUARD STATION
HAPPY JACK
1853
WALDROUP PLACE
7 DRY FALLS
WALDROUP C.
TOP OF COCONINO 1524

N

1883
HOG HILL 2044

DEEP HOLES SWIM OR WADE

BEAVER CREEK

CHAINED AREA 1737

APACHE MAID TANKS 1631

BELL TRAIL #13

WET

LONG CANYON

#13

APACHE MAID #15

SIGN POST TRAIL
1567

WET BEAVER CREEK

BASALT
KAIBAB LS.
TOROWEAP F.
COCONINO S.S.
SUPAI FORMATION

CASNER BUTTE 1564
BELL TRAIL #13
S #

BELL CROSSING 1250

FLAGSTAFF

17

CAMP VERDE
BEAVER CREEK RANGER STATION
EXIT MILE POST 299

RED TANK DRAW

INTERSTATE 17
4 KMS

BELL TRAIL #13

BEAVER CREEK RANCH 1165

ROAD #618

BEAVER CREEK RANGER STATION 1158

C.G.

CAMP VERDE
WEST CLEAR CREEK

ROAD #618

SCALE   0   1   2   3   4   5   KMS

West Clear Creek, Arizona

Location and Access West Clear Creek is located south of Flagstaff, Sedona and Wet Beaver Creek, and east of Camp Verde and Interstate 17. To reach the bottom end of the hike, exit I-17 at Camp Verde at mile posts and Exits 285 or 287, and drive southeast on Highway 260 for about 11 kms. Between mile posts 226 and 227, turn north on Forest Road #618 for 3 kms, then head east for another 3-4 kms to the locked gate leading to the old abandoned Bull Pen Ranch. To get to the upper part of the hike, drive southeast out of Camp Verde on Highway 260. About 38 kms out of town, you'll pass under some large power lines. A little further on and between mile posts 245 and 246, turn left or north on Road #142, and follow this rough dirt track to Calloway Butte. At that point turn left again and follow Road #142A to its end which is located just beyond the same power lines you passed earlier. From the highway to the trailhead is about 8 kms. You'll need two cars, or hitch hike back to the upper canyon.

Trail or Route Conditions From the trailhead, look for a well-used path heading west to an overlook. From there, the trail zig zags down the steep canyon wall to the creek. There are good campsites at the very bottom, but the trail gradually disappears. From there to a point about 10 kms above the Bull Pen Ranch, you'll be walking down the creek bed itself or on either side. Much of your time will be spent rock-hopping half-meter-sized boulders above the small stream. *You also must have an air mattress tied to your pack(take nylon cord) to ferry it across all the deep holes.* The author counted 49 such holes; 28 of which were "swimmers", 21 were "waders". There are no problems getting in or out of the pools; you simply wade in, swim through, and walk out the bottom end. No big jumps either, but walk to the left on a trail around one waterfall. The author did this trip alone as usual and had no problems because of it. The deep holes are all in the upper and middle parts of the canyon and mostly where the stream cuts through Coconino Sandstone. In the lower part of the canyon you'll use Trail #17 to reach the Bull Pen Ranch. Rumors say the upper canyon east of this mapped area is also great(see the book, *Hiker's Guide to Arizona*).

Elevations The trailhead is about 1825 meters; the Bull Pen Ranch about 1097 meters.

Hike Length and Time Needed The National Geographic team which went through the canyon to complete the book, *America's Wild and Scenic Rivers*, took 5 days for apparently this same route(?). The author took 2 days. You'll want 3, maybe 4 days to enjoy the trip. Trailhead to trailhead is about 32-34 kms.

Water Clear, cold and pure water all the way, with many side canyon springs.

Map USGS or BLM map Sedona(1:100,000), or Buckhorn Mtn., Walker Mtn., Calloway Butte(1:24,000), and the Coconino National Forest map.

Main Attractions Deep, cold swimming pools and wild solitude. Also big trout and great fishing.

Ideal Time to Hike In summer with temperatures of 30° C. or higher.

Hiking Boots A pair of tough wading boots or shoes.

Author's Experience The author parked next to Highway 260 in the upper end, then walked all the way to Road #618 and got a ride back to the main highway. Then with a sign reading, *My Car, 20 Miles,* he got another ride back to his car, all in 2 days and with a total walk-time of 17 hours.

An air mattress is an essential for hiking West Clear Creek.

Map 114,   West Clear Creek, Arizona

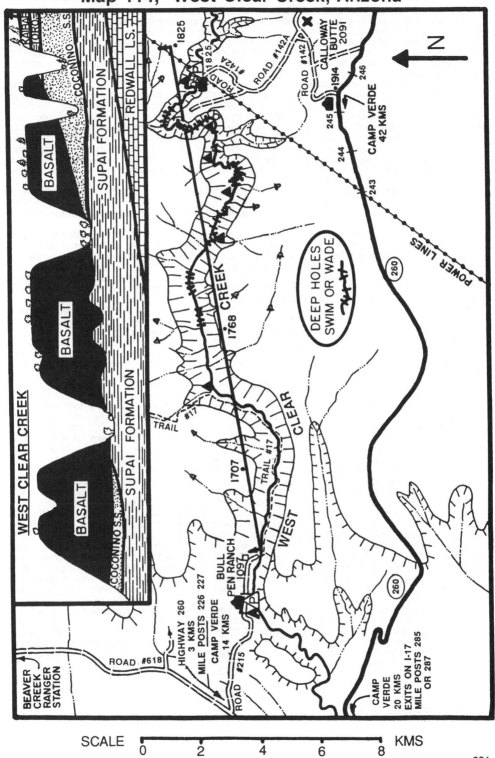

SCALE

0   2   4   6   8   KMS

Native American Civilizations on the Colorado Plateau

By observing the map you can see that in times past there were a number of different groups of people living on the Colorado Plateau. The names applied here are names given them by modern day archaeologists. For the most part these people were all interrelated groups with about the same living standards, the same kind of dwellings, had similar eating habits, etc. The following is an excerpt adapted from the information on the BLM publication, *Grand Gulch Primitive Area*(Map). Grand Gulch is part of the area covered by the Mesa Verde Anasazi, but their way of life was similar to those people in other parts of the Colorado Plateau.

The Basketmakers were the earliest known inhabitants of Grand Gulch. This culture is thought to have derived from an earlier nomadic people whose livelihood was based on hunting and gathering. As yet no artifacts have been discovered in the Gulch which pre-date the Basketmakers. When the nomadic people learned to plant and cultivate corn introduced from the south(as well as squash and beans and the domestication of turkeys), they became more sedentary and the Basketmaker Culture evolved. It flourished here from 200 to 600 A.D. They built pit houses made of mud with caked over stick walls and roofs. Their name was derived from the finely woven baskets they made. They also used flint tools and wooden digging sticks. The most prevalent remains of the Basketmaker Culture in the Gulch are their slab-lined storage bins. These may still be seen on the mesa tops or on high ledges protected from the weather and rodents.

A series of droughts apparently drove the people into the surrounding mountains. When they returned around 1050 A.D., their culture had been influenced by the Mesa Verde people to the east and the Kayenta people from the south. As time passed the Mesa Verde influence predominated in the Grand Gulch area.

The Basketmaker Culture had developed into the Pueblo Culture. The Pueblo Culture is characterized by the making of fine pottery with some highly decorated; the cultivation of cotton and weaving of cotton cloth; and the high degree of architectural and stone masonry skill seen in the cliff dwellings in the Gulch. They also developed irrigation of their crops by building checkdams and diversion canals. These have been found on the mesa tops near Dark Canyon, and tributary canyons to the Colorado River. The kiva, a round ceremonial structure found in Grand Gulch, is still in use by the modern day descendants of the Anasazi, the Hopi and New Mexico Pueblo Indians.

Grand Gulch is also known for the diversity of rock art scattered throughout. The rock art consists of both petroglyphs(pecked into the rock) and pictographs(painted on with pigments). As the figures do not represent a written language, the meaning of the various panels is left to our imaginations.

Periods of drought in the 12th and 13th centuries, plus depletion of natural resources and pressure from nomadic Indians from the north, are thought to be some possible reasons for the abandonment of this region. By the late 1200's, the Anasazi moved south into Arizona and southeast into the Rio Grande Valley of New Mexico.

That part of the Colorado Plateau which has the most Indian ruins, at least the most ruins in wilderness areas which are in a state of good preservation, is the region covered by the Mesa Verde Anasazi. Most of these ruins are located in wild and inaccessible canyons, far away from 4WD's and ORV's and other molesters. Most are cliff dwellings, but many are also on mesa tops. Ruins you must hike to are in Sand and Rock Creek Canyons in Colorado; and Mule, Arch, Fish, Owl, Road, Lime, Johns, Slickhorn, Grand Gulch, Dark and Salt Creek Canyons in Utah. There are also ruins in many other scattered locations, but these just mentioned are the best. Most of these sites are rather small, usually only 2 to 6 structures per site. Sometimes there's only a granary. The places with really large cliff dwellings which are also well preserved, are usually set aside as national monuments. The very best place to view the big cliff pueblos is at Mesa Verde National Park. But you can't do any hiking there except along with the hordes of tourists at selected sites. That's why it's not included in this book. The next best place is probably Navajo National Monument, where you can hike to both of those sites, but practically under armed guard. Canyon De Chelly has some fantastic ruins as well, but outsiders can visit only one set of ruins, unless a Navajo guide is hired as escort. Chaco Canyon has some fine ruins too, but they aren't cliff dwellings and there's not much hiking involved.

The author has ranked the best hikes to Indian ruins in the section *Best Hikes*. See this listing and those concerning Indian art. Since this book specializes in hiking and not Indian history, it's best you buy one or two books on that subject if you plan to spend much time in the region. There are several good history books listed under *Further Reading*, as well as some books on various present-day Native Americans of the southwest and the Colorado Plateau. Look for these books and others in national park or monument visitor centers.

Now a request on the part of all concerned people: when visiting isolated ruins, please leave the artifacts lying around in place for those who will follow in your footsteps. The author is proud to say he has not one potsherd or corn cob from any of his visits to hundreds of ruins. Besides, it's against the law to remove any artifacts. Part of the Federal law states; *No person may excavate, remove, damage, or otherwise alter or deface any archaeological resource located on public lands or Indian lands unless such activity is pursuant to a permit issued........*

Native American Civilizations--Colorado Plateau

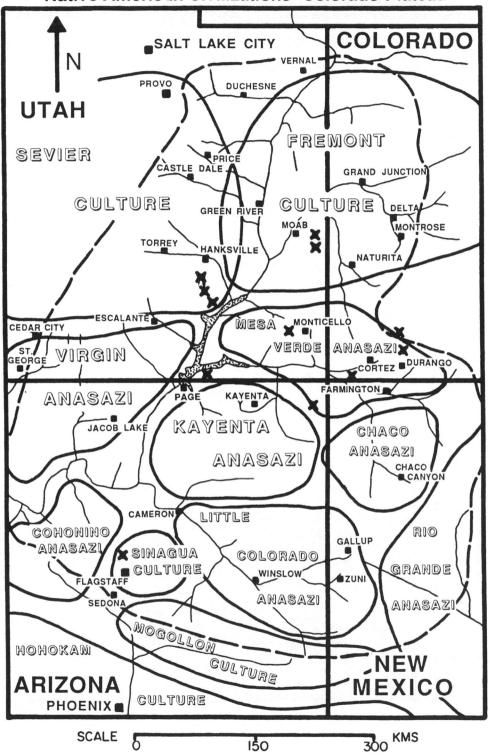

SCALE   0   150   300   KMS

Geology of the Colorado Plateau

In the cross sections below, the author has gone back through time beginning near the head of Desolation Canyon in Utah's Uinta Basin, down along the Green River, through Canyonlands National Park, and along the Colorado River as it winds its way through the Grand Canyon. Perhaps the best source for geology information along the Green and Colorado Rivers are the Powell River Runners Guide series. For information on areas away from the Green and Colorado Rivers, see the geology cross-sections put out by the various natural history associations of the national parks of the southwest.

DESOLATION CANYON CROSS-SECTION

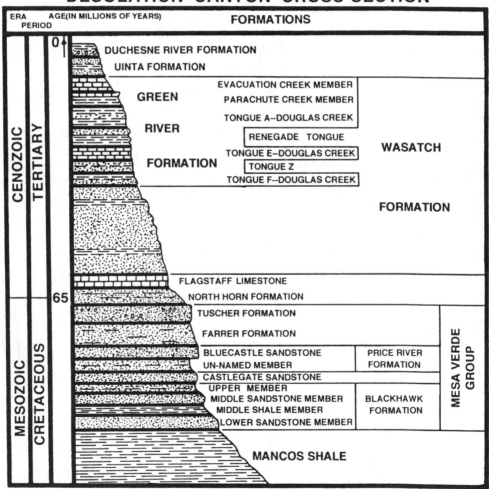

CANYONLANDS CROSS-SECTION

Era	Period	Age	Unit	Group/Formation
MESOZOIC	CRETACEOUS		MANCOS SHALE	
			DAKOTA SANDSTONE	
		135	SHALE MEMBER	CEDAR MOUNTAIN FORMATION
			BUCKHORN CONGLOMERATE	
	JURASSIC		BRUSHY BASIN SHALE	MORRISON FORMATION
			SALT WASH SANDSTONE	
			SUMMERVILLE FORMATION	SAN RAFAEL GROUP
			CURTIS FORMATION	
			ENTRADA SANDSTONE	
			CARMEL FORMATION	
		195	NAVAJO SANDSTONE	GLEN CANYON GROUP
			KAYENTA FORMATION	
			WINGATE SANDSTONE	
	TRIASSIC		MOSSBACK MEMBER	CHINLE FORMATION
		225	MOENKOPI FORMATION	
PALEOZOIC	PERMIAN		WHITE RIM SANDSTONE	CUTLER FORMATION
			ORGAN ROCK TONGUE	
			CEDAR MESA SANDSTONE	
			HALGAITO FORMATION	
		280	RICO FORMATION	
	PENN.		HONAKER TRAIL FORMATION	HERMOSA FORMATION
			PARADOX SALT FORMATION	

GRAND CANYON CROSS-SECTION

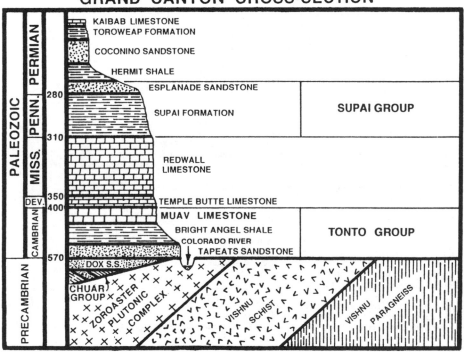

Era	Period	Age	Unit	Group
PALEOZOIC	PERMIAN		KAIBAB LIMESTONE	
			TOROWEAP FORMATION	
			COCONINO SANDSTONE	
			HERMIT SHALE	
		280	ESPLANADE SANDSTONE	SUPAI GROUP
	PENN.		SUPAI FORMATION	
		310		
	MISS.		REDWALL LIMESTONE	
		350		
	DEV.	400	TEMPLE BUTTE LIMESTONE	
	CAMBRIAN		MUAV LIMESTONE	TONTO GROUP
			BRIGHT ANGEL SHALE	
			COLORADO RIVER	
		570	TAPEATS SANDSTONE	
PRECAMBRIAN			DOX S.S.	
			CHUAR GROUP	
			ZOROASTER PLUTONIC COMPLEX	
			VISHNU SCHIST	
			VISHNU PARAGNEISS	

Best Hikes

The Best Hikes as judged by the author. The order of listing will vary between people, but these are all great hikes. Based on scenic beauty, lots of deep and dark narrows, challenges of passage, and interesting geology. Included is the geologic formation most prominently featured.

1. **Paria River and Buckskin Gulch,** Navajo Sandstone, Utah-Arizona--longest slot canyon.
2. **Lower White Canyon**(Black Hole), Cedar Mesa Sandstone, Utah--challenging, cold water.
3. **Upper Kaibito Creek,** Navajo Sandstone, Navajo Nation(Arizona)--very narrow, dark and fotogenic.
4. **Zion Narrows**(Kolob C., long ropes needed, difficult), Navajo Sandstone, Utah--most popular.
5. **Antelope Canyon,** Navajo Sandstone, Navajo Nation(Arizona)--most fotogenic.
6. **Coyote Gulch**(Escalante River), Navajo Sandstone, Utah--arches and huge undercuts,
7. **West Canyon,** Navajo Sandstone, Navajo Nation(Arizona)--challenging upper end, ropes needed.
8. **Little Wildhorse Canyon**(San Rafael Swell), Navajo Sandstone, Utah--fotogenic.
9. **West Clear Creek,** Coconino Sandstone, Arizona--lots of swimming.
10. **Lower Black Box**(San Rafael River), Coconino Sandstone, Utah--swimming.
11. **The Chute, Muddy Creek**(San Rafael Swell), Coconino Sandstone, Utah--easy walking.
12. **Upper Black Box**(San Rafael River), Coconino Sandstone, Utah--can be challenging.
13. **Little Death Hollow**(Escalante River), Wingate Sandstone, Utah--best Wingate narrows.
14. **Peek-a-boo, Spooky & Brimstone Gulches**(Escalante R.), Navajo S.S., Utah--narrowest slots.
15. **Starting Water & M. Kaibito,** Navajo S.S., Navajo N.(Arizona), tight slot, ropes needed, waterfall.
16. **Death Hollow**(Escalante River), Navajo Sandstone, Utah--challenging narrows.
17. **Round Valley Draw**(Upper Paria River), Navajo Sandstone, Utah--fotogenic.
18. **Bull Valley Gorge**(Upper Paria River), Navajo Sandstone, Utah--scene of pickup wreck.
19. **Great West Canyon,** Navajo Sandstone, Utah--watery, deep, scenic, ropes needed.
20. **Water Holes Canyon,** Navajo Sandstone, Navajo Nation(Arizona)--short, sweet & fotogenic.
21. **Wet Beaver Creek,** Coconino Sandstone, Arizona--lots of swimming.
22. **Orderville Canyon,** Navajo Sandstone, Utah--deep narrows, waterfalls.
23. **Parunuweap Canyon**(East Fork of Virgin River), Navajo Sandstone, Utah--deep, challenging.
24. **Lower Halls Creek Narrows,** Navajo Sandstone, Utah--maybe deepest Navajo narrows.
25. **National Canyon,** Muav & Temple Butte Limestones, Hualapai N.(Arizona)--very deep and scenic.

Best Hikes to see Anasazi Indian Ruins
1. **Navajo National Monument,** Navajo Sandstone, Navajo Nation(Arizona).
2. **Grand Gulch,** Cedar Mesa Sandstone, Utah.
3. **Inscription House Canyon,** Navajo Sandstone, Navajo Nation(Arizona).
4. **Mule Canyon,** Cedar Mesa Sandstone, Utah.
5. **Upper Salt Creek,** Cedar Mesa Sandstone, Canyonlands N.P., Utah.
6. **Sand and Rock Creek Canyons,** Entrada Sandstone, Colorado.
7. **Road and Lime Canyons,** Cedar Mesa Sandstone, Utah.
8. **Slickhorn Canyon,** Cedar Mesa Sandstone, Utah
9. **Johns Canyon,** Cedar Mesa Sandstone, Utah.
10. **Fish and Owl Creek Canyons,** Cedar Mesa Sandstone, Utah.
11. **Davis and Lavender Canyons,** Cedar Mesa Sandstone, Utah.
12. **Natural Bridges National Monument,** Cedar Mesa Sandstone, Utah.

Non-Hiking Areas to see Anasazi Indian Ruins
14. **Mesa Verde National Park,** Cliff House Sandstone, Colorado.
15. **Canyon de Chelly National Monument,** De Chelly Sandstone, Arizona.
16. **Walnut Canyon National Monument,** Kaibab Limestone, Arizona.
17. **Wupatki National Monument,** on top of the Moenkopi Formation, Arizona.
18. **Hovenweep National Monument,** on top of various formations, Utah-Colorado.

Best Hikes to see Indian Rock Art--Petroglyphs and Pictographs
1. **Horseshoe Canyon,** Pictographs, Utah.
2. **Snake Gulch,** Pictographs, Arizona.
3. **South(Pictograph) Fork--Horse Canyon,** Pictographs and Petroglyphs, Utah
4. **Desolation Canyon,** Petroglyphs, Utah.
5. **Chaco Canyon National Monument,** Petroglyphs, New Mexico.
6. **Grand Gulch,** Petroglyphs and Pictographs, Utah.
7. **Cottonwood Draw,** Petroglyphs, Utah(see the author's book on the San Rafael Swell).
8. **Jumpup Canyon,** Pictographs, Arizona.
9. **Burnt Canyon,** Petroglyphs, Arizona.

Indian Rock Art Panels with Vehicle Access
8. **Newspaper Rock,** Indian Creek, Utah, Petroglyphs.
9. **Nine Mile Canyon,** Northeast of Price, Utah, Pictographs and Petroglyphs.
10. **Thompson Canyon,** North of I-70 & Thompson, Utah, Pictographs.
11. **Buckhorn Wash,** San Rafael Swell, Utah, Pictographs and Petroglyphs.
12. **Black Dragon Canyon,** San Rafael Swell, Utah, Pictographs.
13. **Three Fingers Canyon,** San Rafael Swell, Utah, Petroglyphs.

Further Reading

History
A Canyon Voyage, Frederick S. Dellenbaugh, University of Arizona Press, Tucson, Arizona.
Canyon Country Geology, Barnes, Canyon Country Publications, P.O. Box 963, Moab, Utah.
Canyonlands National Park--Early History and First Descriptions, Barnes, Canyon Country Publications, Moab, Utah.
Canyon Rims Recreation Area, F.A. & M.M. Barnes, Canyon Country Pub., Moab, Utah.
Desert River Crossing(Lee's Ferry), Rusho & Crampton, Peregrine Smith Inc, Salt Lake City, Utah.
Grand Canyon Rapids, Muriel W. Smith, Carlton Press, Inc., New York.
Horse Thief Ranch--An Oral History, H. Michael Behrendt, self published, Moab, Utah.
Incredible Passage(Through the Hole-in-the-Rock), Lee Reay, Meadow Lane Publications, Provo Utah.
My Canyonlands, Kent Frost, Abelard-Schuman, New York.
One Man's West, David Lavender, U. of Nebraska Press, Lincoln, Nebraska.
San Juan County, Utah, Allan K. Powell, Utah State Historical Society, Salt Lake City, Utah.
The Exploration of the Colorado River and its Canyons, J. W. Powell, Dover Publications, Inc., New York.
Rim Flying Canyonlands, with Jim Hurst, by Pearl Baker, A to Z Printing, Riverside, California.
Robbers Roost Recollections, Pearl B. Baker, Utah State University Press, Logan, Utah.

Native American Books
Ancient Ruins of the Southwest, D. G. Noble, Northland Press, Flagstaff, Arizona.
Canyon Country Prehistoric Indians, Barnes & Pendleton, Wasatch Publishers, Salt Lake City, Utah.
Havasu Canyon, Gem of the Grand Canyon(history and a guide book), Wampler, Joseph Wampler Books, Box 45, Berkeley California.
Pages from Hope History, Harry C. James, University of Arizona Press, Tucson, Arizona.
People of the Blue Water(Hualapai and Havasupai Indians), F. G. Iliff, University of Arizona Press, Tucson, Arizona.
Prehistoric Rock Art, Barnes, Canyon Country Publications, P.O. Box 963, Moab, Utah.
The Navajo, Kluckhohn and Leighton, Harvard University Press, Cambridge, Massachusetts.

Guide Books
A Guide to Hiking the Inner Canyon, Thybony, Grand Canyon Natural History Association.
A Guide to the Trails of Zion National Park, Zion Natural History Association, Springdale, Utah.
A Naturalist's Guide to Hiking the Grand Canyon, Aitchison, Prentice-Hall, Englewood Cliffs, N.J.
Canyoneering-The San Rafael Swell, Steve Allen, University of Utah Press, Salt Lake City, Utah.
Capitol Reef N. P., A Guide to the Road & Trails, Roylance, Wasatch Publishers, Salt Lake City, Utah.
Exploring the Backcountry of Zion National Park--Off-Trail Routes, Brereton & Dunaway, Zion Natural History Association, Springdale, Utah.
Grand Canyon Treks, I, II, and III(3 volumes), Butchart, La Siesta Press, Glendale, California.
Hiking the Escalante, Lambrechtse, Wasatch Publishers, Salt Lake City.
Hiker's Guide to Arizona, Aitchison & Grubbs, Falcon Press, Billings and Helena, Montana.
Hiker's Guide to Colorado, Boddie, Falcon Press Publishing Co., Billings and Helena, Montana.
Hiker's Guide to Utah, Hall, Falcon Press Publishing Co., Billings and Helena Montana.
Labyrinth, Stillwater, Cataract Canyons, Mutschler, Powell Society Ltd., 777 Vine St., Denver, Colorado.
Marble Gorge & Grand Canyon, Simmons & Gaskill, Powell Society Ltd., 777 Vine St., Denver, Colorado.
River Runners Guide to Canyonlands National Park and Vicinity(Emphasis on Geologic Features), Mutschler, Powell Society Ltd., 777 Vine Street, Denver, Colorado.
River Runners Guide to the Canyons of the Green and Colorado Rivers, Desolation and Gray Canyons(Emphasis on Geologic Features), Mutschler, Powell Society Ltd., 777 Vine Street, Denver, Colorado.
Utah's Scenic San Rafael, Owen McClenahan-self published, Box 892, Castle Dale, Utah.
50 Hikes in Arizona, Nelsons, Gem Guides, 3677 San Gabriel Parkway, Pico Rivera, California.
50 Hikes in New Mexico, Evans, Gem Guides, 3677 San Gabriel Parkway, Pico Rivera, California.

Other Guide Books by the Author

(Prices as of January, 1995. Prices may change without notice)

Climber's and Hiker's Guide to the World's Mountains(3rd Ed.), Kelsey, 928 pages, 447 maps, 451 fotos, ISBN 0-944510-02-7. US$34.95 (Mail orders US$37.00).

Utah Mountaineering Guide, and the Best Canyon Hikes(2nd Ed.), Kelsey, 192 pages, 105 fotos, ISBN 0-9605824-5-2. US$9.95 (Mail orders US$11.50).

Canyon Hiking Guide to the Colorado Plateau(3nd Edition), Kelsey, 288 pages, 116 maps, 159 fotos, ISBN 0-9605824-1-5. US $12.95 (Mail orders US$14.50).

Hiking and Exploring Utah's San Rafael Swell(2nd Ed.), Kelsey, 160 pages, 35 mapped hikes, plus lots of history, 104 fotos, ISBN 0-944510-01-9. US$8.95 (Mail orders US$10.50).

Hiking and Exploring Utah's Henry Mountains and Robbers Roost, Including The Life and Legend of Butch Cassidy, Revised Edition, Kelsey, 224 pages, 38 hikes or climbs, 158 fotos, ISBN 0-944510-4-3, US$9.95 (Mail orders US$11.50).

Hiking and Exploring the Paria River, Kelsey, 208 pages, 30 different hikes from Bryce Canyon to Lee's Ferry, 155 fotos, ISBN 0-944510-4-3. US$10.95 (Mail Orders US$12.50).

Hiking and Climbing in the Great Basin National Park--*A Guide to Nevada's Wheeler Peak, Mt. Moriah, and the Snake Range,* Kelsey, 192 pages, 47 hikes or climbs, 125 fotos, ISBN 0-9605824-8-7. US$9.95 (Mail Orders US$11.50).

Boater's Guide to Lake Powell(Updated Edition), with hiking emphasised, Kelsey, 288 pages, 256 fotos, ISBN 0-9605824-9-5. US$12.95 (Mail Orders US$14.50).

Climbing and Exploring Utah's Mt. Timpanogos, Kelsey, 208 pages, 170 fotos, ISBN 0-944510-00-0. US$9.95 (Mail Orders US$11.50).

River Guide to Canyonlands National Park & Vicinity, Kelsey, 256 pages, 151 fotos, ISBN 0-944510-07-8. US$11.95(Mail Orders US$13.50).

Hiking, Biking and Exploring Canyonlands National Park & Vicinity, Kelsey, 320 pages, 227 fotos, ISBN 944510-08-6. US$14.95(Mail Orders US$16.50).

Life on the Black Rock Desert, *A History of Clear Lake, Utah,* Venetta B. Kelsey, 192 pages, 123 fotos, ISBN 0-944510-03-5. US$9.95(Mail Orders US$11.50).

China on Your Own, *and The Hiking Guide to China's Nine Sacred Mountains*(3rd and Revised Ed.), Jennings/Kelsey, 240 pages, 110 maps, 16 hikes or climbs, ISBN 0-9691363-1-5. US$9.95 (Mail Orders US$11.50). This is out of print, but I have a few copies left.

Distributors for Kelsey Publishing

Primary Distributor All of Michael R. Kelsey's books are sold by this company. If you'd like to order any book, please call or write to the following address.

Wasatch Book Distribution, P.O. Box 11776, 268 S., 200 E., Salt Lake City, Utah, USA, 84111, Tele. 1-801-575-6735, or for bookstores, 1-800-786-6715.

<u>Some of Kelsey's books are sold by the following companies.</u>

Alpenbooks, 3616 South Road, Building C, Suite 1, Mukilteo, Washington, 98275, Tele. 1-206-290-8587.

Canyon Country Publications, P. O. Box 963, Moab, Utah, 84532, Tele. 1-801-259-6700.

Canyonlands Publications, 4999 East Empire, Unit A, Flagstaff, Arizona, 86004, 1-602-527-0730.

Crown West Books(Library Service), 575 E. 1000 S., Orem, Utah, 84058, Tele. 1-801-224-1455.

Ingram Books, 2323 Delgany, Denver, Colorado, 80216, Tele. 1-800-876-1830.

Northern Arizona News, 1709 North, East Street, Flagstaff, Arizona, 86001, Tele. 1-602-774-6171.

Many Feathers, 2626 West, Indian School Road, Phoenix, Arizona, 85012, Tele. 1-602-266-1043.

Nevada Publications, 4135 Badger Circle, Reno, Nevada, 89509, Tele. 1-702-747-0800.

Mountain 'N Air Books, 7251 Foothill Blvd., Tujunga, California, 91042, Tele. 1-818-951-4150, or for bookstores, 1-800-446-9696

Recreational Equipment, Inc.(R.E.I.), P.O. Box C-88126, Seattle, Washington, 98188, Tele. 1-800-426-4840(or check at any of their local stores).

For the UK and Europe, and the rest of the world contact:

CORDEE, 3a De Montfort Street, Leicester, England, UK, LE1 7HD, Tele. 0533-543579, Fax 0533-471176.